Prognostics and Health Management

Wiley Series in Quality & Reliability Engineering

Dr. Andre Kleyner
Series Editor

The Wiley Series in Quality & Reliability Engineering aims to provide a solid educational foundation for both practitioners and researchers in the Q&R field and to expand the reader's knowledge base to include the latest developments in this field. The series will provide a lasting and positive contribution to the teaching and practice of engineering.

The series coverage will contain, but is not exclusive to,

- Statistical methods
- Physics of failure
- Reliability modeling
- Functional safety
- Six-sigma methods
- Lead-free electronics
- Warranty analysis/management
- Risk and safety analysis

Prognostics and Health Management

A Practical Approach to Improving System Reliability Using Condition-Based Data

Douglas Goodman, James P. Hofmeister and Ferenc Szidarovszky
Ridgetop Group, Inc., Arizona, USA

This edition first published 2019
© 2019 John Wiley & Sons Ltd

The right of Douglas Goodman, James P. Hofmeister and Ferenc Szidarovszky to be identified as the authors of this work has been asserted in accordance with law.

Registered Offices
John Wiley & Sons, Inc., 111 River Street, Hoboken, NJ 07030, USA
John Wiley & Sons Ltd, The Atrium, Southern Gate, Chichester, West Sussex, PO19 8SQ, UK

Editorial Office
The Atrium, Southern Gate, Chichester, West Sussex, PO19 8SQ, UK

For details of our global editorial offices, customer services, and more information about Wiley products visit us at www.wiley.com.

Wiley also publishes its books in a variety of electronic formats and by print-on-demand. Some content that appears in standard print versions of this book may not be available in other formats.

Library of Congress Cataloging-in-Publication Data

Names: Goodman, Douglas (Industrial engineer), author. | Hofmeister, James
   P., author. | Szidarovszky, Ferenc, author.
Title: Prognostics and health management : a practical approach to improving
   system reliability using condition-based data / Douglas Goodman, Chief
   Engineer, Ridgetop Group, Inc., James P. Hofmeister, Distinguished
   Engineer, Ridgetop Group, Inc., Ferenc Szidarovszky, Senior
   Researcher, Ridgetop Group, Inc.
Description: Hoboken, NJ, USA : Wiley, [2019] | Includes bibliographical
   references and index. |
Identifiers: LCCN 2018060348 (print) | LCCN 2019000608 (ebook) | ISBN
   9781119356691 (AdobePDF) | ISBN 9781119356707 (ePub) | ISBN 9781119356653
   (hardcover)
Subjects: LCSH: Machinery–Reliability. | Equipment health monitoring. |
   Machinery–Maintenance and repair–Planning. | Structural
   failures–Mathematical models.
Classification: LCC TJ174 (ebook) | LCC TJ174 .G66 2019 (print) | DDC
   621.8/16–dc23
LC record available at https://lccn.loc.gov/2018060348

Cover Design: Wiley
Cover Image: © Sergey Nivens/Shutterstock

Set in 10/12pt WarnockPro by SPi Global, Chennai, India

Printed in Great Britain by TJ International Ltd, Padstow, Cornwall

10 9 8 7 6 5 4 3 2 1

Contents

List of Figures

Series Editor's Foreword

As quality and reliability science evolves, it reflects the trends and transformations of the technologies it supports. A device utilizing a new technology, whether it be a solar power panel, a stealth aircraft, or a state-of-the-art medical device, needs to function properly and without failure throughout its mission life.

In addition to addressing the reliability of new technology, the field of quality and reliability engineering has been going through its own evolution, developing new techniques and methodologies aimed at process improvement and reduction in the number of design- and manufacturing-related failures. One of these disciplines is prognostics and health management/monitoring (PHM), a fast-growing field intended to ensure safety and provide the state of health and estimate remaining useful life (RUL) of components and systems. PHM injects a more proactive approach into system reliability, where application of physics of failure (PoF), degradation analysis, and modern algorithms allow the prediction of failure time and, consequently, the ability to take actions preventing failures from happening.

The advancement and growing application of functional safety standards, along with the fast development of autonomous vehicles, increases the pressure to achieve exceptionally high system reliability, thus making PHM an indispensable tool to meet these expectations.

PHM can provide many advantages to users and maintainers, including financial benefits such as operational and maintenance cost reductions and extended lifetime. Despite the additional cost required to facilitate prognostics (monitoring systems, algorithms development, etc.), PHM has positive effects on the overall lifecycle cost of the system by avoiding costly and sometimes catastrophic failures.

This book has been written by the leading experts and state-of-the-art practitioners in the field of prognostics and health management/monitoring. It discusses the many technical aspects of PHM along with its cost benefits and will be an excellent addition to this book series. The Wiley Series in Quality & Reliability Engineering aims to provide a solid educational foundation for researchers and practitioners in the field of quality and reliability engineering and to expand the knowledge base by including the latest developments in these disciplines.

Despite its obvious importance, quality and reliability education is paradoxically lacking in today's engineering curriculum. Few engineering schools offer degree programs or even a sufficient variety of courses in quality or reliability methods. Therefore, the majority of quality and reliability practitioners receive their professional training from colleagues, professional seminars, publications, and technical books. The lack of formal education opportunities in this field greatly emphasizes the importance of technical publications for professional development.

We hope that this book, as well as the whole series, will continue Wiley's tradition of excellence in technical publishing and provide a lasting and positive contribution to the teaching and practice of engineering.

Dr. Andre Kleyner
Editor of the Wiley Series in Quality & Reliability Engineering

Preface

A prognostics and health management/monitoring (PHM) system can be thought of as consisting of three major systems: a sensing system consisting of a sensor and a feature vector framework, a prognosis system comprising a prediction framework and a performance-validation framework, and a health-management framework. Although health management is probably the most complex and most expensive, this book presents topics related to sensing systems and prognosis. An important goal of those systems is to provide accurate prognostic information regarding the prognosis of the health of the system being monitored. This book begins by presenting approaches to reliability predictions based on traditional model-driven, data-driven, and hybrid-driven approaches as necessary background to understanding the rationale for the signature-driven approaches presented later in the book. Those traditional, or handbook, methods are evaluated as inaccurate and misleading when used for prognostic estimation of a future failure.

This book then develops approaches to modeling and data handling that take into account failure modes and operational environment and conditions, and presents an approach using signatures created by extracting leading indicators of failure/condition indicators as feature data that forms condition-based data (CBD) signatures; such signatures can be normalized and converted into dimensionless ratios called fault-to-failure progression (FFP) signatures. FFP signatures form families of characteristic curves, and each family of such curves is dependent on the totality of the mechanisms of failures causing degradation, resulting in changes in monitored signals.

A set of degradation signature models is developed, each model representing a single mode of degradation, and it is shown how those models can be used to transform FFP signatures into degradation-progression signature (DPS) data that can then be transformed into functional failure signature (FFS) data that is particularly amenable to processing by prediction algorithms. FFS data forms linearized transfer curves with values of 0 or less in the absence of degradation and 100 or larger when the component, assembly, or system being monitored (a prognostic-enabled target) reaches or exceeds a level of degradation such that the prognostic target is no longer capable of operating within specifications. When all noise – defined as any change in data not due to degradation – is removed from monitored signals, an FFP signature can be transformed into DPS data that forms a linear straight-line transfer curve, which can be converted into FFS data by defining and using a threshold value that defines when functional failure occurs.

However, since it is not practical to remove all noise, this book also presents some useful signal-conditioning techniques to ameliorate and/or mitigate the effects of noise, including, but not limited to, the following: data fusion, data transforms, domain transforms, filtering, threshold margins, data smoothing, and data trending. Those techniques are illustrated using example sets of noisy data, including those that exhibit signal variations and changes in the shapes of curves caused by the effects of temperature and feedback.

This book also presents the effects of nonlinear rates of degradation and multiple modes of degradation, and methods and techniques for handling those effects. A heuristic-based approach to using degradation models and data conditioning is presented. It is shown that despite not knowing, for example, the exact value of the power for a power-function type of degradation or life value for an exponential-function type of degradation, or not knowing that the CBD signature is the result of an exponential rate of change in the degradation rate of an underlying power-function type of degradation, the FFS data becomes sufficiently linear to result in very accurate prognostic information. Insufficient accuracy (errors) of prognostic information is caused by and/or due to insufficiency in the sensing system, primarily due to two factors: (i) the signal-conditioning and data-handling routines to fuse, transform, condition, and process data; and (ii) the use of insufficient rates of system sampling and data sampling (inexact modeling). It is asserted that, given sufficiently high system-sampling rates and data-sampling rates, the accuracy of prognostic information approaches a level limited by the system. System limits on prognostic accuracy are due to measurement uncertainty and errors: for example, digitization, quantization errors, mathematical rounding, manufacturing tolerances, variations in measurement values of environmental effects such as temperature and pressure, variations in values due to manufacturing, and so on. An often-ignored contributor to prognostic accuracy/inaccuracy is the sampling method employed by the sensing system: it does not make sense to require a minimum prediction horizon for functional failure of 20 weeks within 10% accuracy and a time-of-failure precision of ± 1 week from a sensing system that samples a node to acquire data once every 2 weeks.

It is shown that despite the effects of noise-induced nonlinearity of the transfer curves, the design and development of prediction algorithms for PHM systems is fairly straightforward and not very complex, and easily produces very accurate prognostic information such as estimates of remaining useful life (RUL) and state of health (SoH). For illustration, the authors use computational prediction algorithms incorporated in Adaptive Remaining Useful Life Estimator™ (ARULE™) programs.

A sensing system should acquire and manipulate CBD to sufficiently ameliorate and mitigate all noise using methods and means including sampling, filtering, masking, fusing, transforming of data types and domains such as time and frequency, smoothing, and so on. The ultimate goal of a sensing system used in a PHM system should be that of a transducer that transforms monitored signals into a sufficiently linear transfer curve to support producing accurate prognostic information. Prediction information should minimally contain the following:

- If the amplitude of the input is at or below a degradation threshold, health is 100%.
- If the amplitude of the input is at or above a failure threshold, health is 0%.
- Otherwise, the prognostic target is degraded: health is between 0% and 100%.
- RUL estimates: estimated time remaining before end of life (EOL) occurs.
- Prognostic horizon (PH) estimates: current time plus estimated RUL.

Included in this book is a design for a robust prototype of an exemplary PHM system comprising a framework architecture and control/data flow that supports prognostic enabling of six assemblies (line-replaceable units): two switch-mode power supplies and four electro-mechanical actuators. Each power supply is prognostic enabled using a single sensor for a single failure mode: loss of filtering capacitance. Each actuator is prognostic enabled using a set of phase current sensors for three different failure modes: increased on-resistor of power-switching transistor in an H bridge controller for the actuator; degradation in the windings of a brushless DC motor of the actuator; and excessive loading on the shaft of that DC motor. The system comprises two subsystems, each having three prognostic-enabled units: a power supply and two electro-mechanical actuators. A design is presented to do the following: (i) control polling and sampling of the units to acquire, condition, and process CBD to produce prediction information; (ii) support checkpoint/restart in which the system can be stopped, paused, or ended and then restarted; and (iii) show how alerts could be used to support event-triggered maintenance. A chapter on prognostic enabling – selection, evaluation, and other considerations – concludes this book.

Acknowledgments

The authors would like to express our thanks for the understanding, patience, and support given to us by our families, especially our wives, and by our colleagues. We would also like to thank the numerous government agencies and technical points of contact behind the grants that supported the research results reflected in this book: Department of Defense – NAVAIR, NAVSEA, Air Force, and Army, NASA, and the US Department of Energy. We also acknowledge the support, understanding, and encouragement we received from the staff of Wiley books. Last, but not least, we acknowledge the support and assistance of our staff editor, Richard Thompson, provided in the way of logistics, editing, coordination, and liaison-especially with the Wiley staff.

1

Introduction to Prognostics

1.1 What Is Prognostics?

Prognostics is predictive diagnostics and provides the state of degraded health and makes an accurate prediction of when a resulting future failure in the system is likely to occur. The purpose of prognostics is to detect degradation and create predictive information such as estimates of state of health (SoH) and remaining useful life (RUL) for systems. Doing so yields the following benefits: (i) provides advance warning of failures; (ii) minimizes unscheduled maintenance; (iii) predicts the time to perform preventive replacement; (iv) increases maintenance cycles and operational readiness; (v) reduces system sustainment costs by decreasing inspection, inventory, and downtime costs; and (vi) increases reliability by improving the design and logistic support of existing systems (Pecht 2008; Kumar and Pecht 2010; O'Connor and Kleyner 2012).

Prognostics, as defined and used in this book, includes data acquisition (DA) and data manipulation (DM) by sensors (S) and processing within a sensor framework; DA, DM, and state detection (SD) employing processing and computational routines within a feature-vector framework to produce feature data (FD) consisting of condition indicators that are leading indicators of failure (signatures); and health assessment (HA) and prognostic assessment (PA) within a prediction framework/prognostic-information framework. The sensor framework, feature-vector framework, prediction framework, and control and data flow framework form a prognosis subsystem within a prognostics and health management/monitoring (PHM) system (see Figure 1.1). Health management in a PHM system includes the generation of advisory information and health management/monitoring (HM): a health management framework (CAVE3 2015; IEEE 2017).

The focus of this book is prognostics although there is a strong linkage with System Health Management. Some aspects of health management are referenced and/or alluded to at times in the book. Included in this book is a framework for producing performance metrics to evaluate the accuracy of prognostic information. Figure 1.2 is a framework diagram for the exemplary PHM system used in this book.

1.1.1 Chapter Objectives

The scientific field encompassing the sensing, collecting, and processing of data to diagnose and provide a prognosis of the health of a system and estimates of future failure events is complex. Several prognostics concepts and solutions use a

Prognostics and Health Management: A Practical Approach to Improving System Reliability Using Condition-Based Data,
First Edition. Douglas Goodman, James P. Hofmeister and Ferenc Szidarovszky.
© 2019 John Wiley & Sons Ltd. Published 2019 by John Wiley & Sons Ltd.

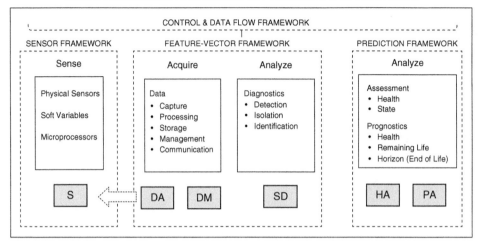

Figure 1.1 Core prognostic frameworks in a PHM system. Source: based on IEEE (2017).

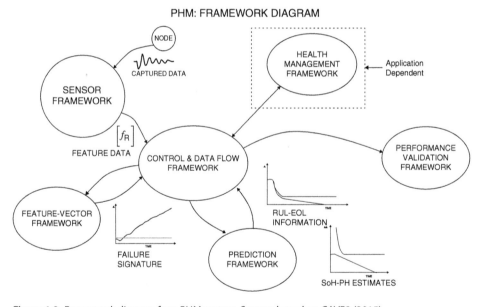

Figure 1.2 Framework diagram for a PHM system. Source: based on CAVE3 (2015).

classical approach based on reliability theory and statistics: the fundamental ideas and methods of those classical approaches will be presented and discussed in this chapter. First, the basic ideas and methods of reliability engineering will be outlined, followed by brief discussions of the main ideas of the different elements of prognostic health management. This chapter includes some cost-minimizing and cost-benefit models.

The primary objective of this chapter is to provide a general overview of this complex scientific field. Detailed descriptions and model development for a heuristic approach to modeling condition-based data (CBD) to support prognosis in a PHM system are presented in the later chapters.

1.1.2 Chapter Organization

The remainder of this chapter is organized to present and discuss a heuristic-based approach to modeling CBD signatures as follows:

1.2 Foundation of Reliability Theory
This section presents the foundation upon which reliability is built: failure distributions, probability and reliability, probability density functions and relationships, failure rate, and expected value and variance.
1.3 Failure Distributions Under Extreme Stress Levels
This section presents models for failure and extreme stress, cumulative damage, and exponential and Weibull failures and distribution.
1.4 Uncertainty Measures in Parameter Estimation
This section presents an introduction to measurement metrics dealing with uncertainty, such as likelihood, variance, and covariance.
1.5 Expected Number of Failures
This section presents the effects and interaction of repair, replacement, and partial repair activities on assessing the number of failures and decrease in effective age due to partial repairs.
1.6 System Reliability and Prognosis and Health Management
This section presents a framework for a PHM system based on condition-based maintenance (CBM) and introduces the reader to such a framework, CBD and signatures, and CBM, which can transform signatures into curves useful for processing to produce prognostic information.
1.7 Prognostic Information
This section introduces basic concepts and measures that are applicable to prognostic information: RUL, SoH, prognostic horizon (PH), prognostic distance (PD), and convergence.
1.8 Decisions on Cost and Benefits
This section presents a brief overview of decision problems and their mathematical modeling during different stages of the lifetime of equipment, with some illustrative mathematical models.
1.9 Introduction to PHM: Summary
This section summarizes the material presented in this chapter.

1.2 Foundation of Reliability Theory

Engineering systems and their components and elements are subject to degradation that sooner or later results in failure. A component that fails sooner compared to other components is said to be less reliable. Failures of like objects, such as components and assemblies, do not all occur at the same point in time; failures are distributed over time.

1.2.1 Time-to-Failure Distributions

The time to failure (TTF) of a particular object is usually modeled using probability theory, which represents the average behavior of a large set of identical objects and not the behavior of a particular object. It is considered as a random variable, denoted by X. The cumulative distribution function (CDF) of X is the probability that failure occurs at or before time t:

$$F(t) = P(X \leq t) \tag{1.1}$$

It is usually assumed that the object starts its operation at $t = 0$, so $F(t) = 0$ as $t \leq 0$. Furthermore, $F(t)$ tends to unity as $t \to \infty$: the function is non-decreasing. In reliability engineering, the most frequently used distribution types are as follows:

1. *Exponential* distribution:

$$F(t) = 1 - e^{-\lambda t} \qquad t \geq 0 \tag{1.2}$$

where the parameter $\lambda > 0$ determines the lifetime.
2. *Weibull* distribution:

$$F(t) = 1 - e^{-\left(\frac{t}{\eta}\right)^{\beta}} \qquad t \geq 0 \tag{1.3}$$

where $\eta > 0$ is referred to as the *scale* and $\beta > 0$ is referred to as the *shape*.
3. *Gamma* distribution:

$$F(t) = \int_0^t \frac{x^{k-1} e^{-\frac{x}{\theta}}}{\theta^k \Gamma(k)} dx \qquad t \geq 0 \tag{1.4}$$

where k represents a shape parameter and θ represents a scale parameter. Furthermore, $\Gamma(k)$ is the gamma function value at k. Very often, variable $\lambda = \frac{1}{\theta}$ is used in applications.
4. *Normal* distribution:

$$F(t) = \phi\left(\frac{t - \mu}{\sigma}\right) \qquad -\infty < t < \infty \tag{1.5}$$

where μ is a real parameter and $\sigma > 0$.
5. *Lognormal* distribution:

$$F(t) = \phi\left(\frac{\ln(t) - \mu}{\sigma}\right) \qquad t > 0 \tag{1.6}$$

where μ is a real value and $\sigma > 0$. Here $\phi(t)$ is the standard normal distribution function defined in Eq. (1.7):

$$\phi(t) = \int_{-\infty}^t \frac{1}{\sqrt{2\pi}} e^{-\frac{x^2}{2}} dx \tag{1.7}$$

This does not have an exact simple formula; only function approximations are available, or the value of $\phi(t)$ for any particular value of t can be obtained from the corresponding function table.
6. *Logistic* distribution:

$$F(t) = \frac{1}{1 + e^{-z}} = \frac{e^z}{1 + e^z} \quad \text{with } z = \frac{t - \mu}{\sigma} \qquad -\infty < t < \infty \tag{1.8}$$

where μ and $\sigma > 0$ are the model parameters.

7. *Gumbel* distribution:

$$F(t) = 1 - e^{-e^z} \text{ with } z = \frac{t - \mu}{\sigma} \qquad -\infty < t < \infty \qquad (1.9)$$

where the meanings of μ and σ are the same as before. This distribution is often called the *log-Weibull* distribution.

8. *Log-logistic* distribution:

$$F(t) = \frac{e^x}{1 + e^x} \text{ with } x = \frac{\ln(t) - \mu}{\sigma} \qquad t > 0 \qquad (1.10)$$

where Eq. (1.10) also has an alternative formulation:

$$F(t) = \frac{t^\beta}{\alpha^\beta + t^\beta} \qquad (1.11)$$

with $\alpha, \beta > 0$ and $\alpha = e^\mu$ and $\beta = \frac{1}{\sigma}$

9. *Log-Gumbel* distribution:

$$F(t) = 1 - e^{-e^x} \text{ with } x = \frac{\ln(t) - \mu}{\sigma} \qquad t > 0 \qquad (1.12)$$

10. *Log-gamma* distribution:

$$F(t) = \int_0^t \frac{\lambda(\lambda \ln(x))^{\alpha-1} e^{-\lambda \ln(x)}}{\Gamma(\alpha)x} dx \qquad t > 0 \qquad (1.13)$$

where λ and α are positive parameters.

The CDFs of exponential and Weibull distributions are illustrated in Figures 1.3 and 1.4.

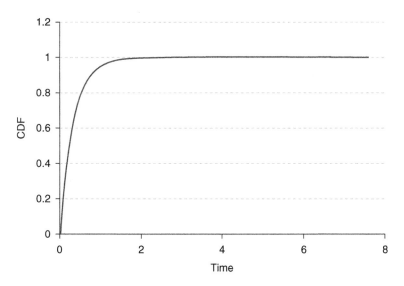

Figure 1.3 Graph of the exponential CDF with $\lambda = 3$.

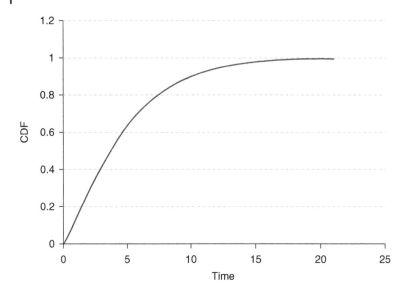

Figure 1.4 Graph of the Weibull CDF with $\beta = 1.2$ and $\eta = 5$.

1.2.2 Probability and Reliability

Most probability values can be determined by using the CDF of X. The probability that failure occurs at or before time t is given as $F(t)$, and the probability that the object is still working at time t is the complement of $F(t)$,

$$P(X > t) = 1 - F(t) = R(t) \tag{1.14}$$

which is called the *reliability function* of X. Sometimes it is also denoted by $\overline{F}(t)$. The probability that failure occurs during time interval $(t_1, t_2]$ is given as

$$P(t_1 < X \le t_2) = F(t_2) - F(t_1) \tag{1.15}$$

Example 1.1 Assume the random variable X is exponential with parameter $\lambda = 1$. Then

$$F(t) = 1 - e^{-t}$$

$$P(X > 2) = 1 - (1 - e^{-2}) = 0.1353$$

$$P(1 < X < 2) = (1 - e^{-2}) - (1 - e^{-1}) = e^{-1} - e^{-2} = 0.3679 - 0.1353 = 0.2326$$

Example 1.2 Assume next that random variable X is Weibull with $\beta = 1.5$ and $\eta = 5$. Then $F(t) = 1 - e^{-\left(\frac{t}{5}\right)^{1.5}}$, and therefore

$$P(X > 2) = 1 - \left(1 - e^{-\left(\frac{2}{5}\right)^{1.5}}\right) = e^{-0.2530} \approx 0.7765$$

or

$$P(1 < X < 2) = \left(1 - e^{-\left(\frac{2}{5}\right)^{1.5}}\right) - \left(1 - e^{-\left(\frac{1}{5}\right)^{1.5}}\right) = e^{-0.2^{1.5}} - e^{-0.4^{1.5}}$$

$$= 0.9144 - 0.7765 \approx 0.1379$$

1.2.3 Probability Density Function

The probability density function (PDF) of X is the derivative of $F(t)$:

$$f(t) = F'(t) \tag{1.16}$$

1. *Exponential* PDF:
   It is easy to see that in the exponential case

   $$f(t) = \lambda e^{-\lambda t} \tag{1.17}$$

2. *Weibull* PDF:

   $$f(t) = \frac{\beta}{\eta^\beta} t^{\beta-1} e^{-\left(\frac{t}{\eta}\right)^\beta} \tag{1.18}$$

3. *Gamma* PDF:
   The gamma PDF is the integrand of Eq. (1.4):

   $$f(t) = \frac{t^{k-1} e^{-\frac{t}{\theta}}}{\theta^k \Gamma(k)} \tag{1.19}$$

4. *Normal* PDF:
   The normal PDF has a closed form representation:

   $$f(t) = \frac{1}{\sigma\sqrt{2\pi}} e^{-\frac{(t-\mu)^2}{2\sigma^2}} \tag{1.20}$$

5. *Lognormal* PDF:

   $$f(t) = \frac{1}{t\sigma\sqrt{2\pi}} e^{-\frac{(\ln(t)-\mu)^2}{2\sigma^2}} \tag{1.21}$$

6. *Logistic* PDF:

   $$f(t) = \frac{e^z}{\sigma(1 + e^z)^2} \quad \text{with } z = \frac{t - \mu}{\sigma} \tag{1.22}$$

7. *Gumbel* PDF:

   $$f(t) = \frac{1}{\sigma} e^{z - e^z} \quad \text{with } z = \frac{t - \mu}{\sigma} \tag{1.23}$$

8. *Log-logistic* PDF:

   $$f(t) = \frac{e^x}{\sigma t(1 + e^x)^2} \quad \text{with } x = \frac{\ln(t) - \mu}{\sigma} \tag{1.24}$$

9. *Log-Gumbel* PDF:

   $$f(t) = e^{x - e^x} \cdot \frac{1}{\sigma t} \quad \text{with } x = \frac{\ln(t) - \mu}{\sigma} \tag{1.25}$$

10. *Log-gamma* PDF:

    $$f(t) = \frac{\lambda(\lambda(\ln(t))^{\alpha-1} e^{-\lambda \ln(t)}}{\Gamma(\alpha) t} \tag{1.26}$$

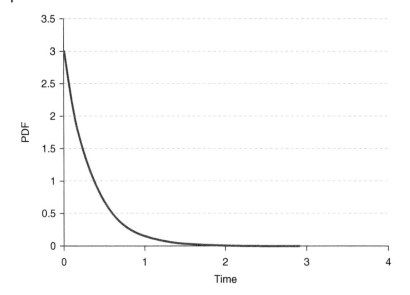

Figure 1.5 Graph of the exponential PDF with $\lambda = 3$.

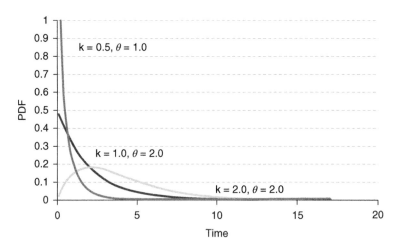

Figure 1.6 Graphs of gamma PDFs.

The PDFs of exponential, gamma, and Weibull distributions are plotted in Figures 1.5–1.7.

Similar to the CDF of any random variable, most probability values can be obtained by using the corresponding PDF:

$$F(t) = P(X \le t) = \int_0^t f(x)dx \tag{1.27}$$

$$R(t) = P(X \ge t) = \int_t^\infty f(x)dx \tag{1.28}$$

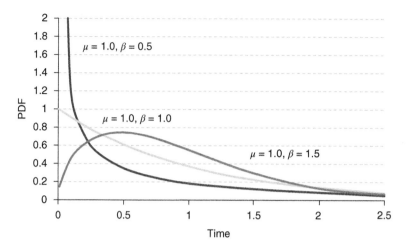

Figure 1.7 Graphs of Weibull PDFs.

and

$$P(t_1 < X \le t_2) = \int_{t_1}^{t_2} f(x)dx \qquad (1.29)$$

Example 1.3 Assume now that random variable X is exponential with $\lambda = 1$. Then

$$f(t) = 1 \cdot e^{-1 \cdot t} = e^{-t}$$

so

$$P(X > 2) = \int_2^\infty e^{-t}dt = [-e^{-t}]_2^\infty = 0 + e^{-2} = 0.1353$$

and

$$P(1 < X < 2) = \int_1^2 e^{-t}dt = [-e^{-t}]_1^2 = -e^{-2} + e^{-1} = 0.2326$$

as in Example 1.1.

Example 1.4 Assume next that X is Weibull with parameter values $\beta = 1.5$ and $\eta = 5$. Then

$$f(t) = \frac{1.5}{5^{1.5}}t^{0.5}e^{-\left(\frac{t}{5}\right)^{1.5}}$$

therefore

$$P(X > 2) = \int_2^\infty \frac{1.5}{5^{1.5}}t^{0.5}e^{-\left(\frac{t}{5}\right)^{1.5}}dt = \left[-e^{-\left(\frac{t}{5}\right)^{1.5}}\right]_2^\infty = 0 + e^{-0.4^{1.5}} = 0.7765$$

and

$$P(1 < X < 2) = \int_1^2 \frac{1.5}{5^{1.5}}t^{0.5}e^{-\left(\frac{t}{5}\right)^{1.5}}dt = \left[-e^{-\left(\frac{t}{5}\right)^{1.5}}\right]_1^2 = e^{-0.4^{1.5}} + e^{-0.2^{1.5}} = 0.1379$$

We get the same answers as in Example 1.2.

1.2.4 Relationships of Distributions

There is a strong relation between normal and lognormal, logistic and log-logistic, gamma and log-gamma, and Gumbel and log-Gumbel distributions. If X is normal, then e^X is lognormal; if X is logistic, then e^X is log-logistic; if X is Gumbel, then e^X is log-Gumbel and if X is gamma than e^X is log-gamma.

The domain of exponential, Weibull, gamma, and lognormal distribution functions is the $[0, \infty\}$ interval: that is, they are defined for positive arguments. However, in the normal case, t can be any real value. If the value of X can only be positive and we still want to use a normal distribution, then Eq. (1.20) must be modified:

$$f_+(t) = \frac{1}{A\sigma\sqrt{2\pi}} e^{-\frac{(t-\mu)^2}{2\sigma^2}} \tag{1.30}$$

where

$$A = \int_0^\infty f(t)dt = 1 - F(0) = 1 - \phi\left(-\frac{\mu}{\sigma}\right) = \phi\left(\frac{\mu}{\sigma}\right) \tag{1.31}$$

And, therefore, the corresponding CDF is the following:

$$F_+(t) = \int_0^t f_+(x)dx = \frac{1}{A}(F(t) - F(0)) = \frac{\phi\left(\frac{t-\mu}{\sigma}\right) - \phi\left(\frac{-\mu}{\sigma}\right)}{\phi\left(\frac{\mu}{\sigma}\right)} \tag{1.32}$$

An exponential variable is very seldom used in reliability analysis since it doesn't model degradation or the aging of an object. In order to show this disadvantage, consider an object of age T, and find the probability that it will work for an additional t units of time. It is a conditional probability:

$$P(X > T + t | X > T) = \frac{P(X > T + t)}{P(X > T)} = \frac{R(T + t)}{R(T)} = \frac{e^{-\lambda(T+t)}}{e^{-\lambda T}} = e^{-\lambda t} = R(t)$$

showing that this probability does not depend on the age T of the object. That is, the survival probability remains the same regardless of the age of the object.

Other important relations between different distribution types can be summarized as follows. If $\beta = 1$ in a Weibull variable, then it becomes exponential with $\lambda = \frac{1}{\eta}$. Assume next that $X_1, X_2, ..., X_k$ are independent exponential variables with an identical parameter λ; then their sum $X_1 + X_2 + ... + X_k$ is a gamma variable with parameters k and $\theta = \frac{1}{\lambda}$. Similarly, if X is exponential with parameter λ, then X^α is Weibull with parameters $\beta = \frac{1}{\alpha}$ and $\eta = \frac{1}{\lambda^\alpha}$. Normal variables are characterized based on the *central limit theorem*, which asserts the following: Let $X_1, X_2, ..., X_N, ...$ be a sequence of independent, identically distributed random variables; then their partial sum $\sum_{k=1}^N X_k$ converges to a normal distribution as $N \to \infty$.

1.2.5 Failure Rate

The *failure rate* (or hazard rate) is defined mathematically as follows:

$$\rho(t) = \lim_{\Delta t \to 0} \frac{P(t < X < t + \Delta t | t < X)}{\Delta t} \tag{1.33}$$

The numerator is the probability that the object will break down in the next Δt time periods given that it is working at time t. If $\Delta t = 1$, then the fraction gives the probability that the object will break down during the next unit time interval. By simple calculation,

$$\frac{P(t < X < t + \Delta t | t < X)}{\Delta t} = \frac{P(t < X < t + \Delta t)}{\Delta t P(t < X)} = \frac{F(t + \Delta t) - F(t)}{\Delta t P(X > t)}$$

which converges to the following:

$$\rho(t) = \frac{f(t)}{R(t)} \tag{1.34}$$

That is, $\rho(t)$ can be easily computed based on the PDF and reliability function of X. TTF distributions are characterized by their PDFs ($f(t)$), CDFs ($F(t)$), reliability functions ($R(t)$), and failure rates ($\rho(t)$). If any of these functions is known, then the other three can be easily determined as shown next.

If $f(t)$ is known, then $F(t) = \int_0^t f(t)dt$, $R(t) = 1 - F(t)$, and $\rho(t) = \frac{f(t)}{R(t)}$.

If $F(t)$ is known, then $f(t) = F'(t)$, $R(t) = 1 - F(t)$, and $\rho(t) = \frac{f(t)}{R(t)}$.

Assume next that $R(t)$ is known. Then $F(t) = 1 - R(t)$, $f(t) = F'(t)$, and $\rho(t) = \frac{f(t)}{R(t)}$.

And finally, assume that $\rho(t)$ is given. From its definition

$$\rho(t) = \frac{f(t)}{R(t)} = \frac{f(t)}{1 - F(t)} = -\frac{-f(t)}{1 - F(t)}$$

and by integration

$$\int_0^t \rho(x)dx = -\int_0^t \frac{-f(x)}{1 - F(x)}dx = [-\ln(1 - F(x))]_{x=0}^t = -\ln(1 - F(t)) + \ln(1)$$

so

$$\ln(1 - F(t)) = -\int_0^t \rho(x)dx$$

implying that

$$F(t) = 1 - e^{-\int_0^t \rho(x)dx} \tag{1.35}$$

$$f(t) = F'(t)$$

$$R(t) = 1 - F(t) = e^{-\int_0^t \rho(x)dx}$$

Exponential Failure

In the exponential case,

$$\rho(t) = \frac{\lambda e^{-\lambda t}}{e^{-\lambda t}} = \lambda \tag{1.36}$$

which is a constant and does not depend on the age of the object. This relation also shows that the exponential variable does not include degradation.

Weibull Failure

In the Weibull case,

$$\rho(t) = \frac{\beta}{\eta^\beta} t^{\beta - 1} \tag{1.37}$$

which has different shapes depending on the value of β, as shown in Figure 1.8.

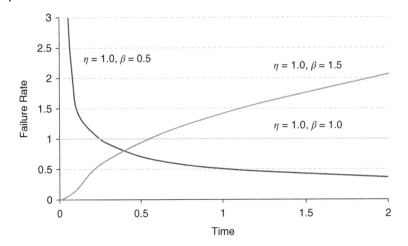

Figure 1.8 Failure rates of Weibull variables.

If $\beta < 1$, then the exponent of t is negative, so $\rho(t)$ strictly decreases from the infinite limit at $t = 0$ to the zero limit at $t = \infty$. If $\beta = 1$, then $\rho(t)$ is constant; and if $1 < \beta < 2$, then the exponent of t is positive and below unity, so $\rho(t)$ strictly increases and is concave in t. In the case of $\beta = 2$, the exponent of t equals 1, so $\rho(t)$ is a linear function; and finally, as $\beta > 2$, the exponent of t is greater than 1, so $\rho(t)$ strictly increases and is convex.

Gamma Failure
If X is gamma, then:

$$\rho(t) = \dfrac{\dfrac{t^{k-1}e^{-\frac{t}{\theta}}}{\theta^k \Gamma(k)}}{\int_t^\infty \dfrac{x^{k-1}e^{-\frac{x}{\theta}}}{\theta^k \Gamma(k)}dx} = \dfrac{1}{\int_t^\infty \left(\dfrac{x}{t}\right)^{k-1} e^{-\frac{x-t}{\theta}}dx}$$

And by introducing the new integration variable $u = x - t$, we have:

$$\rho(t) = \dfrac{1}{\int_0^\infty \left(\dfrac{u}{t}+1\right)^{k-1} e^{-\frac{u}{\theta}}du} \tag{1.38}$$

If $k = 1$, then this is a constant since

$$\int_0^\infty e^{-\frac{u}{\theta}}du = \left[-\theta e^{-\frac{u}{\theta}}\right]_0^\infty = \theta$$

and so $\rho(t) = \frac{1}{\theta}$. If <1, then $\rho(t)$ decreases; and if $k > 1$, then $\rho(t)$ increases in t. Figure 1.9 shows the possible shapes of $\rho(t)$, which always converge to $\frac{1}{\theta}$ as $t \to \infty$.

Standard-Normal Failure
Assume next that X is standard normal:

$$\rho_n(t) = \dfrac{\dfrac{1}{\sqrt{2\pi}}e^{-\frac{t^2}{2}}}{\int_t^\infty \dfrac{1}{\sqrt{2\pi}}e^{-\frac{\tau^2}{2}}d\tau} = \dfrac{1}{\int_0^\infty e^{-\frac{x^2+2xt}{2}}dx} \tag{1.39}$$

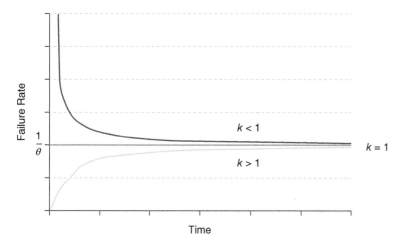

Figure 1.9 Failure rates of gamma variables.

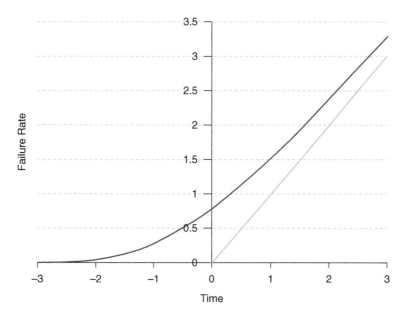

Figure 1.10 Failure rate of the standard normal variable.

It is easy to show that $\rho_n(t)$ strictly increases and is convex in t, $\rho_n(0) = \sqrt{\frac{2}{\pi}}$, $\lim_{t \to -\infty} \rho_n(t) = 0$, and $\lim_{t \to \infty} \rho_n(t) = \infty$. Furthermore, the $45°$ line is its asymptote as $t \to \infty$. The shape of $\rho_n(t)$ is shown in Figure 1.10.

Normal Failure

If X is a normal variable with parameters μ and σ, then $\rho(t) = \frac{1}{\sigma}\rho_n\left(\frac{t-\mu}{\sigma}\right)$, $\lim_{t \to -\infty} \rho(t) = 0$, and $\lim_{t \to \infty} \rho(t) = \infty$. Furthermore, its asymptote is $\frac{t-\mu}{\sigma^2}$ as $t \to \infty$.

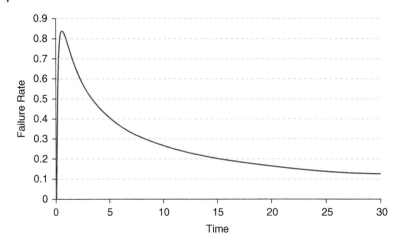

Figure 1.11 Failure rate of the lognormal variable with $\mu = 0$ and $\sigma = 1$.

Lognormal Failure

The failure rate of a lognormal variable X is shown in Figure 1.11. It is a mound-shaped curve with a maximum at t^*, which can be determined as follows. There is a unique value $\bar{t}$ such that $\rho_n(\bar{t}) = \bar{t} + \sigma$, and then $t^* = \exp(\mu + \sigma\bar{t})$. Clearly

$$\lim_{t \to 0+} \rho(t) = \lim_{t \to \infty} \rho(t) = 0$$

Logistic Failure

The failure rate of the logistic variable is

$$\rho(t) = \frac{1}{\sigma(1 + e^{-z})} \quad \text{with} \quad z = \frac{t - \mu}{\sigma} \qquad (1.40)$$

Gumbel Failure

In the case of the Gumbel distribution,

$$\rho(t) = \frac{1}{\sigma} e^z \quad \text{with} \quad z = \frac{t - \mu}{\sigma} \qquad (1.41)$$

which is a simple exponential function.

Log-Logistic Failure

Similarly, in the case of a log-logistic variable:

$$\rho(t) = \frac{e^x}{\sigma t(1 + e^x)} \quad \text{with} \quad x = \frac{\ln(t) - \mu}{\sigma} \qquad (1.42)$$

Log-Gumbel Failure

For the log-Gumbel distribution,

$$\rho(t) = \frac{e^x}{\sigma t} = \frac{1}{\sigma} e^{x(1-\sigma)-\mu} \quad \text{with} \quad x = \frac{\ln(t) - \mu}{\sigma} \qquad (1.43)$$

as before.

Log-Gamma Failure

The failure rate of the log-gamma variable is

$$\rho(t) = \frac{y^{a-1}e^{-(\lambda+1)y}}{\int_y^\infty u^{a-1}e^{-\lambda u}du} \quad \text{with} \quad y = \ln(t) \tag{1.44}$$

Figures 1.12–1.14 show the graphs of some of these failure rates.

Example 1.5 Assume that the failure rate of a variable is $\rho(t) = t^2$. Then, from Eq. (1.35),

$$F(t) = 1 - e^{-\int_0^t \tau^2 d\tau} = 1 - e^{-\frac{t^3}{3}}$$

showing that this is a Weibull variable with $\beta = 3$ and $\eta = 3^{\frac{1}{3}}$.

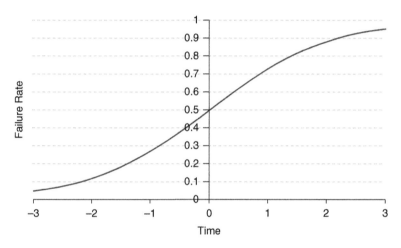

Figure 1.12 Logistic failure rate with $\mu = 0$ and $\sigma = 1$.

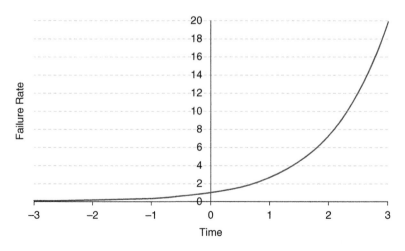

Figure 1.13 Gumbel failure rate.

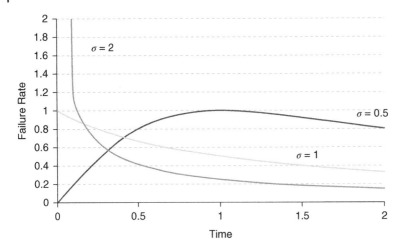

Figure 1.14 Log-logistic failure rate with $\mu = 0$.

Example 1.6 Assume next that for a variable

$$F(t) = 1 - e^{-t^2}$$

Then its PDF is

$$f(t) = 2te^{-t^2}$$

and since the reliability function is

$$R(t) = e^{-t^2}$$

the failure rate becomes

$$\rho(t) = \frac{2te^{-t^2}}{e^{-t^2}} = 2t$$

1.2.6 Expected Value and Variance

The *expected value* and *variance* of any continuous random variable X can be obtained as follows:

$$\mu = E(X) = \int_{-\infty}^{+\infty} xf(x)dx \tag{1.45}$$

where $f(x)$ is the PDF of X:

$$\sigma^2 = \text{Var}(X) = \int_{-\infty}^{+\infty} (x - \mu)^2 f(x)dx = E(X^2) - E^2(X) \tag{1.46}$$

The expectations and variances of the most frequently used random variables are summarized in Table 1.1, where $\gamma \approx 0.57721566$ is the Euler constant.

Example 1.7 Assume that the CDF of a variable X is given as:

$$F(t) = 1 - e^{-2t} \qquad\qquad (0 < t < \infty)$$

Then

$$f(t) = 2e^{-2t}$$

Table 1.1 Expectations and variances.

Type of distributions	Parameters	Domain	$E(X)$	$Var(X)$
Exponential	λ	$[0, \infty)$	$1/\lambda$	$1/\lambda^2$
Weibull	β, η	$[0, \infty)$	$\eta \Gamma\left(1 + \frac{1}{\beta}\right)$	$\eta^2 \left[\Gamma\left(1 + \frac{2}{\beta}\right) - \Gamma^2\left(1 + \frac{1}{\beta}\right)\right]$
Gamma	k, θ	$[0, \infty)$	$k\theta$	$k\theta^2$
Normal	μ, σ^2	$(-\infty, \infty)$	μ	σ^2
Lognormal	μ, σ^2	$[0, \infty)$	$\exp\left(\mu + \frac{\sigma^2}{2}\right)$	$(e^{\sigma^2} - 1)\exp(2\mu + \sigma^2)$
Logistic	μ, σ	$(-\infty, \infty)$	μ	$\sigma^2 \pi^2 / 3$
Gumbel	μ, σ	$(-\infty, \infty)$	$\mu - \sigma\gamma$	$\pi^2 \sigma$
Log-logistic	μ, σ	$(0, \infty)$	$e^\mu \Gamma(1 + \sigma)\Gamma(1 - \sigma)$	$e^{2\mu}[\Gamma(1 + 2\sigma)\Gamma(1 - 2\sigma)$ $- (\Gamma(1 + \sigma)\Gamma(1 - \sigma))^2]$

The expected value is given by the integral

$$E(X) = \int_0^\infty 2te^{-2t}\,dt = \left[2t\frac{e^{-2t}}{-2}\right]_0^\infty - \int_0^\infty 2\frac{e^{-2t}}{-2}\,dt$$

where integration by parts is used with $u = 2t$, $v' = e^{-2t}$.
Therefore,

$$E(X) = \int_0^\infty e^{-2t}\,dt = \left[\frac{e^{-2t}}{-2}\right]_0^\infty = 0 + \frac{1}{2} = \frac{1}{2}$$

The second moment of X is the following:

$$E(X^2) = \int_0^\infty 2t^2 e^{-2t}\,dt = \left[2t^2\frac{e^{-2t}}{-2}\right]_0^\infty - \int_0^\infty 4t\frac{e^{-2t}}{-2}\,dt$$

where integration by parts is used again with $u = 2t^2$ and $v = e^{-2t}$. From $E(X)$,

$$E(X^2) = \int_0^\infty 2te^{-2t}\,dt = \frac{1}{2}$$

and therefore

$$Var(X) = \frac{1}{2} - \left(\frac{1}{2}\right)^2 = \frac{1}{4}$$

Notice that X is exponential with $\lambda = 2$, and the results we obtained by integration match the formulas of the first line in Table 1.1.

The material of this section can be found in most introductory books on probability and statistics, as well as those on reliability engineering. See, for example, Ayyub and McCuen (2003), Ross (1987, 2000), Milton and Arnold (2003), Elsayed (2012), Finkelstein (2008), or Nakagawa (2008).

1.3 Failure Distributions Under Extreme Stress Levels

Traditional life data analysis assumes normal stress levels and other normal conditions. During operations, extreme stress levels may occur that significantly alter life characteristics. In this section, some of the most frequently used models are outlined. They play significant roles in reliability engineering, especially in designing accelerated testing.

1.3.1 Basic Models

We first discuss some of the most popular models.

Arrhenius Model

The *Arrhenius life stress model* (Arrhenius 1889) is based on the relationship

$$L(V) = Ce^{\frac{B}{V}} \tag{1.47}$$

where L is a quantifiable life measure such as median life, mean life, and so on; V represents the stress level, formulated in temperature given in absolute units as Kelvin degrees; and B *and* C are model parameters determined by using sample elements.

This model is used mainly if the stress is thermal (i.e. temperature). We will next show how this rule is applied in cases of exponential and Weibull TTF distributions. Other distribution types can be treated similarly. Notice that $L(V)$ is decreasing in V, meaning higher stress level results in shorter useful life.

The exponential PDF is known to be

$$f(t) = \frac{1}{m}e^{-\frac{t}{m}}$$

where m is referred to as the *expected TTF*. Combining this form with Eq. (1.47) leads to a modified form of the PDF, which depends on time as well as on the stress level:

$$f(t) = \frac{1}{Ce^{\frac{B}{V}}}e^{-\frac{t}{Ce^{\frac{B}{V}}}} \tag{1.48}$$

It can be shown that the expectation is $Ce^{\frac{B}{V}}$, the median is $0.693Ce^{\frac{B}{V}}$, the mode is zero, and the standard deviation is $Ce^{\frac{B}{V}}$. The Arrhenius-exponential reliability function is $e^{-\frac{t}{Ce^{\frac{B}{V}}}}$.

The Weibull PDF has the following form:

$$f(t) = \frac{\beta}{\eta}\left(\frac{t}{\eta}\right)^{\beta-1}e^{-\left(\frac{t}{\eta}\right)^{\beta}}$$

where the scale parameter is η. And the Arrhenius-Weibull PDF can be derived by setting $\eta = L(V)$, which gives the form

$$f(t) = \frac{\beta}{Ce^{\frac{B}{V}}}\left(\frac{t}{Ce^{\frac{B}{V}}}\right)^{\beta-1}e^{-\left(\frac{t}{Ce^{\frac{B}{V}}}\right)^{\beta}} \tag{1.49}$$

The expected value is $Ce^{\frac{B}{V}}\Gamma\left(\frac{1}{\beta}+1\right)$, the median is $Ce^{\frac{B}{V}}(\ln(2))^{\frac{1}{\beta}}$, the mode is $Ce^{\frac{B}{V}}\left(1-\frac{1}{\beta}\right)^{\frac{1}{\beta}}$, and the standard deviation is $Ce^{\frac{B}{V}}\sqrt{\Gamma\left(\frac{2}{\beta}+1\right)-\Gamma\left(\frac{1}{\beta}+1\right)^2}$.

Eyring Model

The *Eyring formula* (Eyring 1935) is used mainly for thermal stress, but it is also often used for other types of stresses. It is given as

$$L(V) = \frac{1}{V}e^{-\left(A-\frac{B}{V}\right)} \tag{1.50}$$

where L, V, A and B have the same meanings as before.

In comparing the Eyring formula with the Arrhenius model, notice the appearance of a hyperbolic factor $\frac{1}{V}$ in the Eyring formula, which decreases faster in V than the Arrhenius rule.

The Eyring-exponential PDF can be written as follows:

$$f(t) = Ve^{\left(A-\frac{B}{V}\right)}e^{-Vte^{A-\frac{B}{V}}}$$

since we chose $m = L(V)$ as before. The mean is $\frac{1}{V}e^{-\left(A-\frac{B}{V}\right)}$, the median is $0.693\,\frac{1}{V}e^{-\left(A-\frac{B}{V}\right)}$, the mode is zero, and the standard deviation is $\frac{1}{V}e^{-\left(A-\frac{B}{V}\right)}$.

The Eyring-Weibull PDF has a much more complicated form:

$$f(t) = \beta Ve^{\left(A-\frac{B}{V}\right)}\left(tVe^{\left(A-\frac{B}{V}\right)}\right)^{\beta-1}e^{-\left(tVe^{\left(A-\frac{B}{V}\right)}\right)^{\beta}} \tag{1.51}$$

where we again selected $\eta = L(V)$. It can be shown that the mean is $\frac{1}{V}e^{-\left(A-\frac{B}{V}\right)}\Gamma\left(\frac{1}{\beta}+1\right)$, the median is $\frac{1}{V}e^{-\left(A-\frac{B}{V}\right)}(\ln(2))^{\frac{1}{\beta}}$, and the mode is $\frac{1}{V}e^{-\left(A-\frac{B}{V}\right)}\left(1-\frac{1}{\beta}\right)^{\frac{1}{\beta}}$, with standard deviation

$$\frac{1}{V}e^{-\left(A-\frac{B}{V}\right)}\sqrt{\Gamma\left(\frac{2}{\beta}+1\right)-\Gamma\left(\frac{1}{\beta}+1\right)^{2}}$$

Generalized Eyring Model

The *generalized Eyring relationship* (Nelson 1980) is given by the more complex formula

$$L(V,U) = \frac{1}{V}e^{A+\frac{B}{V}+CU+D\frac{U}{V}} \tag{1.52}$$

where V is the thermal stress (in absolute units); U is the non-thermal stress (i.e. voltage, vibration, and so on); and A, B, C, *and* D are model parameters.

In the case of a negative A value and $C = D = 0$, model Eq. (1.52) reduces to Eq. (1.50). Similar to previous cases, the modified generalized Eyring-exponential PDF can be written as

$$f(t) = \frac{1}{L(V,U)}e^{-\frac{t}{L(V,U)}} \tag{1.53}$$

where $L(V,U)$ is given in Eq. (1.52). The mean is $L(V,U)$, the median is $L(U,V)\cdot0.693$, the mode is zero, and the standard deviation is again $L(V,U)$.

The generalized Eyring-Weibull PDF can be written as

$$f(t) = \frac{\beta}{L(V,U)}\left(\frac{t}{L(V,U)}\right)^{\beta-1}e^{-\left(\frac{t}{L(V,U)}\right)^{\beta}} \tag{1.54}$$

with mean $L(V, U)\Gamma\left(\frac{1}{\beta} + 1\right)$, median $L(V, U)(\ln(2))^{\frac{1}{\beta}}$, mode $L(V, U)\left(1 - \frac{1}{\beta}\right)^{\frac{1}{\beta}}$, and

standard deviation $L(V, U)\sqrt{\Gamma\left(\frac{2}{\beta} + 1\right) - \Gamma\left(\frac{1}{\beta} + 1\right)^2}$.

Inverse Power Law

The *inverse power law* (Kececioglu and Jacks 1984) is commonly used for non-thermal stresses:

$$L(V) = \frac{1}{KV^n} \tag{1.55}$$

where V represents the stress level, K and n are model parameters to be determined based on sample elements, and L is the quantifiable life measure. The power-exponential and power-Weibull probability density functions are given by Eqs. (1.53) and (1.54), where $L(V, U)$ is replaced by Eq. (1.55).

Temperature-Humidity Model

The most frequently used *temperature-humidity relationship* is given as

$$L(V, U) = Ae^{\frac{\alpha}{V} + \frac{\beta}{U}} \tag{1.56}$$

The temperature-humidity exponential and Weibull density functions are given by Eqs. (1.53) and (1.54), respectively.

Temperature-Only Model

The *temperature-only relationship* is based on the equation

$$L(V, U) = \frac{C}{U^n e^{-\frac{\beta}{V}}} \tag{1.57}$$

which is a combination of the Arrhenius and inverse power laws. The corresponding exponential and Weibull PDFs are obtained again from Eqs. (1.53) and (1.54). Here, U is the non-thermal stress (e.g. voltage, vibration, etc.), V is the temperature (in absolute units), and n *and* β are model parameters.

General Log-Linear Model

In the case of multiple stresses, the function $L(V)$ is usually substituted with a *general log-linear function*:

$$L(V) = e^{a_0 + \sum_{j=1}^{n} a_j V_j} \tag{1.58}$$

where V_1, V_2, ..., V_n are the different stress components, $V = (V_1, V_2, ..., V_n)$ is an n-element vector, and a_0, a_1, ..., a_n are model variables. The corresponding PDFs are obtained as before.

Time-Varying Stress Model

Time-varying stress models use different $L(V)$ functions in the different sections of a time interval:

$$L(V) = L_i(V) \text{ if } t_i < t \leq t_{i+1} \quad (i = 0, 1, ..., N-1) \tag{1.59}$$

where $t_0 = 0 < t_1 < t_2 < ... < t_N = T$, with T being the endpoint of the time span being considered.

Therefore, the corresponding PDFs are time-varying: in each subinterval (t_i, t_{i+1}), the corresponding $L_i(V)$ function represents the life measure under consideration.

1.3.2   Cumulative Damage Models

Finally, we briefly discuss the *cumulative damage relationship*, where first the inverse power law is assumed. Let $V(t)$ denote the time-varying stress; then the life-stress relationship is

$$L(V(t)) = \left(\frac{A}{V(t)}\right)^n \tag{1.60}$$

where $A = K^{-\frac{1}{n}}$ in comparison to Eq. (1.55).

Exponential Failure Distribution
If the TTF distribution is exponential, then $m(t, V) = L(V(t))$. Defining

$$I(t, V) = \int_0^t \left(\frac{V(\tau)}{A}\right)^n d\tau \tag{1.61}$$

as the cumulative damage, the reliability function becomes $e^{-I(t, V)}$ and therefore the PDF has the form

$$f(t, V) = \left(\frac{V(t)}{A}\right)^n e^{-I(t,V)} \tag{1.62}$$

Weibull Failure Distribution
If the TTF distribution is Weibull, then the reliability function is $e^{-(I(t,V))^\beta}$, where $I(t, V)$ is given by Eq. (1.61), and the corresponding PDF becomes

$$f(t, V) = \beta \left(\frac{V(t)}{A}\right)^n (I(t, V))^{\beta-1} e^{-(I(t,V))^\beta} \tag{1.63}$$

Exponential Failure Distribution and Arrhenius Life Stress
Assume next that the Arrhenius life stress model Eq. (1.47) is used. If the TTF is exponentially distributed, then similar to Eq. (1.61), we have

$$I(t, V) = \int_0^t \frac{1}{C} e^{-\frac{B}{V(\tau)}} d\tau \tag{1.64}$$

and the associated PDF becomes

$$f(t, V) = \left(\frac{1}{C} e^{-\frac{B}{V(t)}}\right) e^{-I(t,V)} \tag{1.65}$$

Weibull Failure Distribution and Arrhenius Life Stress
If the TTF distribution is Weibull, then analogous to earlier cases $\eta(t, V) = L(t, V)$, and therefore similar to Eq. (1.62) we get the associated PDF:

$$f(t, V) = \beta \left(\frac{1}{C} e^{-\frac{B}{V(t)}}\right) (I(t, V))^{\beta-1} e^{-(I(t,V))^\beta} \tag{1.66}$$

1.3.3   General Exponential Models

Next, we show the case of a general exponential relationship,

$$L(V(t)) = C e^{BV(t)} \tag{1.67}$$

in which case

$$I(t, V) = \int_0^t \frac{1}{C} e^{-BV(\tau)} d\tau \tag{1.68}$$

Failure Distributions

With exponential TTF distributions, the PDF becomes

$$f(t, V) = \left(\frac{1}{C}e^{-BV(t)}\right)e^{-I(t,V)} \tag{1.69}$$

And in the Weibull case,

$$f(t, V) = \beta\left(\frac{1}{C}e^{-BV(t)}\right)(I(t, V))^{\beta-1}e^{-(I(t,V))^{\beta}} \tag{1.70}$$

If the general log-linear function Eq. (1.58) is assumed, then

$$I(t, V) = \int_0^t e^{-\left(a_0 + \sum_{j=1}^n a_j V_j(\tau)\right)} d\tau \tag{1.71}$$

And in the exponential case, the corresponding PDF has the form

$$f(t, V) = e^{-\left(a_0 + \sum_{j=1}^n a_j V_j(t)\right)}e^{-I(t,V)} \tag{1.72}$$

It is modified in the Weibull case as follows:

$$f(t, V) = \beta\left(e^{-\left(a_0 + \sum_{j=1}^n a_j V_j(t)\right)}\right)(I(t, V))^{\beta-1}e^{-(I(t,V))^{\beta}} \tag{1.73}$$

Reliability Function

The reliability function is

$$R(t, V) = e^{-I(t,V)} \text{ or } R(t, V) = e^{-(I(t,V))^{\beta}} \tag{1.74}$$

in the exponential or Weibull case. The CDF is clearly

$$F(t, V) = 1 - R(t, V)$$

The mean, median, mode, and variance can be obtained from the corresponding PDF or CDF by using standard methods known from probability theory.

Example 1.8 As mentioned earlier, the model parameters can be obtained based on measured sample elements. As an example, consider model Eq. (1.48), and assume that $t_1, t_2, \ldots, t_N$ are the measured times of failures. Then the likelihood function has the form

$$L(C, B) = \prod_{k=1}^N \frac{1}{Ce^{\frac{B}{V}}}e^{-\frac{t_k}{Ce^{B/V}}} = \frac{1}{\left(Ce^{\frac{B}{V}}\right)^N}e^{-\frac{1}{Ce^{\frac{B}{V}}}\sum_{k=1}^N t_k}$$

with logarithm

$$l(C, B) = -N\left(\ln(C) + \frac{B}{V}\right) - \frac{1}{Ce^{\frac{B}{V}}}\sum_{k=1}^N t_k$$

which is then maximized with respect to model parameters B and C, when stress level V is considered a given constant. If the value of V changes form from one sample to the next, then the likelihood function becomes more complex:

$$\prod_{k=1}^N \frac{1}{Ce^{B/V_k}}e^{-\frac{t_k}{Ce^{B/V_k}}} = \frac{1}{C^N e^{B\sum_{k=1}^N \frac{1}{V_k}}}e^{-\frac{1}{C}\sum_{k=1}^N \frac{t_k}{e^{B/V_k}}}$$

with logarithm

$$l(C, B) = -N \ln(C) - B \sum_{k=1}^{N} \frac{1}{V_k} - \frac{1}{C} \sum_{k=1}^{N} \frac{t_k}{e^{B/V_k}}$$

which is then maximized.

Several books and articles have been written about accelerated testing, including Nelson (2004).

1.4 Uncertainty Measures in Parameter Estimation

Consider first a PDF that depends on only one parameter, θ, which is determined by using a finite sample. Let $f(t|\theta)$ denote the PDF, and let $t_1, t_2, \ldots, t_N$ be the sample elements. The likelihood function is

$$L(\theta) = \prod_{k=1}^{N} f(t_k|\theta)$$

and its logarithm is written as

$$l(\theta) = \sum_{k=1}^{N} \ln(f(t_k|\theta)) \tag{1.75}$$

which is then maximized to get the best estimate for the unknown parameter θ. By sampling repeatedly, we obtain different sample elements and therefore different estimates for θ. The value of θ depends on the random selection of the sample elements, so it is also a random variable. Its uncertainty can be characterized by its variance. Let the maximizer of Eq. (1.75) be denoted by θ^*.

The Fisher information is defined as the negative of the second derivative of $l(\theta)$ with respect to θ at $\theta = \theta^*$:

$$I = -\frac{\partial^2 l(\theta)}{\partial \theta^2}\Big|_{\theta=\theta^*} \tag{1.76}$$

And it can be proved that the variance of θ is the reciprocal of I:

$$\mathrm{Var}(\theta) = \frac{1}{I} \tag{1.77}$$

(Frieden 2004; Pratt et al. 1976).

Example 1.9 Assume X is normal with a known variance σ^2 but an unknown expectation μ. In this case,

$$L(\mu) = \prod_{k=1}^{N} \frac{1}{\sigma\sqrt{2\pi}} e^{\frac{-(t_k-\mu)^2}{2\sigma^2}} = \frac{1}{(\sigma\sqrt{2\pi})^N} e^{-\frac{1}{2\sigma^2}\sum_{k=1}^{N}(t_k-\mu)^2}$$

and so

$$l(\mu) = -N \ln(\sigma\sqrt{2\pi}) - \frac{1}{2\sigma^2} \sum_{k=1}^{N} (t_k - \mu)^2$$

which must be maximized. Notice first that as $\mu \to -\infty$ or $\mu \to +\infty$, since $l(\mu)$ tends to $-\infty$, the maximum is interior. By differentiation:

$$\frac{\partial l(\mu)}{\partial \mu} = \frac{1}{\sigma^2} \sum_{k=1}^{N} (t_k - \mu) = \frac{1}{\sigma^2} \left(\sum_{k=1}^{N} t_k - N\mu \right) = 0$$

implying that the maximum likelihood estimate is $\mu^* = \frac{1}{N} \sum_{k=1}^{N} t_k = \bar{t}$, which is the sample mean.

It gives the expectation of parameter μ. In order to obtain the variance of μ, we compute the Fisher information first:

$$I = -\frac{\partial^2 l(\mu)}{\partial \mu^2} = \frac{N}{\sigma^2}$$

showing that the variance of μ is $\frac{\sigma^2}{N}$, so μ can be approximated with large N values with a normal distribution with mean μ^* and standard deviation $\frac{\sigma}{\sqrt{N}}$.

Example 1.10 Assume next that X is exponential with an unknown parameter λ. If t_1, $t_2, \dots, t_N$ are the sample elements, then

$$L(\lambda) = \prod_{k=1}^{N} \lambda e^{-\lambda t_k} = \lambda^N e^{-\lambda \sum_{k=1}^{N} t_k}$$

with logarithm

$$l(\lambda) = N \ln(\lambda) - \lambda \sum_{k=1}^{N} t_k$$

As $\lambda \to 0$ or $\lambda \to \infty$, the value of $l(\lambda)$ tends to negative infinity, so the optimum is local. Since

$$\frac{\partial l(\lambda)}{\partial \lambda} = \frac{N}{\lambda} - \sum_{k=1}^{N} t_k$$

the likelihood estimate is

$$\lambda^* = \frac{N}{\sum_{k=1}^{N} t_k} = \frac{1}{\bar{t}}$$

The Fisher information is clearly

$$I = -\frac{\partial^2 l(\lambda)}{\partial \lambda^2} = \frac{N}{\lambda^2}$$

and the variance of λ becomes $\frac{\lambda^{*2}}{N}$.

If the distribution depends on more than one parameter – say, its PDF is $f(t|\theta_1, \dots, \theta_n)$ – then the likelihood function is

$$L(\theta_1, \dots, \theta_n) = \prod_{k=1}^{N} f(t_k|\theta_1, \dots, \theta_n)$$

and its logarithm is

$$l(\theta_1, \ldots, \theta_n) = \sum_{k=1}^{N} \ln(f(t_k|\theta_1, \ldots, \theta_n))$$

The Fisher information is now a matrix, the negative Hessian of $l(\theta_1, \ldots, \theta_n)$:

$$I(\theta_1, \ldots, \theta_n) = \begin{pmatrix} -\dfrac{\partial^2 l}{\partial \theta_1^2} & -\dfrac{\partial^2 l}{\partial \theta_1 \partial \theta_2} & \cdots & -\dfrac{\partial^2 l}{\partial \theta_1 \partial \theta_n} \\[2mm] -\dfrac{\partial^2 l}{\partial \theta_2 \partial \theta_1} & -\dfrac{\partial^2 l}{\partial \theta_2^2} & \cdots & -\dfrac{\partial^2 l}{\partial \theta_2 \partial \theta_n} \\[2mm] \vdots & \vdots & \ddots & \vdots \\[2mm] -\dfrac{\partial^2 l}{\partial \theta_n \partial \theta_1} & -\dfrac{\partial^2 l}{\partial \theta_n \partial \theta_2} & \cdots & -\dfrac{\partial^2 l}{\partial \theta_n^2} \end{pmatrix}$$

It can be shown that the inverse matrix of $I(\theta_1, \ldots, \theta_n)$ gives the covariance matrix of $(\theta_1, \ldots, \theta_n)$.

Example 1.11 Let X now be a normal variable with unknown parameters μ and σ^2. From Example 1.9, we know that

$$l(\mu, V) = -\frac{N}{2} \ln(V) - N \ln\left(\sqrt{2\pi}\right) - \frac{1}{2V} \sum_{k=1}^{N} (t_k - \mu)^2$$

where we use the simplifying notation $V = \sigma^2$. Notice that

$$\frac{\partial l}{\partial \mu} = -\frac{1}{2V} \sum_{k=1}^{N} 2(t_k - \mu)(-1) = \frac{1}{V} \sum_{k=1}^{N} (t_k - N\mu)$$

The first order condition implies that the likelihood estimate of μ is

$$\mu \mathbin{*}= \frac{1}{N} \sum_{k=1}^{N} t_k = \bar{t}$$

Similarly

$$\frac{\partial l}{\partial V} = -\frac{N}{2V} + \frac{1}{2V^2} \sum_{k=1}^{N} (t_k - \mu)^2$$

showing that the estimate of V is

$$V \mathbin{*}= \frac{1}{N} \sum_{k=1}^{N} (t_k - \mu \mathbin{*})^2$$

In order to determine the Fisher information matrix, we need to find the Hessian of $l(\mu, V)$. By differentiation,

$$\frac{\partial^2 l}{\partial \mu^2} = -\frac{N}{V}$$

$$\frac{\partial^2 l}{\partial \mu \partial V} = -\frac{1}{V^2} \left(\sum_{k=1}^{N} t_k - N\mu\right)$$

$$\frac{\partial^2 l}{\partial V^2} = \frac{N}{2V^2} - \frac{1}{2} \cdot \frac{2}{V^3} \sum_{k=1}^{N} (t_k - \mu)^2$$

At the point (μ^*, V^*),

$$\frac{\partial^2 l}{\partial \mu^2} = -\frac{N}{V}, \frac{\partial^2 l}{\partial \mu \partial V} = 0, \frac{\partial^2 l}{\partial V^2} = \frac{N}{2V^2} - \frac{1}{V^3} NV = -\frac{N}{2V^2}$$

So

$$I(\mu, V) = \begin{pmatrix} \frac{N}{V} & 0 \\ 0 & \frac{N}{2V^2} \end{pmatrix}$$

with inverse

$$Cov(\mu, V) = \begin{pmatrix} \frac{V}{N} & 0 \\ 0 & \frac{2V^2}{N} \end{pmatrix} = \begin{pmatrix} \frac{\sigma^2}{N} & 0 \\ 0 & \frac{2\sigma^4}{N} \end{pmatrix}$$

Therefore

$$Var(\mu) = \frac{\sigma^2}{N}, Var(\sigma^2) = \frac{2\sigma^4}{N}, Cov(\mu, \sigma^2) = 0$$

1.5 Expected Number of Failures

The expected number of failures in a given time period $[0, t]$ is an important characteristic of the reliable operation of an object, which may be an element, a unit, a block, or even an entire system. Clearly, this expectation depends on the way we react to failures. Three cases will be discussed in this section: minimal repair, partial repair, and failure replacement. We will also include lengths of time needed to perform repairs or replacements.

The material in this section is based on renewal theory, the fundamentals of which can be found, for example, in Cox (1970) and Ross (1996).

1.5.1 Minimal Repair

Consider first the case of minimal repairs, where each repair requires T time when the object is not operational. Let S_1 denote the time of the first failure, $M(t)$ the expected number of failures in the interval $[0, t]$, and X_t the actual number of failures in this interval. Then

$$M(t) = E(X_t) = E_{S_1}(E(X_t|S_1)) \tag{1.78}$$

where

$$E(X_t|S_1) = \begin{cases} 0 & \text{if } t < S_1 \\ 1 & \text{if } S_1 < t < S_1 + T \\ 1 + M(t - T) - M(S_1) & \text{if } t > S_1 + T \end{cases} \tag{1.79}$$

Therefore

$$M(t) = \int_0^{t-T} (1 + M(t - T) - M(s))f(s)ds + \int_{t-T}^{t} f(s)ds$$

where $f(t)$ is the PDF of the TTF so that

$$M(t) = F(t - T)(1 + M(t - T)) - \int_0^{t-T} M(s)f(s)ds + F(t) - F(t - T)$$

$$= M(t - T)F(t - T) + F(t) - \int_0^{t-T} M(s)f(s)ds \qquad (1.80)$$

In the special case when $T = 0$, this integral equation is modified as

$$M(t) = M(t)F(t) + F(t) - \int_0^t M(s)f(s)ds \qquad (1.81)$$

where $M(0) = 0$. We will show that with the failure rate $\rho(\tau)$,

$$M(t) = \int_0^t \rho(\tau)d\tau \qquad (1.82)$$

solves this equation by substitution. Notice first that based on the integration domain shown in Figure 1.15,

$$\int_0^t M(s)f(s)ds = \int_0^t \int_0^s \rho(\tau)d\tau f(s)ds = \int_0^t \int_\tau^t \rho(\tau)f(s)dsd\tau$$

$$= \int_0^t \rho(\tau)[F(t) - F(\tau)]d\tau = \int_0^t \rho(\tau)d\tau F(t) - \int_0^t \rho(\tau)(1 - R(\tau))d\tau$$

$$= M(t)F(t) - M(t) + \int_0^t \frac{f(\tau)}{R(\tau)}R(\tau)d\tau = M(t)F(t) - M(t) + F(t)$$

so the right side of Eq. (1.81) becomes

$$M(t)F(t) + F(t) - [M(t)F(t) - M(t) + F(t)] = M(t)$$

The simple formula in Eq. (1.82) shows that $\{X_t\}$ is a non-homogenous Poisson process with arrival rate $\rho(t)$, which has several important consequences. If the failures are

Figure 1.15 Integration domain.

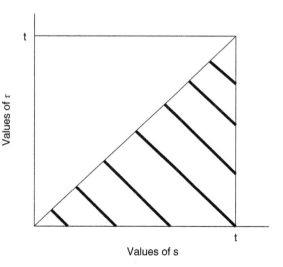

t

Values of τ

t

Values of s

independent, then X_t is a Poisson variable with $\lambda = M(t)$ expectation, so

$$P(X_t = n) = \frac{M(t)^n}{n!}e^{-M(t)} \tag{1.83}$$

Assume that the repairable failures occur at times $0 < t_1 < t_2 < \ldots$ and repairs are done instantly. Let $F_k(t)$ denote the CDF of t_k. Then

$$F_k(t) = P(t_k \le t) = P(X_t \ge k) = \sum_{n=k}^{\infty} \frac{M(t)^n}{n!}e^{-M(t)} \tag{1.84}$$

and the corresponding PDF is its derivative,

$$
\begin{aligned}
f_k(t) = F_k'(t) &= \sum_{n=k}^{\infty} \left(\frac{nM(t)^{n-1}}{n!}\rho(t)e^{-M(t)} + \frac{M(t)^n}{n!}e^{-M(t)}(-\rho(t)) \right) \\
&= \sum_{n=k}^{\infty} \left(\frac{M(t)^{n-1}}{(n-1)!} - \frac{M(t)^n}{n!} \right) e^{-M(t)}\rho(t) = \frac{M(t)^{k-1}}{(k-1)!}e^{-M(t)}\rho(t)
\end{aligned}
\tag{1.85}
$$

Example 1.12 In the special case when X is exponential, $\rho(t) = \lambda$ and $M(t) = \lambda t$, so

$$f_k(t) = \frac{(\lambda t)^{k-1}}{(k-1)!}\lambda e^{-\lambda t} = \frac{\lambda^k t^{k-1} e^{-\lambda t}}{(k-1)!} \tag{1.86}$$

which is a gamma PDF with parameters k and $\theta = \frac{1}{\lambda}$.

Example 1.13 If X is Weibull, then $\rho(t) = \frac{\beta}{\eta^\beta}t^{\beta-1}$ and $M(t) = \left(\frac{t}{\eta}\right)^\beta$, so

$$f_k(t) = \frac{\left(\frac{t}{\eta}\right)^{\beta(k-1)}}{(k-1)!}\frac{\beta}{\eta^\beta}t^{\beta-1}e^{-\left(\frac{t}{\eta}\right)^\beta} = \frac{\beta t^{\beta k-1}e^{-\left(\frac{t}{\eta}\right)^\beta}}{(k-1)!\eta^{\beta k}} \tag{1.87}$$

1.5.2 Failure Replacement

Assume next that upon failures, failure replacement is performed, which requires T time to complete. Then, similar to the previous case,

$$
E(X_t|S_1) = \begin{cases}
0 & \text{if } t < S_1 \\
1 & \text{if } S_1 < t < S_1 + T \\
1 + E(X_{t-S_1-T}) & \text{if } t > S_1 + T
\end{cases}
\tag{1.88}
$$

Therefore, Eq. (1.78) implies that

$$
\begin{aligned}
M(t) &= \int_{t-T}^{t} f(s)ds + \int_0^{t-T} (1 + M(t-s-T))f(s)ds \\
&= F(t) - F(t-T) + F(t-T) + \int_0^{t-T} M(t-s-T)f(s)ds \\
&= F(t) + \int_0^{t-T} M(t-s-T)f(s)ds
\end{aligned}
$$

If $t \leq T$, then the integral is zero, so

$$M(t) = \begin{cases} F(t) & \text{if } t \leq T \\ F(t) + \int_0^{t-T} M(t-T-s)f(s)ds & \text{if } t > T \end{cases} \tag{1.89}$$

If replacement is instantaneous, then $T = 0$, so

$$M(t) = F(t) + \int_0^t M(t-s)f(s)ds \tag{1.90}$$

gives the equation for $M(t)$. There is no simple formula for the solution of this integral equation. However, the use of Laplace transforms provides an interesting insight. Let $M^*(s)$, $F^*(s)$, and $f^*(s)$ denote the Laplace transforms of $M(t)$, $F(t)$, and $f(t)$, respectively. Then from Eq. (1.90), we have

$$M^*(s) = F^*(s) + M^*(s)f^*(s)$$

implying that

$$M^*(s) = \frac{F^*(s)}{1-f^*(s)} = \frac{\frac{1}{s}f^*(s)}{1-f^*(s)}$$

If $g(t)$ denotes the inverse Laplace transform of $f^*(s)/(1-f^*(s))$, then

$$M(t) = \int_0^t g(\tau)d\tau \tag{1.91}$$

showing that $\{X_t\}$ is again a non-homogenous Poisson process with arrival rate $g(t)$.

Let $0 < t_1 < t_2 < \ldots$ denote times of failures. Then $z_k = t_{k+1} - t_k (k = 0, 1, \ldots$ and $t_0 = 0)$ are independent random variables with PDF $f(t)$. Then the density $f_k(t)$ of t_k is the k-fold convolution of $f(t)$:

$$f_k(t) = f(t) * f(t) * \ldots * f(t)$$

Notice that

$$E(t_k) = E(z_1) + E(z_2) + \ldots + E(z_k) = E(z_1)k = \mu k,$$

where μ is the TTF mean value. Similarly

$$\text{Var}(t_k) = \text{Var}(z_1) + \text{Var}(z_2) + \ldots + \text{Var}(z_k) = \text{Var}(z_1)k = \sigma^2 k$$

where σ^2 is the TTF variance value. If z_k is exponential, then t_k is gamma; if z_k is gamma, then t_k is also gamma; and if z_k is normal, then t_k is also a normal variable.

Based on these values, in other cases the PDF of t_k can be estimated by a lognormal or gamma variable with a given expectation and variance; and confidence intervals can be constructed for any significance level selected by the user.

If repairs or replacements need T time to perform, then the time of the kth failure is $\bar{t}_k = t_k + (k-1)T$, with the CDF

$$\bar{F}_k(t) = P(t_k + (k-1)T \leq t) = P(t_k \leq t - (k-1)T)$$

$$= \begin{cases} 0 & \text{if } t \leq (k-1)T \\ F_k(t - (k-1)T) & \text{otherwise} \end{cases} \tag{1.92}$$

where $F_k(t)$ is the CDF of $f_k(t)$.

1.5.3 Decreased Number of Failures Due to Partial Repairs

Assume next that partial repairs are performed, resulting in the number of failures decreasing by a factor of $\alpha < 1$. Then, similar to the previous cases,

$$E(X_t|S_1) = \begin{cases} 0 & \text{if } t < S_1 \\ 1 & \text{if } S_1 < t < S_1 + T \\ 1 + \alpha(M(t-T) - M(S_1)) & \text{if } t > S_1 + T \end{cases} \tag{1.93}$$

So

$$M(t) = \int_{t-T}^{t} f(s)ds + \int_{0}^{t-T} [1 + \alpha(M(t-T) - M(s))]f(s)ds$$

$$= F(t) - F(t-T) + F(t-T) + \alpha M(t-T)F(t-T) - \alpha \int_{0}^{t-T} M(s)f(s)ds$$

$$= \begin{cases} F(t) & \text{if } t \le T \\ F(t) + \alpha M(t-T)F(t-T) - \alpha \int_{0}^{t-T} M(s)f(s)ds & \text{if } t > T \end{cases} \tag{1.94}$$

Assume again instantaneous repairs with $T = 0$. Then (1.94) is modified as

$$M(t) = F(t) + \alpha M(t)F(t) - \alpha \int_{0}^{t} M(s)f(s)ds \tag{1.95}$$

1.5.4 Decreased Age Due to Partial Repairs

Consider finally the case where partial repairs decrease the effective age of the object by a factor $\alpha < 1$. Then we have

$$E(X_t|S_1) = \begin{cases} 0 & \text{if } t < S_1 \\ 1 & \text{if } S_1 < t < S_1 + T \\ 1 + M((t - S_1 - T + \alpha S_1)) - M(\alpha S_1) & \text{if } t > S_1 + T \end{cases} \tag{1.96}$$

since at time t the effective age is $(t - S_1 - T) + \alpha S_1$, where the first term is the length of interval $[S_1 + T, t]$ and the second term is the effective age at time $S_1 + T$ after repair. Therefore

$$M(t) = \int_{t-T}^{t} f(s)ds + \int_{0}^{t-T} [1 + M((t - s - T + \alpha s)) - M(\alpha s)]f(s)ds$$

$$= \begin{cases} F(t) & \text{if } t \le T \\ F(t) + \int_{0}^{t-T} [M(t - s - T + \alpha s) - M(\alpha s)]f(s)ds & \text{if } t > T \end{cases} \tag{1.97}$$

If $T = 0$ then

$$M(t) = F(t) + \int_{0}^{t} [M(t - s + \alpha s) - M(\alpha s)]f(s)ds \tag{1.98}$$

Notice that the case $\alpha = 1$ corresponds to minimal repair, and Eqs. (1.95) and (1.98) reduce to Eq. (1.81) as they should. Equations (1.81), (1.90), and (1.95) are usually called the *renewal equations* (Mitov and Omey 2014).

The value of $M(t)$ for given values of t usually requires the solution of integral equations; however, in the case of model Eq. (1.81), the solution was given in a simple formula. Assume first that the cost of each repair is c_r and no discounting of money is considered. The total repair cost in interval $[0, t]$ is clearly $c_r M(t)$ in the average.

Let r denote the discount factor, and assume minimal repair each time. Then the expected discounted cost in interval $[0, t]$ is the following:

$$c_r \sum_{k=1}^{\infty} \int_0^t e^{-r\tau} \frac{M(\tau)^{k-1}}{(k-1)!} \rho(\tau) e^{-M(\tau)} d\tau = c_r \int_0^t \left(\sum_{k=1}^{\infty} \frac{M(\tau)^{k-1}}{(k-1)!} \right) e^{-r\tau} \rho(\tau) e^{-M(\tau)} d\tau$$

$$= c_r \int_0^t e^{M(\tau)} \rho(\tau) e^{-r\tau} e^{-M(\tau)} d\tau = c_r \int_0^t e^{-r\tau} \rho(\tau) d\tau \tag{1.99}$$

Example 1.14 In the exponential case, $\rho(\tau) = \lambda$, so this expression is simplified as

$$c_r \int_0^t e^{-r\tau} \lambda d\tau = c_r \lambda \left[\frac{e^{-r\tau}}{-r} \right]_0^t = c_r \lambda \frac{1 - e^{-rt}}{r}$$

Example 1.15 In the Weibull case, $\rho(\tau) = \frac{\beta}{\eta^\beta} \tau^{\beta-1}$, so the expected cost is

$$c_r \int_0^t e^{-r\tau} \frac{\beta}{\eta^\beta} \tau^{\beta-1} d\tau = c_r \int_0^{rt} e^{-z} \frac{\beta}{\eta^\beta} \frac{z^{\beta-1}}{r^{\beta-1}} \frac{dz}{r}$$

where the new integration variable $z = r\tau$ is introduced with $d\tau = \frac{dz}{r}$ and $\tau = \frac{z}{r}$. By simplification, it becomes

$$\frac{c_r \beta}{(r\eta)^\beta} \int_0^{rt} z^{\beta-1} e^{-z} dz = \frac{c_r \beta}{(r\eta)^\beta} \gamma(\beta, rt) \tag{1.100}$$

where $\gamma(\beta, s)$ is

$$\gamma(\beta, s) = \int_0^s x^{\beta-1} e^{-x} dx$$

The values of $\gamma(\beta, s)$ with any particular values of β and s can be found in the appropriate function tables.

1.6   System Reliability and Prognosis and Health Management

A PHM system is complex and comprises several subsystems. In the classical approach, this complex system is based on probabilistic considerations that have many disadvantages in comparison to the new approach that will be introduced and discussed in this book. First, the probability distributions reflect the average behavior of many identical items, not the behavior of one particular item under consideration. Second, the parameters of the distribution functions used in the process are not exact; their parameters are uncertain, as demonstrated earlier. A more practical approach is to use CBD, from which you extract feature data (FD) consisting of leading indicators of failure; then, as necessary, perform signal conditioning, data transforms, and domain transforms to create data that forms a fault-to-failure progression (FFP) signature.

FFP signatures have been successfully processed by prediction algorithms to produce accurate estimates of system health and remaining life. Increased accuracy is achieved by further transforming FFP signature data in degradation progression signature (DPS) data and then transforming DPS data into functional failure signature (FFS) data. FFS data processed by predication algorithms employing Kalman-like filtering, random-walk, and/or other trending algorithms results in very accurate prognostic estimates and rapid convergence from, for example, an initial error of 50% to 10% within fewer than 20 data samples. This rapid, accurate convergence is very useful for CBM in PHM systems (Hofmeister et al. 2016, 2017a,b).

1.6.1 General Framework for a CBM-Based PHM System

Figure 1.16 illustrates a PHM system forming a framework for CBM. The system monitors, captures, and processes CBD to extract signatures indicative of the state of health (SoH) of devices, components, assemblies, and subsystems in systems. Extracted signatures are processed to produce prognostic information, such as SoH and RUL, which are further processed to manage maintenance and logistics of the system (CAVE3 2015; Hofmeister et al. 2016).

Figure 1.17 is an exemplary illustration of a framework for CBM. The framework consists of (i) a sensor framework, (ii) a feature-vector framework, (iii) a prediction framework; (iv) a health-management framework, (v) a performance-validation framework, and (vi) a control- and data-flow framework.

A sensor framework comprises sensors to capture signals at nodes within the system. Smart sensors provide initial data conditioning such as analog-to-digital data conversion and digital-signal processing, and produce and transmit FD such as scalar values. A feature-vector framework provides additional data conditioning, including data fusion and transformation to create signatures, such as FFS that are used as input data to a prediction framework that produces prognostic information. A health-management framework processes prognostic information to create diagnostic, prognostic, and logistics directives and actions to maintain the health and reliability

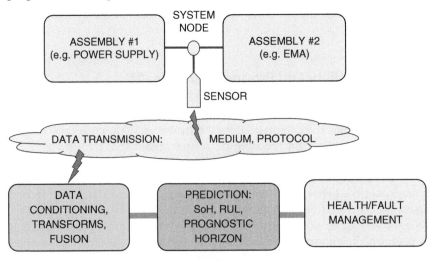

Figure 1.16 High-level block diagram of a PHM system.

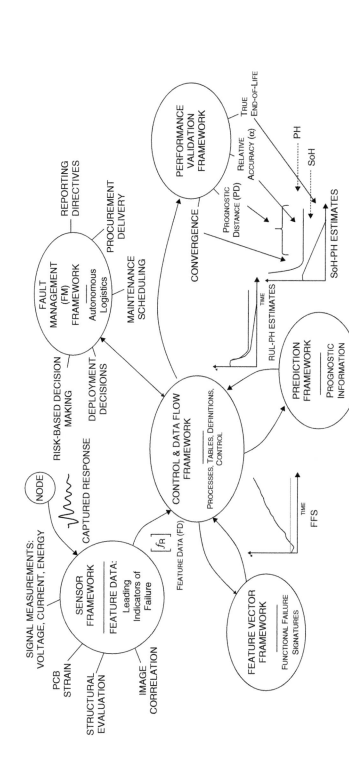

Figure 1.17 A framework for CBM for PHM. Source: after CAVE3 (2015).

of the system. A performance-validation framework provides means and methods to produce prognostic-performance metrics for evaluation of the accuracy of prognostic information. A control- and data-flow framework manages PHM functions and actions (CAVE3 2015; Pecht 2008; Kumar and Pecht 2010).

Reliability $R(t)$ (refer back to Eq. (1.14) of a prognostic target is the probability that the prognostic target will operate satisfactorily for a required period of time. Reliability is related to lifetime and mean time before failure (MTBF) by the following (Speaks 2005). Theoretically, the mean time before failure is obtained from probabilistic reasoning as the expectation of TTF:

$$\mathrm{MTBF} = \int_0^\infty tf(t)dt$$

where $f(t)$ is a PDF. By examining a number of test units, it can be obtained practically as

$$\mathrm{MTBF} = (NT)/F$$

where N = number of test units, T = test time, and F = number of test failures.

The relative frequency of number the of failures until time T is clearly given as F/N, which is an estimator of the CDF of TTF at time $t = T$. If the distribution is exponential, then its parameter can be estimated from

$$1 - e^{-\lambda T} = F/N \tag{1.101}$$

so an estimate of λ is

$$\lambda = -\frac{1}{T} \ln\left(1 - \frac{F}{N}\right) \tag{1.102}$$

Then the CDF is given as $F(t) = 1 - e^{-\lambda t}$, the reliability function is $R(t) = e^{-\lambda t}$, and the constant failure rate is λ. We previously derived these results; refer back to Eq. (1.2), for example. If the distribution of TTF has more than one parameter, then the information provided by λ is not sufficient to estimate $R(t)$. For example, the variance can be estimated by repeated testing, so two equations would be obtained for the usual two parameters (such as Weibull, gamma, normal, and so on).

MTBF is a statistical measure that applies to a set of prognostic targets N and is not applicable for estimating the lifetime or RUL of a specific prognostic target in a system.

Example 1.16   As an example of how MTBF is used, suppose that MTBF is determined to be 500 days. Then the probability that a specific prognostic target will continue to operate within specifications for 500 days = 0.368 (36.8%): the exponential value of (−1). In particular, notice that MTBF is not used to estimate the time of failure of the prognostic target.

1.6.2 Relationship of PHM to System Reliability

What is the relationship of PHM to system reliability? As an example, suppose a PHM system is designed to accurately predict the time when functional failure will occur for every prognostic target in the system; further suppose that the fault/health management framework of that system enables replacement and/or repair of every prognostic target prior to the actual time of functional failure; also suppose that the time when maintenance is performed occurs after the time when degradation is detected for each

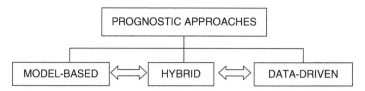

Figure 1.18 Taxonomy of prognostic approaches.

prognostic target; and also suppose that the time to replace and/or repair prognostic targets in a PHM-enabled system is less than the time to replace and/or repair unexpected outages. In such a PHM-enabled system, the effective MTBF of the system is increased, and therefore the reliability of the system is also increased. In addition to increasing the reliability of the system, PHM reduces the cost of maintaining the system because the operating lives of the prognostic targets are increased and because it is generally less expensive, in both time and materials, to handle planned versus unexpected outages.

Any number of traditional modeling approaches lend themselves as base platforms for PHM-enabling systems that can be classified as model-based prognostics, data-driven prognostics, or hybrid-driven prognostics (see Figure 1.18). Model-based approaches often offer potentially greater precision and accuracy in prognostic estimations, which can be difficult to apply in complex systems. Data-driven approaches are simpler to apply but can produce less precision and less accuracy in prognostic estimations. Hybrid-driven approaches offer a high degree of precision and accuracy and are applicable to complex systems, including on-vehicle and off-vehicle, taking advantage of both approaches (Pecht 2008; Medjaher and Zerhouni 2013).

In Chapter 2, we will return to these concepts and approaches in more detail.

1.6.3 Degradation Progression Signature (DPS) and Prognostics

A CBM system, such as that presented in this book, uses a CBD approach. As a prognostic target in a system progresses from a state of no damage (zero degradation) to a state of damage (degraded) and then to a failed state, one or more signals at one or more nodes in the system will change. Such signal changes can be sensed, measured, collected, and processed by a PHM system to produce highly accurate prognostic information for use in fault and/or health management. As an example, suppose a power supply in a system becomes damaged, and the ripple voltage of the output changes and increases as the amount of damage (accumulated degradation) in the power supply increases. Further, suppose the amplitude of the ripple voltage can be sensed, measured, and collected by a sensor framework in a PHM system. The collection of CBD amplitudes forms an FFP signature, such as that shown in Figure 1.19.

Clearly the FFP curve is not smooth, due to the effects of measurement uncertainty and noise. Now suppose a feature-vector framework in that PHM system consists of data-conditioning algorithms that filter, smooth, fuse, and transform CBD-based FFP signatures into DPS data: a transfer curve. An ideal DPS has a constant steady-state value in the absence of degradation; it linearly increases in correlation to the level of degradation; and it reaches a maximum amplitude when the level of degradation is at its maximum: physical failure occurs. For this example, such an ideal DPS is shown as the line with positive slope in Figure 1.20; also shown are the points at which the onset of degradation occurs and when physical failure occurs.

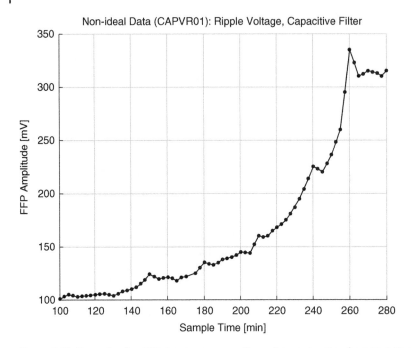

Figure 1.19 Example of an FFP signature – a curvilinear (convex), noisy characteristic curve.

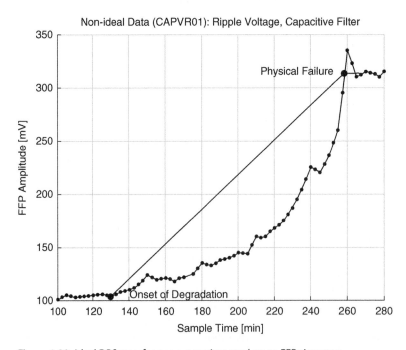

Figure 1.20 Ideal DPS transfer curve superimposed on an FFP signature.

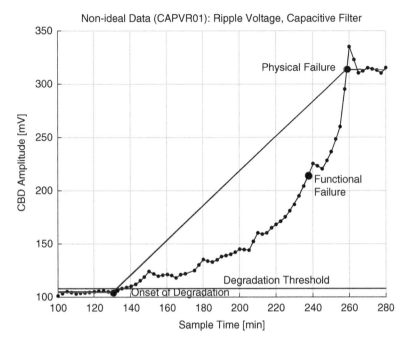

Figure 1.21 Ideal DPS, degradation threshold, and functional failure.

Further suppose a PHM system allows a reliability engineer to specify a noise margin to mask signal variations due to, for example, environmental and operational noise. An example of a noise margin is that indicated by the lower horizontal line on the plot in Figure 1.21. Also suppose a PHM system allows a reliability engineer to specify a level of degradation that defines a threshold at which a prognostic target is no longer capable of satisfactorily operating within specifications: functional failure occurs, as indicated by the circled Functional Failure point in the plot shown in Figure 1.21. A PHM system can be designed to support reliability $R(t)$ with emphasis on operating satisfactorily, meaning to operate within a required set of specifications: a functional-failure threshold used for transforming a physical-failure-based DPS into an FFS, such as that shown in Figure 1.22.

1.6.4 Ideal Functional Failure Signature (FFS) and Prognostics

Now suppose the PHM system transforms FFS amplitudes into percent values, as shown in Figure 1.23. Transforming CBD into a DPS and then into an FFS creates useful input data for prediction algorithms to produce accurate prognostic information: (i) when FFS = 0, no degradation; (ii) when FSS > 0 and < 100, degradation but operating within specifications; and (iii) when FFS = 100, functional failure has occurred.

To further illustrate the usefulness of FFS transfer curves, suppose a PHM-enabled system has three similar prognostic targets for which ideal DPS transfer curves are shown in Figure 1.24. The plots exhibit four significant variabilities: (i) each prognostic target starts to degrade at a different sample time; (ii) each prognostic target has a different nominal value in the absence of degradation; (iii) each prognostic target has a

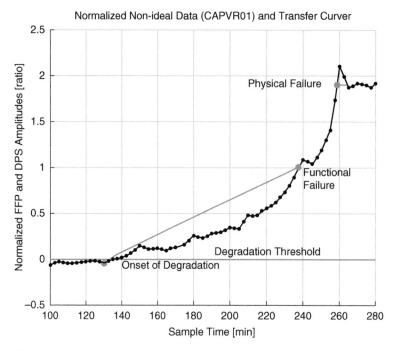

Figure 1.22 Normalized and transformed FFP and DPS transformed into FFS.

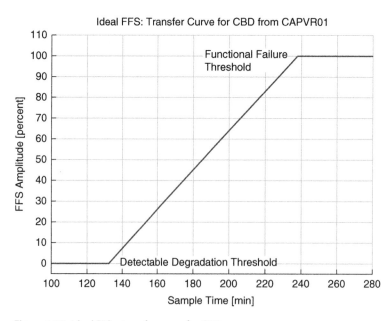

Figure 1.23 Ideal FFS – transfer curve for CBD.

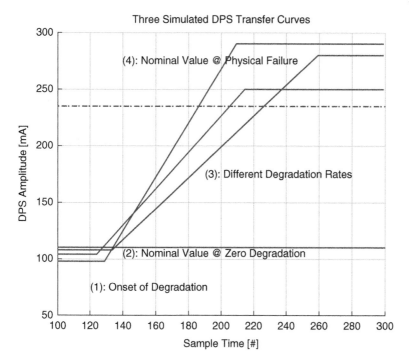

Figure 1.24 Variability in DPS transfer curves.

different rate of degradation; and (iv) each prognostic target exhibits a different amplitude at physical failure. Included in the plots are the horizontal lines representing the thresholds specified by a reliability engineer for noise (below) and functional failure (above).

Now suppose the PHM system and its prediction algorithms are designed to translate clock time to relative time with respect to the detection of degradation during prognostic processing; and also suppose the PHM system and its prediction algorithms truncate FFS values to 0 and 100. Then the plots of the FFS transforms are shown in Figure 1.25.

The design of a prediction algorithm to produce accurate prognostic information is greatly facilitated when FFS data is the input. For example, at any relative sample time (ST), the corresponding sample amplitude (SA) is used to calculate an estimated failure time (FT), RUL, and SoH using the following equations:

$$FT = 100\, ST/SA = \text{failure time} \tag{1.103}$$

$$RUL = FT - ST = \text{remaining useful life} \tag{1.104}$$

$$SoH = 100\, RUL/FT = \text{state of health} \tag{1.105}$$

Example 1.17 An example of how a prediction algorithm might produce prognostic information from data in an FFS is illustrated in Figure 1.26. For any given FFS data point, such as that for amplitude SA = 40.2 and ST = 36, compute the estimated failure time FT = 89.5 using Eq. (1.103), then compute the estimated RUL = 53.5 using Eq. (1.104), and then compute the estimated SoH = 59.8% using Eq. (1.105).

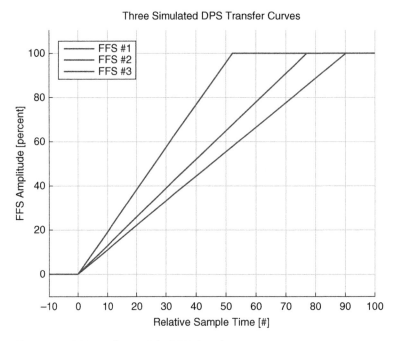

Figure 1.25 FFS transforms of the DPS plots shown in Figure 1.23.

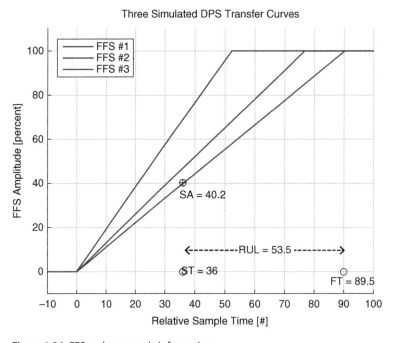

Figure 1.26 FFS and prognostic information.

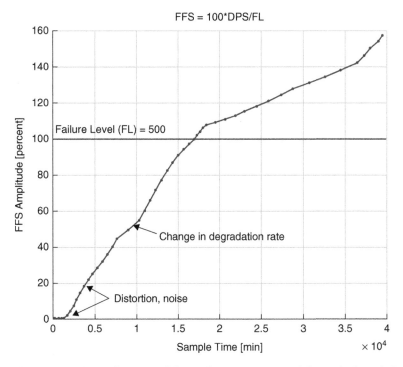

Figure 1.27 FFS transfer curve exhibiting distortion, noise, and change in degradation rate.

1.6.5 Non-ideal FFS and Prognostics

In practice, it generally is not possible to obtain ideal FFS transfer curves. Instead, the curves will be noisy and distorted, and will exhibit changes associated with changes in the degradation rate – similar to what is shown in Figure 1.27. Small perturbations may be due to measurement uncertainty and have no relationship or significance with respect to the actual SoH and/or RUL of a prognostic object: in such cases, prediction algorithms should employ methods to filter or otherwise mitigate these variations. On the other hand, an FFS change might be caused by a real, significant change in the rate of accumulated damage in a prognostic target: in such cases, prediction algorithms should employ methods to recognize and take into account these changes in degradation rates, which should modify the estimates of SoH and RUL. Mitigation of and accounting for both causes of signal variability are competing objectives in the design of prediction algorithms: in practice, to design of prediction algorithms that satisfy both objectives, the algorithms need to incorporate appropriate design compromises. In subsequent chapters of this book, you will see examples of how FFS variabilities are successfully handled.

1.7 Prognostic Information

Prognostic information includes the following significant metrics: RUL, SoH, and PH. Figure 1.28 shows example plots of an ideal RUL and ideal PH: (i) the estimated time of

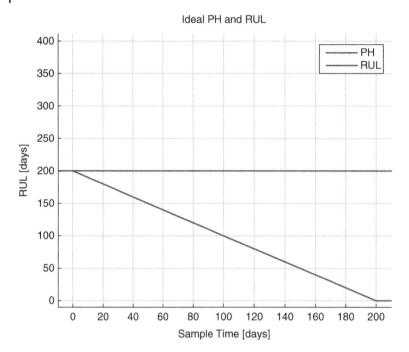

Figure 1.28 Example plots of an ideal RUL and ideal PH.

failure relative to the onset of degradation is PH, and the initial value specified by a reliability engineer is exactly correct; (ii) the onset of degradation is detected exactly at the time the prognostic target begins to degrade; (iii) the time when degradation reaches the specified level for failure is exactly detected; and (iv) the FFS input to a prediction algorithm is ideal. The PH and RUL plots are perfectly linear, with no noise or distortion – the RUL is an exact linear progression from 200 to 0 days.

Similarly, Figure 1.29 shows the plots of an ideal SoH and an ideal PH accuracy. SoH is an exact linear progression from 100% to 0%, and PH accuracy is exactly 100% before and after the onset of degradation.

1.7.1 Non-ideality: Initial-Estimate Error and Remaining Useful Life (RUL)

A major cause of non-ideality in prognostic information is the difference between an estimated TTF of a prognostic target defined by a reliability engineer (an initial RUL estimate) and the actual TTF between the onset of degradation and functional failure. It is extremely unlikely that estimated initial TTF values will exactly equal actual TTF values, and therefore there will be initial-estimate errors. For example, in Figures 1.28 and 1.29, the defined TTF value is 200 days and is exactly equal to the true TTF value at functional failure. But suppose a reliability engineer had estimated the TTF after the onset of degradation to be 300 days. In that case, when degradation is detected, the estimated RUL will be 300 days, and the actual TTF will be 200 days – an initial error of 100 days, as shown in Figure 1.30.

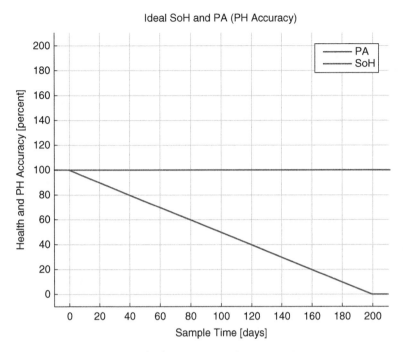

Figure 1.29 Example plots of an ideal SoH transfer curve and PH accuracy.

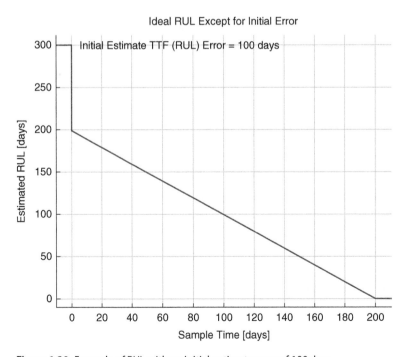

Figure 1.30 Example of RUL with an initial-estimate error of 100 days.

1.7.2   Convergence of RUL Estimates Given an Initial Estimate Error

It is desirable to design prediction algorithms that (i) mitigate FFS variability not caused by degradation and (ii) respond to FFS variability caused by degradation. A design approach is to treat FFS data points as particles having inertia and momentum. Particles do not exhibit rapid changes; instead, they tend to maintain velocity and direction. A satisfactory design objective is to employ a dampening factor for changes in amplitude and velocity. Another design approach is to develop a random-walk solution for particles that progress from a zero-degradation state (lower-left corner) to a maximum-degradation state (upper-right corner) – a solution that employs Kalman-like filtering (Hamilton 1994; Bucy and Joseph 2005): (i) use the previous data point of an FFS model; (ii) calculate the predicted location of the next data point; (iii) adapt the FFS model to an adjusted location between the predicted and the actual location of the next data point; (iv) use dampening factors and coefficients to adjust the FFS model; (v) use the adapted FFS model to estimate when functional failure is likely to occur; and (vi) calculate prognostic information such as RUL, SoH, and PH.

Example 1.18   Figure 1.31 is an example of such a solution applied to an initial-estimate error. In addition to a high-value initial-estimate error, a reliability engineer might specify a TTF value that is low compared to that for a prognostic target. Suppose for a prognostic target a reliability engineer specifies 100 days as the estimated TTF after the onset of degradation, and the target actually takes 200 days to become functionally failed: a random-walk with Kalman-like solution will resemble that shown in Figure 1.32.

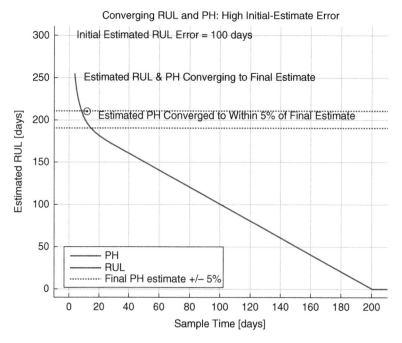

Figure 1.31   Random-walk with Kalman-like filtering solution for a high-value initial-estimate error.

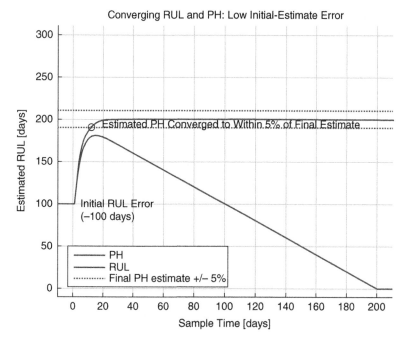

Figure 1.32 Random-walk with Kalman-like filtering solution for a low-value initial-estimate error.

1.7.3 Prognostic Distance (PD) and Convergence

Referring back to Figures 1.31 and 1.32, the dotted horizontal lines represent a desired zone of accuracy called alpha (α). A prognostic parameter of interest is the prognostic distance when prognostic estimates RUL and PH are within an application-specific value α (Saxena et al. 2009):

$$PD_\alpha = \text{prognostic distance when estimated PH within accuracy } \alpha$$

$$PD_{MAX} = \text{maximum prognostic: TTF} - \text{time of onset of degradation}$$

Example 1.19 For this example, the desired accuracy is 5% ($\alpha = 0.05$), the estimated PH is within $+/- \alpha$ when time ≥ 12, and the estimated PD_α equals 188 (200 – 12). Another prognostic parameter of interest is the maximum PD: time of failure minus time of onset of degradation. For this example, the maximum PD (PD_{MAX}) equals 200. We then define a figure of merit for convergence:

$$\chi_\alpha = 100\, PD_\alpha/PD_{MAX} = \text{convergence efficiency at } \alpha \tag{1.106}$$

For this example, χ_α=94.0%.

1.7.4 Convergence: Figure of Merit (χ_α)

The convergence efficiency of a prediction algorithm is dependent on many factors, including the following:

- The dampening used in the design of the prediction algorithm. A large dampening factor increases the number of FFS data points required to converge to a desired level of accuracy.
- The value of the initial-estimated TFF. The larger the initial error, the more FFS data points need to be processed to converge to a desired level of accuracy.
- The CBD rate of sampling. The slower the sampling rate, then even if the amount of data processed remains the same, the time between data points is increased, and the longer it takes to reach a desired level of accuracy.

Failure of a prediction algorithm to meet your requirements for convergence efficiency might be caused by many factors other than design, including (i) errors in specifications, such as the initial-estimated value for TTF; (ii) design or operational specifications of the sensor framework; and (iii) insufficient data conditioning that results in overly detrimental offset errors, distortion, and noise in the FFS input.

1.7.5 Other Sources of Non-ideality in FFS Data

Other sources of non-ideality in FFS data include, but are not limited to, the following: (i) offset errors in detecting the onset of degradation and the time of functional failure; (ii) distortion in FFS data due to the effects of using noise margins; (iii) measurement uncertainty associated with, for example, quantization errors of data converters; (iv) unfiltered variability caused by environmental effects such as voltage and temperature;

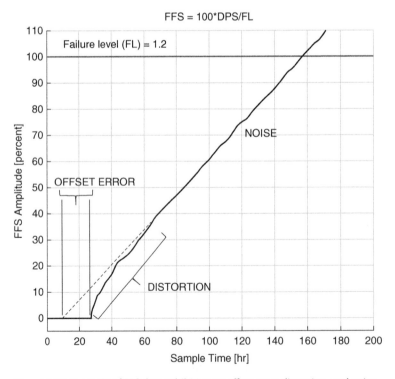

Figure 1.33 Example of FFS data exhibiting an offset error, distortion, and noise.

and (v) the effects of self-healing as the operational and/or environment of a prognostic target changes. Figure 1.33 illustrates a non-ideal FFS and the effects of some of the causes of the non-ideality.

1.8 Decisions on Cost and Benefits

Over the lifetime of any equipment, hard decisions arise regarding the choice of the right or best equipment to purchase; selection, inspection, maintenance, and repair strategies; and finding the optimal time for preventive replacement in order to avoid costly failures that might result in additional damage or accidents. This section gives a brief overview of the decision problems and their mathematical modeling during different stages of the lifetime of equipment, with some illustrative mathematical models.

1.8.1 Product Selection

There are usually several alternative choices of similar equipment available for purchase, and the choice must be based on several criteria. This leads us to use multi-objective optimization techniques. (Szidarovszky et al. 1986) give a summary of the different concepts and algorithms for solving such problems, so we will not give those details; instead, this section of the book will provide an illustrative example to provide an idea of their usage.

Example 1.20 Consider a company that needs a special machine, and assume there are several possible choices. The selection is based on multiple criteria such as the price, expected lifetime, maintainability, ability to fix repairable failures, and worker convenience when working on the machine. Table 1.2 gives the estimated objective values for three alternatives.

Here, Price and Expected Lifetime are known data. In the absence of concrete data, the other three factors are only subjective measures of how good an option is and can only be estimated from the experience of firms using the same equipment; convenience can be further assessed via a demonstration by the selling company. Objectives without hard numerical values are usually characterized by measures of how good they are on a 0–100 scale, when 0 means unacceptable and 100 indicates the most satisfying alternative.

Notice that a lower price is better; however, for all other factors, a larger value is considered better, since a longer lifetime is preferable and the others are satisfaction measures. These objectives cannot be compared as they are given; we need to transform them to a unified common measure. The most logical way is to transform the price and

Table 1.2 Decision table for purchasing a machine.

	Price ($)	Expected lifetime (years)	Maintainability	Repairing repairable failures	Convenience
Machine 1	5000	8	60	50	80
Machine 2	5800	15	70	70	50
Machine 3	6300	20	90	80	60

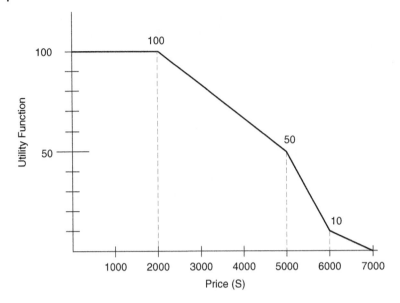

Figure 1.34 Utility function of price.

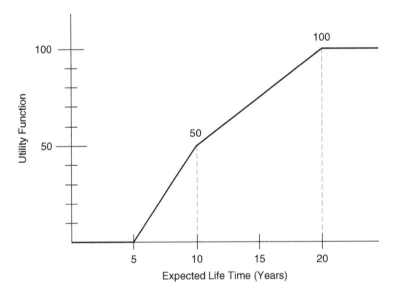

Figure 1.35 Utility function of expected lifetime.

expected lifetime data to measures showing how satisfied the decision-maker is with these values. To do so, the decision-maker constructs utility functions for those two objectives, showing their satisfaction with the different values. Figures 1.34 and 1.35 show the utility functions for these two objectives.

Table 1.3 Utility table for purchasing a machine.

	Price	Expected life time	Maintainability	Repairing repairable failure	Convenience
Machine 1	50	30	60	50	80
Machine 2	18	75	70	70	50
Machine 3	0	100	90	80	60

These functions are piecewise linear functions:

$$U_1(P) = \begin{cases} 100 & \text{if } 0 \le P \le 2000 \\ \dfrac{8000 - P}{60} & \text{if } 2000 \le P \le 5000 \\ \dfrac{6250 - P}{25} & \text{if } 5000 \le P \le 6000 \\ \dfrac{7000 - P}{100} & \text{if } 6000 \le P \le 7000 \\ \dfrac{7000 - P}{100} & \text{if } 6000 \le P \le 7000 \end{cases}$$

$$U_2(LT) = \begin{cases} 0 & \text{if } 0 \le LT \le 5 \\ 10LT - 50 & \text{if } 5 \le LT \le 10 \\ 5LT & \text{if } 10 \le LT \le 20 \\ 100 & \text{if } LT \ge 20 \end{cases}$$

By replacing the Price and Expected Lifetime columns with their satisfaction levels, we get Table 1.3.

In the next step, the decision-maker must decide how important the various objectives are. The easiest method is to assign importance weights. Assume in this case the following weights are given: 30%, 20%, 20%, 20%, and 10%. Then the weighted averages of the objectives (as average satisfaction levels) are calculated for each machine:

$$\text{Machine 1} = 50(0.3) + 30(0.2) + 60(0.2) + 50(0.2) + 80(0.1) = 51$$

$$\text{Machine 2} = 18(0.3) + 75(0.2) + 70(0.2) + 70(0.2) + 50(0.1) = 53.4$$

$$\text{Machine 3} = 0(0.3) + 100(0.2) + 90(0.2) + 80(0.2) + 60(0.1) = 60$$

Since machine 3 has the largest score, it should be chosen for purchase.

1.8.2 Optimal Maintenance Scheduling

The objective of any maintenance action is either to retain the condition of the object under consideration or to restore it to the desired condition. Therefore, we can divide maintenance actions into two major groups: *preventive* and *corrective maintenance* types.

If maintenance is carried out at predetermined time intervals without checking the condition of the item, we are referring to *predetermined maintenance*. If the item is regularly monitored, *condition-based maintenance* can be performed. This can result in either improving the item without changing its required function; or in modifying it, if the required function of the item is altered. If failure is detected, then after the fault is diagnosed, it is corrected. And finally, the operation of the item is checked out to be sure the *repair* was done correctly and successfully.

In this section, some mathematical models for optimal maintenance scheduling will be introduced.

Predetermined Maintenance

First, we introduce a model for scheduling *predetermined maintenance* actions. Consider a piece of equipment that is subject to preventive maintenance actions, which are performed in a uniform time grid. Between maintenance actions, only minimal repairs are performed, in cases of failure. Let h denote the length of time between consecutive maintenance actions. Without maintenance, let $F(t), f(t), R(t)$, and $\rho(t)$ denote the CDF, PDF, reliability function, and failure rate of time to non-repairable failure.

In addition to the value of h, two other decision variables can be introduced. One characterizes the level of maintenance, so as a result of each maintenance action the effective age of the equipment decreases by u, where $0 < u < h$. It must also be determined how long the equipment is scheduled to operate, which can be represented by a positive integer n showing that when the nth maintenance is due, the equipment is replaced. So, maintenance actions are performed at times $h, 2h, \ldots, (n-1)h$, and the equipment is replaced at time nh. For $k = 1, 2, \ldots, n$, let $F_k(t), f_k(t)$, and $\rho_k(t)$ denote the CDF, PDF, and failure rate of the equipment in interval $[(k-1)h, kh]$. Then clearly

$$F_k(t) = F(t - (k-1)u), f_k(t) = f(t - (k-1)u) \text{ and } \rho_k(t) = \rho(t - (k-1)u) \quad (1.107)$$

Before the model of cost minimization is developed, an important constraint must be presented. The reliability of the equipment must not go below a user-selected threshold r_0:

$$R_n(nh) = 1 - F_n(nh) = 1 - F(nh - (n-1)u) \geqq r_0 \quad (1.108)$$

implying that

$$nh - (n-1)u \leqq F^{-1}(1 - r_0) \quad (1.109)$$

For example, in the case of a Weibull distribution, Eq. (1.108) can be written as

$$e^{-\left(\frac{nh-(n-1)u}{\eta}\right)^\beta} \geqq r_0$$

By taking the logarithm on both sides we get

$$\left(\frac{nh - (n-1)u}{\eta}\right)^\beta \leqq -\ln(r_0)$$

or

$$nh - (n-1)u \leqq \eta\{-\ln(r_0)\}^{\frac{1}{\beta}} \quad (1.110)$$

which is a linear constraint.

The expected net cost per unit time is minimized. The following factors must be considered:

- The cost of preventive replacement is c_p, whereas the cost of failure replacement is $c_f > c_p$.
- Repairable failures may occur with failure rate $\rho(t)$. Then, assuming minimal repair, the expected number of such failures in any interval (t_1, t_2) is given as $M(t_1, t_2) = \int_{t_1}^{t_2} \rho(\tau)d\tau$. Each repair costs c_r.
- The cost of each maintenance action depends on the level of maintenance: $a + bu$ by assuming the linear constraint of Eq. (1.110).
- The revenue-generating density of the working equipment is assumed to be $A - BT$, where T is the effective age of the equipment. So, revenue generated between effective ages T_1 and T_2 is the integral

$$\int_{T_1}^{T_2} (A - BT)dT = \left[AT - \frac{BT^2}{2} \right]_{T_1}^{T_2} = (T_2 - T_1)\left(A - \frac{B(T_1 + T_2)}{2} \right) \quad (1.111)$$

- After the equipment is replaced at time t, it is assumed that its salvage value is Ve^{-wt}, where V is the value of a new piece of equipment, and as time progresses the value of the equipment decreases exponentially.

In computing the expected cost per unit time, we have to consider two cases:

Case 1. The equipment does not face non-repairable failure before the scheduled replacement time. Then the net cost per unit of time is given as

$$\frac{1}{nh} \left\{ c_p + (n-1)(a+bu) + c_r \sum_{l=1}^{n}(M(lh - (l-1)u) - M((l-1)h - (l-1)u)) \right.$$

$$\left. -Ve^{-wnh} - \sum_{k=1}^{n} h\left(A - \frac{B((l-1)h - (l-1)u + lh - (l-1)u)}{2} \right) \right\} \quad (1.112)$$

The first term is the cost of preventive replacement, the second term is the overall cost of maintenance actions, and the third term is the cost of repairs. The last two terms represent the salvage value and the generated revenue.

The last term can be further simplified:

$$hAn - \frac{Bh}{2}\left[\sum_{k=1}^{n} l(2h - 2u) + n(-h + 2u) \right]$$

$$= hAn - \frac{Bh}{2}[n(n+1)(h-u) + n(2u-h)] = \frac{2hAn + Bhn(n(u-h)-u)}{2} \quad (1.113)$$

Let's denote Eq. (1.112) as $COST_1 (n, h, u)$.

Case 2. Non-repairable failure occurs at a time $t \in ((k-1)h, kh]$. Then the net cost per unit time is

$$\frac{1}{t}\{c_f + (k-1)(a+bu) + c_r[\sum_{l=1}^{k-1}(M(lh - (l-1)u) - M((l-1)h - (l-1)u))$$

$$+ M(t - (k-1)u) - M((k-1)h - (k-1)u)] - Ve^{-wt}$$

$$- [2hA(k-1) + Bh(k-1)((k-1)(u-h) - u)]/2$$

$$- (t - (k-1)h)[A - (B(t + (k-1)h - 2(k-1)u))/2]\} \tag{1.114}$$

where we used Eq. (1.113) with $k-1$ instead of n and Eq. (1.111) for the interval

$$((k-1)h - (k-1)u, t - (k-1)u)$$

Notice that in computing the repair costs and generated revenue, in addition to the intervals $(0, h], (h, 2h], \ldots, ((k-2)h, (k-1)h]$, we have to take into account the last interval $((k-1)h, t]$ as well. Equation (1.114) can be denoted by $COST_2(t, k, h, u)$.

Then the expected net cost per unit time during a complete cycle (until replacement) is given as

$$COST_1(n, h, u)(1 - F_n(nh)) + \sum_{k=1}^{n} \int_{(k-1)h}^{kh} COST_2(t, k, h, u) f_k(t) dt \tag{1.115}$$

which is minimized with respect to decision variables $0 < u < h$, $n \in \{1, 2, \ldots\}$. We have to assume that the revenue generating density is positive: $A - Bnh \geq 0$; otherwise the machine would produce loss.

The previous formulation considered one type of repairable and non-repairable failures. If I types of repairable failures are considered, the third terms of Eqs. (1.112) and (1.114) have a second summation with respect to i before c_r, which is replaced by c_{ri}; and M must be replaced by M_i. These represent the cost and number of repairable failures of type i. Assume next that there are J types of non-repairable failures with CDFs $F^{(j)}(t)$ for $j = 1, 2, \ldots, J$. Let the failure times be denoted by $X^{(j)}, j = 1, 2, \ldots, J$. The first non-repairable failure occurs at $X = \min\{X^{(1)}, X^{(2)}, \ldots, X^{(J)}\}$, with CDF

$$F(t) = P(\min\{X^{(j)}\} \leq t) = 1 - P(\min\{X^{(j)}\} > t) = 1 - P\left(\bigcap_{j=1}^{J}(X^{(j)} > t)\right)$$

$$= 1 - \prod_{j=1}^{J} P(X^{(j)} > t) = 1 - \prod_{j=1}^{J}(1 - F^{(j)}(t))$$

by assuming that the non-repairable failures are independent of each other.

Optimal Coordination

In the next related model, the *optimal coordination* of the maintenance actions of several machines will be discussed. Consider a machine shop with K machines with equal lifetime of N time periods. The maintenance of these machines must be scheduled optimally in order to reduce setup costs. Let

$$x_{ki} = \begin{cases} 1 & \text{if machine } k \text{ is maintained at time period } i \\ 0 & \text{otherwise} \end{cases}$$

for $i = 1, 2, \ldots, N$ and $k = 1, 2, \ldots, K$

The constraints are as follows:

- For machine k, the time difference between consecutive maintenance actions must not be more than m_k: at least one maintenance must be done during the first m_k time periods:

$$\sum_{j=1}^{m_k} x_{kj} \geq 1 \qquad (k = 1, 2, \ldots, K)$$

and in general

$$\sum_{j=i+1}^{i+m_k} x_{kj} \geq x_{ki} \qquad (i = 1, 2, \ldots, N - m_k,\; ; k = 1, 2, \ldots, K) \qquad (1.116)$$

If $x_{ki} = 0$, then this constraint has no restriction. And if $x_{ki} = 1$, then at least one of $x_{k,i+1}, \ldots, x_{k,i+m_k}$ must be equal to 1.

- Let

$$z_i = \begin{cases} 1 & \text{if at least one maintenance is done at period } i \\ 0 & \text{otherwise} \quad (i = 1, \ldots, N) \end{cases}$$

Then we require that

$$z_i \leq \sum_{k=1}^{K} x_{ki} \leq K z_i \qquad (i = 1, \ldots, N) \qquad (1.117)$$

If no maintenance is made in period i, then all $x_{ki} = 0$, so $z_i = 0$. If at least one maintenance is done in period i, then $\sum_{k=1}^{N} x_{ki} \geq 1$, so z_i must not be 0, and the right-hand side of this inequality allows it to be equal to 1.

- The maximum number of maintenance actions in each time period must not exceed M:

$$\sum_{k=1}^{K} x_{ki} \leq M \qquad (i = 1, \ldots, N) \qquad (1.118)$$

The objective function is to minimize the number of time periods when maintenance is done:

$$\text{Minimize} \sum_{i=1}^{N} z_i \qquad (1.119)$$

This problem is a linear binary optimization problem, which can be solved by routine methods.

We also might minimize the total cost:

$$\text{Minimize} \sum_{i=1}^{N} C z_i + \sum_{i=1}^{N} \sum_{k=1}^{K} c_{ki} x_{ki} \qquad (1.120)$$

where C is the setup cost and c_{ki} is the maintenance cost of machine k in time period i.

A comprehensive summary of the most important maintenance models and methods is presented, for example, in (Valdez-Flores and Feldman 1989), (Nakagawa 2006, 2008), (Jardine 2006), and (Wang 2002).

1.8.3 Condition-Based Maintenance or Replacement

Next, we present a model of CBM or replacement if repair is not feasible or not economical (Hamidi et al. 2017). Consider a machine that produces utility with rate $u(t)$ after t time periods of operation, which includes the value of its produced utility minus all costs of operation. Assume at time T a sensor shows the start of a degradation process, resulting in a decrease in utility production, in addition to increased operation and replacement costs because of the increased level of degradation. In formulating the mathematical model to find the optimal time of repair or replacement, the following notations are used:

$\alpha =$ Degradation coefficient
$\beta =$ Cost of repair or replacement increase coefficient
$x =$ Planned time of repair or replacement
$u(t)e^{-\alpha(t-T)} =$ Density of utility value produced at time $t > T$
$Ke^{\beta(x-T)} =$ Cost of repair or replacement at time x

So, the total net profit per unit time can be given as

$$g(x) = \frac{\int_0^T u(t)dt + \int_T^x u(t)e^{-\alpha(t-T)}dt - Ke^{\beta(x-T)}}{x} \tag{1.121}$$

which is maximized with respect to the decision variable x. Notice first that the derivative of $g(x)$ has the same sign as

$$G(x) = [u(x)e^{-\alpha(x-T)} - \beta Ke^{\beta(x-T)}]x - \left[\int_0^T u(t)dt + \int_T^x u(t)e^{-\alpha(t-T)}dt - Ke^{\beta(x-T)}\right]$$

where

$$G(T) = (u(T) - \beta K)T - \int_0^T u(t)dt + K$$

and $G(x)$ tends to negative infinity as x converges to infinity.

Notice that

$$G'(x) = u'(x)e^{-\alpha(x-T)}x - \alpha u(x)e^{-\alpha(x-T)}x - \beta^2 Ke^{\beta(x-T)}x + u(x)e^{-\alpha(x-T)} - \beta Ke^{\beta(x-T)}$$

$$-u(x)e^{-\alpha(x-T)} + K\beta e^{\beta(x-T)} = u'(x)e^{-\alpha(x-T)}x - \alpha u(x)e^{-\alpha(x-T)}x - \beta^2 Ke^{\beta(x-T)}x < 0$$

since $u(x)$ is nonincreasing. So $G(x)$ is a strictly decreasing function of x. Let $T^* > T$ be the latest time period when the repair or replacement must be performed.

So, we have the following three possibilities:

- $G(T) \leq 0$. Then $G(x) < 0$ for all $x > T$, so $g(x)$ decreases, and therefore T is the optimal x value.
- $G(T^*) \geq 0$. Then $G(x) > 0$ for all $x < T^*$, so $g(x)$ increases, and therefore T^* is the optimum x value.
- Otherwise, $G(T) > 0$ and $G(T^*) < 0$, and there is a unique $\overline{T} \in (T, T^*)$ such that $G(\overline{T}) = 0$. This $\overline{T}$ value is the optimum scheduled repair or replacement time.

In cases where there are several machines with different parameters, and a common time x is to be determined to repair or replace them together in order to decrease setup costs, then Eq. (1.121) is generalized as

$$g(x) = \sum_i \frac{\left\{ \int_0^{T_i} u_i(t)dt + \int_{T_i}^x u_i(t)e^{-\alpha_i(t-T_i)}dt - K_i e^{\beta_i(x-T_i)} \right\}}{x} \tag{1.122}$$

where the summation is made for all machines involved. Similar to the previous case, the derivative of $g(x)$ has the same sign as

$$G(x) = \left\{ \sum_i u_i(x)e^{-\alpha_i(x-T_i)} - K_i\beta_i e^{\beta_i(x-T_i)} \right\} x$$

$$- \sum_i \left\{ \int_0^{T_i} u_i(t)dt + \int_{T_i}^x u_i(t)e^{-\alpha_i(t-T_i)}dt - K_i e^{\beta_i(x-T_i)} \right\}$$

and furthermore

$$G'(x) = \sum_i \{ u_i'(x)e^{-\alpha_i(x-T_i)}x - \alpha_i u_i(x)e^{-\alpha_i(x-T_i)}x - K_i\beta_i^2 e^{\beta_i(x-T_i)}x + u_i(x)e^{-\alpha_i(x-T_i)}$$

$$- K_i\beta_i e^{\beta_i(x-T_i)} - u_i(x)e^{-\alpha_i(x-T_i)} + K_i\beta_i e^{\beta_i(x-T_i)} \}$$

$$= \sum_i \{ u_i'(x)e^{-\alpha_i(x-T_i)}x - \alpha_i u_i(x)e^{-\alpha_i(x-T_i)}x - K_i\beta_i^2 e^{\beta_i(x-T_i)}x \} < 0$$

So, the optimum depends on the signs of $G(T)$ and $G(T^*)$, as in the previous case.

Consider next the special case of one machine with $u(t) \equiv u$ and $\beta = 0$: that is, a constant production rate and no increase in repair/replacement costs. Then

$$G(T) = uT - \beta KT - uT + K = K(1 - \beta T)$$

If $\beta T < 1$, then the optimal time is larger than T.

Example 1.21 As an illustrative example, let $\alpha = 0.1$, $\beta = 0.05$, $K = 3.5$, $T = 2$, and $T^* = \infty$, and the utility function can be defined as $u(t) = 2.5$. It can be seen that $G(T) > 0$ and the unique optimum is $\overline{T} = 5.82$. The corresponding total net utility value per unit time is $g(\overline{T}) = 1.50$. Figure 1.36 shows the objective function with different values of repair/replacement time. The figure illustrates that after a failure is detected at $T = 2$, it is best to keep the machine operating until $\overline{T} = 5.82$.

1.8.4 Preventive Replacement Scheduling

The optimum timing of *preventive replacement* is next discussed. In the case of a failure, the unexpected timing and the possible additional damages make the replacement more expensive than a preventive replacement. If replacement is done early, then a useful time interval is lost; and if it is scheduled late, then failure might occur, with a high replacement cost and possible damages and accidents. In scheduling preventive replacement, several factors should be considered.

Let X denote the random variable representing the time to unrepairable failure, when the object must be replaced. The CDF, PDF, reliability function, and failure rate of X are denoted by $F(t)$, $f(t)$, $R(t)$, and $\rho(t)$, respectively. Let c_p and c_f denote the cost of preventive and failure replacements, respectively. Assume that the preventive replacement is scheduled at time T; then the expected replacement cost is $c_p R(T) + c_f F(T)$,

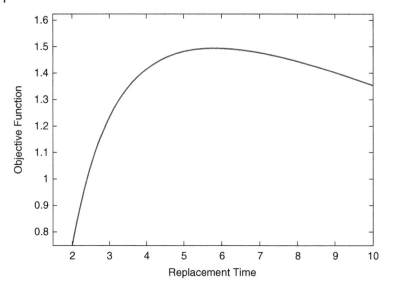

Figure 1.36 Shape of the objective function.

and the expected cycle length is given as $TR(T) + \int_0^T tf(t)dt$, which can be rewritten as $\int_0^T R(t)dt$. If identical cycles repeat indefinitely, then the expected cost per unit time can be obtained based on the renewal theory as

$$\frac{c_p R(T) + c_f F(T)}{\int_0^T R(t)dt} \tag{1.123}$$

which is minimized (Barlow and Proschan 1965; Aven and Jensen 1999). In many cases – for example, in fast-changing technologies – the optimum must be obtained based on the single-cycle criteria. The cost per unit time is given as

$$\begin{cases} \frac{c_f}{X} & \text{if } X \le T \\ \frac{c_p}{T} & \text{if } X \ge T \end{cases}$$

with expected value

$$\frac{c_p}{T}R(T) + \int_0^T \frac{c_f}{t}f(t)dt \tag{1.124}$$

which is then minimized. In the case of several distribution types, the optimal solution can be obtained easily.

Consider first the model in Eq. (1.123). Its derivative has the same sign as

$$(-c_p f(T) + c_f f(T)) \int_0^T R(t)dt - (c_p R(T) + c_f F(T))R(T)$$

or

$$(c_f - c_p)\left\{ \int_0^T R(t)dt \rho(T) + R(T) \right\} - c_f \tag{1.125}$$

The derivative of the bracketed expression is

$$R(T)\rho(T) + \int_0^T R(t)dt\rho'(T) - f(T) = \rho'(T) \int_0^T R(t)dt > 0$$

if $\rho(T)$ is increasing. So Eq. (1.125) is an increasing function, and its value at $T = 0$ is $-c_p < 0$. As $T \to \infty$, it converges to infinity if $\rho(T) \to \infty$ as $T \to \infty$; otherwise it converges to a positive limit if

$$(c_f - c_p) \int_0^\infty R(t)dt\rho(\infty) - c_f > 0$$

Therefore, in these cases there is a unique $\overline{T} > 0$ that makes Eq. (1.125) 0, and this is the unique optimum. If the limit of Eq. (1.125) is non-positive at infinity, then there is no finite optimum, and the equipment must be replaced when it fails.

The derivative of Eq. (1.124) has the form

$$-\frac{c_p}{T^2}R(T) - \frac{c_p}{T}f(T) + \frac{c_f}{T}f(T)$$

with the same sign as

$$(c_f - c_p)\rho(T) - \frac{c_p}{T} \tag{1.126}$$

If $\rho(T)$ increases, then this function also increases; at $T = 0$, its value converges to $-\infty$; and as $T \to \infty$, its limit is positive. So, there is again a unique solution that is the unique optimizer. The solution of monotonic equations can be obtained by standard computer methods (Yakowitz and Szidarovszky 1989).

Example 1.22 Assume a Weibull distribution with parameters $\beta = 1.5$ and $\eta = 5$, and $c_f = 200$ and $c_p = 100$. We showed earlier that the failure rate of a Weibull distribution is

$$\rho(t) = \frac{\beta}{\eta^\beta}t^{\beta-1}$$

and the first-order condition from Eq. (1.126) can be written as

$$(c_f - c_p)\frac{\beta}{\eta^\beta}T^{\beta-1} - \frac{c_p}{T} = 0$$

implying that

$$T = \left\{ \frac{c_p\eta^\beta}{\beta(c_f - c_p)} \right\}^{\frac{1}{\beta}}$$

In this case

$$T = \left\{ \frac{100(5)^{1.5}}{1.5(100)} \right\}^{\frac{1}{1.5}} \approx 3.816$$

1.8.5 Model Variants and Extensions

There are many variants and extensions of these models. Some of them will be briefly introduced.

Assume first that preventive replacements need T_p time to perform and T_f is the same for failure replacements. If we consider a cycle until a replacement starts operating, then T_p or T_f must be included in the cycle length. Eq. (1.123) is modified as

$$\frac{c_p R(T) + c_f F(T)}{T_p R(T) + T_f F(T) + \int_0^T R(t)dt} \tag{1.127}$$

and Eq. (1.124) becomes

$$\frac{c_p}{T + T_p} R(T) + \int_0^T \frac{c_f}{t + T_f} f(t)dt \tag{1.128}$$

These simple models do not take any additional elements into account: the cost of repairs of repairable failures, revenues generated by the product made by the machine, the salvage value after replacement, the optimal timing of the arrival of the spare parts, as well as the length of time required to perform repairs and the replacement. For the development of more advanced models, we introduce the following notations:

$c_r =$ Cost of each repair
$M(T) =$ Expected number of repairable failures in interval $[0, T]$
$u(T) =$ Generated revenue in interval $[0, T]$
$S(T) =$ Salvage value of the equipment after lifetime T
$\alpha =$ Unit inventory cost of spare part
$\beta =$ Unit loss of production delay

A new decision variable must be also included:

$W =$ Arrival schedule of spare part

For the sake of mathematical simplicity, we introduce the following function. If the equipment must be scheduled to be replaced at time t and the spare part arrives at time W, then the additional cost is

$$\gamma(t, W) = \begin{cases} (t - W)\alpha & \text{if } t \geqq W \\ (W - t)\beta & \text{if } t < W \end{cases}$$

In the first case, an inventory cost occurs; and in the second case, a production delay causes losses to the company.

If long-term expected costs are considered, then Eq. (1.124) is generalized as

$$\frac{(c_r M(T) + c_p - S(T) - u(T) + \gamma(T, W))R(T) + \int_0^T (c_r M(t) + c_p - S(t) - u(t) + \gamma(t, W))f(t)dt}{T_p R(T) + T_f F(t) + \int_0^T R(t)dt}$$

$$\tag{1.129}$$

and the corresponding one-cycle model is extended to a more complicated objective function:

$$\frac{c_p + c_r M(T) - S(T) - u(T) + \gamma(T, W)}{T + T_p} R(T)$$

$$+ \int_0^t \frac{c_f + c_r M(t) - S(t) - u(t) + \gamma(t, W)}{t + T_f} f(t) dt \qquad (1.130)$$

Both models have two decision variables, so for their optimization, commercial software is the best option. A complicated analytical approach is offered by the following idea. Consider first the value of T given, and optimize Eqs. (1.129) and (1.130) with respect to W. Clearly, the optimal W value depends on $T : W = W(T)$. Then substitute $W(T)$ into the objective functions, which now will depend only on T, so they become simple, single-variable search algorithms that can be used to find the optimal solutions.

If short-term and long-term objectives are considered simultaneously, then a multi-objective optimization method must be used with these two objective functions.

We can also incorporate the possibility of non-repairable failures in Eqs. (1.121) and (1.122). In this case, the conditional CDF of TTF after time T must be used, which can be obtained as follows:

$$P(X < t | X > T) = \frac{P(T < X < t)}{P(X > T)} = \frac{F(t) - F(T)}{1 - F(T)} \qquad (1.131)$$

Notice that these objectives represent expected costs per unit time without considering their uncertainties, which are characterized by the variances of the costs per unit time. Both expected costs and their variances should be minimized, leading again to the application of multi-objective optimization procedures.

A simple model is finally introduced to find the needed number of repair kits in any interval during the lifecycle of a piece of equipment. A similar model can be developed to plan the necessary amount of labor as well. For the sake of simplicity, assume that in cases of repairable failures, minimal repairs are performed. It is known from Section 1.5.1 that the number of failures in any interval $[t, T]$ follows a Poisson distribution with expectation

$$\lambda = M(T) - M(t) = \int_t^T \rho(\tau) d\tau \qquad (1.132)$$

meaning the times of failures form a non-homogenous Poisson process. Assume that each repair requires one unit of repair kit. It is a very important problem to determine the number of units that guarantees that all failure repairs can be performed with at least a given probability value $1 - p$. Let Q denote the required available number of kits; then with the notation

$$X_{[t,T]} = \text{actual number of failures in interval } [t, T]$$

we require that

$$P(X_{[t,T]} \leq Q) = \sum_{k=0}^{Q} \frac{\lambda^k}{k!} e^{-\lambda} = e^{-\lambda} \sum_{k=0}^{Q} \frac{\lambda^k}{k!} \geq 1 - p$$

implying that

$$\sum_{k=0}^{Q} \frac{\lambda^k}{k!} \geq (1 - p) e^{\lambda} \qquad (1.133)$$

The required value of Q can be obtained by repeatedly adding the terms of the left-hand side, starting with $k = 0$, until we reach or exceed $(1 - p)e^{\lambda}$.

In more complicated cases, the same method as shown in Section 1.5.1 can be used to find the second moment of the number of failures; then its variance can be easily obtained. Let μ and σ^2 denote the expectation and variance of the number of failures in interval $[t, T]$; then its distribution can be approximated by a normal distribution, and the condition from Eq. (1.133) is replaced by the following:

$$P(X_{[t,T]} \leq Q) = \phi\left(\frac{Q - \mu}{\sigma}\right) \geq 1 - p$$

where $\phi(t)$ is a CDF. Therefore

$$Q \geq \sigma\phi^{-1}(1 - p) + \mu \tag{1.134}$$

where the smallest integer exceeding the right-hand side gives the required number of repair kits.

1.9 Introduction to PHM: Summary

This chapter first gave a brief summary of reliability engineering and a probabilistic approach. It provided some statistical information about a large collection of identical items, which gives no information about a particular item. The different types of distributions of time to failure under normal or extreme stress conditions, the uncertainty of model parameters, and the expected number of failures with different types of repair or with failure replacement were discussed.

This introduction also provided a high-level introduction to PHM, including the following topics: (i) traditional approaches to PHM-enabling a system; (ii) DPS transfer curve and failure, and ideal and non-ideal FFS transfer curves; (iii) data conditioning; and (iv) prognostic information. The chapter introduced some important prognostic parameters and a new performance metric: convergence efficiency. These topics are discussed in more detail in the following chapters of this book. This introductory chapter presented only a brief outline of material related to signatures, data conditioning, and prognostic information; that material will be elaborated on further in later chapters. In addition, a brief overview of the cost-benefit analysis through the entire lifetime of equipment was provided.

References

Arrhenius, S.A. (1889). Über die Dissociationswärme und den Einfluß der Temperatur auf den Dissociationsgrad der Elektrolyte. *Zeitschrift für Physikalische Chemie* 4: 96–116.

Aven, T. and Jensen, U. (1999). *Stochastic Models in Reliability*. New York: Springer.

Ayyub, B.M. and McCuen, R.H. (2003). *Probability, Statistics, and Reliability for Engineers and Scientists*, 2e. Boca Raton: CRC Press.

Barlow, R.E. and Proschan, F. (1965). *Mathematical Theory of Reliability*. New York: Wiley.

Bucy, R.S. and Joseph, P.D. (2005). *Filtering for Stochastic Processes with Applications to Guidance*. Providence, RI: AMS Chelsea Publishing.

Cox, D. (1970). *Renewal Theory*. London: Methuen and Co.

Elsayed, E.A. (2012). *Reliability Engineering*. Hoboken, NJ: Wiley.

Eyring, H. (1935). The activated complex in chemical reactions. *The Journal of Chemical Physics* 3 (2): 107–115.

Finkelstein, M. (2008). *Failure Rate Modeling for Reliability and Risk*. London: Springer.

Frieden, B.R. (2004). *Science from Fisher Information: A Unification*. Cambridge, UK: Cambridge University Press.

Hamidi, M., Matsumoto, A., and Szidarovszky, F. (2017). Optimal schedule of repair or replacement after degradation is noticed. 46th Annual Meeting of the Western Decision Science Institute, Vancouver, Canada, 4–8 April.

Hamilton, J. (1994). *Time Series Analysis*. Princeton, NJ: Princeton University Press.

Hofmeister, J., Goodman, D., and Wagoner, R. (2016). Advanced anomaly detection method for condition monitoring of complex equipment and systems. 2016 Machine Failure Prevention Technology, Dayton, Ohio, US, 24–26 May.

Hofmeister, J., Goodman, D., and Szidarovszky, F. (2017a). PHM/IVHM: checkpoint, restart, and other considerations. 2017 Machine Failure Prevention Technology, Virginia Beach, Virginia, US, 15–18 May.

Hofmeister, J., Szidarovszky, F., and Goodman, D. (2017b). An approach to processing condition-based data for use in prognostic algorithms. 2017 Machine Failure Prevention Technology, Virginia Beach, Virginia, US, 15–18 May.

IEEE. (2017). Draft standard framework for prognosis and health management (PHM) of electronic systems. IEEE 1856/D33.

Jardine, A. (2006). Optimizing maintenance and replacement decisions. Annual Reliability and Maintainability Symposium (RAMS), Newport Beach, California, US, 23–26 January.

Kececioglu, D. and Jacks, J.A. (1984). The Arrhenius, Eyring, inverse power law and combination models in accelerated life testing. *Reliability Engineering* 8 (1): 1–9.

Kumar, S. and Pecht, M. (2010). Modeling approaches for prognostics and health management of electronics. *International Journal of Performability Engineering* 6 (5): 467–476.

Medjaher, K. and Zerhouni, N. (2013). Framework for a hybrid prognostics. *Chemical Engineering Transactions* 33: 91–96. https://doi.org/10.3303/CET1333016.

Milton, J.S. and Arnold, J.C. (2003). *Introduction to Probability and Statistics*. Boston, MA: McGraw Hill.

Mitov, K.V. and Omey, E. (2014). *Renewal Processes*. New York: Springer.

Nakagawa, T. (2006). *Maintenance Theory and Reliability*. Berlin/Tokyo: Springer.

Nakagawa, T. (2008). *Advanced Reliability Models and Maintenance Policies*. London: Springer.

National Science Foundation Center for Advanced Vehicle and Extreme Environment Electronics at Auburn University (CAVE3). (2015). Prognostics health management for electronics. http://cave.auburn.edu/rsrch-thrusts/prognostic-health-management-for-electronics.html (accessed November 2015).

Nelson, W. (1980). Accelerated life testing – step stress models and data analysis. *IEEE Transactions on Reliability* R-29 (2): 103–108.

Nelson, W. (2004). *Accelerated Testing: Statistical Models, Test Plans, and Data Analysis*. New York: Wiley.

O'Connor, P. and Kleyner, A. (2012). *Practical Reliability Engineering*. Chichester: Wiley.

Pecht, M. (2008). *Prognostics and Health Management of Electronics*. Hoboken, NJ: Wiley.

Pratt, J.W., Edgeworth, F.Y., and Fisher, R.A. (1976). On the efficiency of maximum likelihood estimation. *Annals of Statistics* 4 (3): 501–514.

Ross, M.S. (1987). *Introduction to Probability and Statistics for Engineers and Scientists*. New York: Wiley.

Ross, M.S. (1996). *Stochastic Processes*, 2e. New York: Wiley.

Ross, M.S. (2000). *Probability Models*, 7e. San Diego, CA: Academic Press.

Saxena, A., Celaya, J., Saha, B. et al. (2009). On applying the prognostic performance metrics. Annual Conference of the Prognostics and Health Management Society (PHM09), San Diego, California, US, 27 Sep.-1 Oct.

Speaks, S. (2005). Reliability and MTBF overview. Vicor Reliability Engineering. http://www.vicorpower.com/documents/quality/Rel_MTBF.pdf (accessed August 2015).

Szidarovszky, F., Gershon, M., and Duckstein, L. (1986). *Techniques of Multiobjective Decision Making in Systems Management*. Amsterdam: Elsevier.

Valdez-Flores, C. and Feldman, R.M. (1989). A survey of preventive maintenance models for stochastically deteriorating single-unit systems. *Naval Research Logistics* 36: 419–446.

Wang, H.Z. (2002). A survey of maintenance policies of deteriorating systems. *European Journal of Operational Research* 139 (3): 469–489.

Yakowitz, S. and Szidarovszky, F. (1989). *An Introduction to Numerical Computations*. New York: Macmillan.

Further Reading

Filliben, J. and Heckert, A. (2003). Probability distributions. In: *Engineering Statistics Handbook*. National Institute of Standards and Technology. http://www.itl.nist.gov/div898/handbook/eda/section3/eda36.htm.

Hofmeister, J., Wagoner, R., and Goodman, D. (2013). Prognostic health management (PHM) of electrical systems using conditioned-based data for anomaly and prognostic reasoning. *Chemical Engineering Transactions* 33: 992–996.

Tobias, P. (2003a). Extreme value distributions. In: *Engineering Statistics Handbook*. National Institute of Standards and Technology. https://www.itl.nist.gov/div898/handbook/apr/section1/apr163.htm.

Tobias, P. (2003b). How do you project reliability at use conditions? In: *Engineering Statistics Handbook*. National Institute of Standards and Technology. https://www.itl.nist.gov/div898/handbook/apr/section4/apr43.htm.

2

Approaches for Prognosis and Health Management/Monitoring (PHM)

2.1 Introduction to Approaches for Prognosis and Health Management/Monitoring (PHM)

You learned in Chapter 1 that the purpose of prognostics is to be able to accurately detect and report a future failure in systems – to predict failure progression. Prognostic approaches in prognostics and health management/monitoring (PHM) to accomplish that purpose can be grouped into broad categories: classical, usage-based, and condition-based (Hofmeister et al. 2017; Pecht 2008; Kumar and Pecht 2010; O'Connor and Kleyner 2012; Sheppard and Wilmering 2009). Classical prognostic approaches can be categorized as model-based, data-driven, or hybrid-driven, as shown in Figure 2.1.

2.1.1 Model-Based Prognostic Approaches

Model-based prognostic approaches include the modeling and use of expressions related to reliability, probability, and physics of failure (PoF) models. Such models are used to study and compare, for example, the relationships of materials, manufacturing, and utilization of the reliability, robustness, and strength of a product, often in structured, designed, controlled experiments and life tests. Such modeling offers potentially good accuracy, but it is difficult to apply and use in complex, fielded systems (Speaks 2005). Those models include distributions and probability models; fundamentals of reliability theory; and models based on reliability testing, such as acceleration factors (AFs), presented in Chapter 1. Other methods will be discussed later.

2.1.2 Data-Driven Prognostic Approaches

Data-driven prognostic approaches include statistical and machine learning (ML) methods; are generally simpler to apply, compared to model-based prognostic approaches; but can produce less precision and less accuracy in prognostic estimations. Statistical methods include both parametric and nonparametric models, such as those shown in (Ross 1987) and (Hollander et al. 2014); and K-nearest neighbor (KNN), a nonparametric method for classification or regression of an object with respect to its neighbors (Medjaher and Zerhouni 2013). Machine learning includes examples such as linear discriminant analysis (LDA) to characterize or separate multiple objects, hidden Markov modeling (HMM) to model a system having hidden states, and principal component analysis (PCA) to convert observations into linearly uncorrelated variables.

Prognostics and Health Management: A Practical Approach to Improving System Reliability Using Condition-Based Data,
First Edition. Douglas Goodman, James P. Hofmeister and Ferenc Szidarovszky.
© 2019 John Wiley & Sons Ltd. Published 2019 by John Wiley & Sons Ltd.

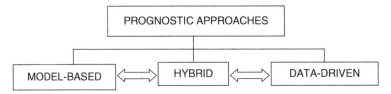

Figure 2.1 Block diagram showing three approaches to PHM. Source: based on Pecht (2008).

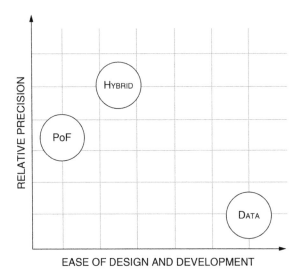

EASE OF DESIGN AND DEVELOPMENT

Figure 2.2 Precision and complexity: relative comparison of classical PHM approaches.

2.1.3 Hybrid Prognostic Approaches

Hybrid approaches employ both model-driven and data-driven approaches to further improve the accuracy of and/or to better understand the relationships of parameters and objects (Medjaher and Zerhouni 2013). Drawbacks are increased computational processing and complexity (see Figure 2.2). We will return to these and other methods later in this chapter.

2.1.4 Chapter Objectives

The objectives of this chapter are to present classical methodologies to support prognostics for PHM and then to present an approach to condition-based maintenance (CBM) for PHM: an approach based on CBD signatures that lays the foundation for Chapter 3.

2.1.5 Chapter Organization

The remainder of this chapter is organized to present and discuss classical approaches to modeling to support prognostics for PHM and to introduce our approach to CBM:

2.2 Model-Based Prognostics
  This section presents approaches to model-based prognostics, including topics on analytical modeling, distribution modeling, PoF and reliability modeling, acceleration

factors, complexity related to reliability modeling, failure distribution, failure rate and failures in time, and advantages and disadvantages of model-based prognostics.

2.3 Data-Driven Prognostics

This section presents approaches to data-driven prognostics including topics on statistical methods and machine learning – classification and clustering.

2.4 Hybrid-Driven Prognostics

This section presents approaches to hybrid-driven prognostics: model-based combined with data-driven prognostics.

2.5 An Approach to Condition-Based Maintenance (CBM)

This section presents an approach to CBM, including topics on modeling CBD signatures, comparing life consumption and PoF, and CBD signature methodologies. An illustration of CBD-signature modeling is included.

2.6 Approaches to PHM: Summary

This section summarizes the material presented in this chapter.

2.2   Model-Based Prognostics

Model-based approaches use analytical and PoF models. Analytical models include usage, statistical, and probabilistic models; and they may be validated by other models, such as PoF models and/or reliability-based models. Reliability-based models are associated with testing, such as accelerated life tests (ALTs) and regression analysis: Pecht favors PoF, in which life-cycle loading and failure mechanisms are modeled and applied to assess reliability and evaluate new materials, structures, and technologies (Pecht 2008). It should be noted that, in general, PoF modeling tends to be computationally prohibitive when applied to systems (Sheppard and Wilmering 2009). A simplified approach to model-based prognostics, shown in Figure 2.3, includes model development and model use.

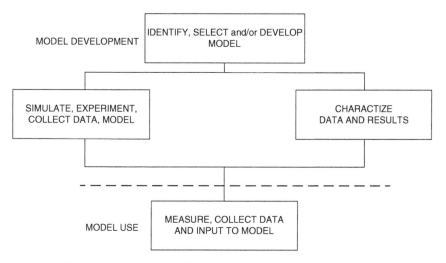

Figure 2.3 Model-based approach to development and use.

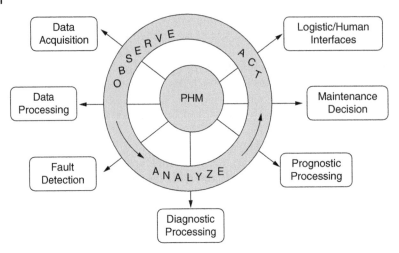

Figure 2.4 Model-use diagram. Source: based on Medjaher and Zerhouni (2013).

Model development includes the following: (i) identification, selection, and/or development of a model; (ii) simulation and/or experimentation to produce data to evaluate and verify the model; and (iii) characterization of the data for subsequent data measurement, collection, and inputting into the model to produce prognostic information when the model is used. Model use includes the following (see Figure 2.4): (i) acquire data, (ii) process data, (iii) detect fault(s), (iv) perform diagnostics, (v) perform prognostics, (vi) make decisions, and (vii) issue maintenance and logistic directives (Medjaher and Zerhouni 2013).

Referring to Figures 2.4 and 2.5, the functionality of the blocks labeled Data Acquisition, Data Processing, Fault Detection, Diagnostic Processing, and Prognostic Processing are embodied in the Sensor Framework, the Feature Vector Framework, the Prediction Framework, and the Performance Validation Framework. Prognostic information is passed to a Fault Management (FM) Framework and/or written to output files for deferred decisions and actions related to maintenance, logistics, and graphical-user interfaces (CAVE3 2015; Hofmeister et al. 2017).

2.2.1 Analytical Modeling

Analytical models, also referred to as physical models, employ load parameters such as those shown in Table 2.1 to estimate how a particular prognostic target in a system changes from a state of 100% healthy (not damaged) to zero health (failed) as damage accumulates (Pecht 2008; Hofmeister et al. 2016, 2017; Vichare 2006; Vichare et al. 2007). The PHM system performs health monitoring, detects an unhealthy condition, and uses, for example, fault-tree or state-diagram analysis to identify and determine the location(s) of the most likely prognostic target(s) causing the fault. Analytical approaches can be divided into two major groups: *inductive* and *deductive* approach.

Analytical Modeling: Inductive Approach

The *inductive* approach is based on reasoning, using qualitative data, from individual case to general conclusions. For example, such an approach might be used to determine

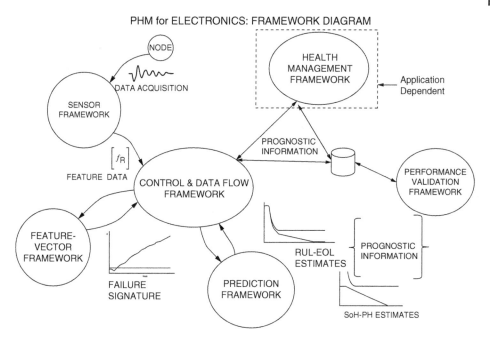

Figure 2.5 A framework for CBM for PHM (CAVE3 2015).

Table 2.1 Load types and examples.

Load	Examples of load type
Electrical	Current, voltage, power, energy
Thermal	Ambient temperature, temperature cycles, gradients, ramp rates
Mechanical	Pressure, vibration, shock load, stress/strain, vibration rate
Chemical	Humidity, reactivity – inert, active, acid, base – reaction rate
Physical	Radiation, magnetic and electrical fields, altitude

how the elimination or reordering of components in a design affects the overall operation, or how the elimination of a sensor affects the possible observation of a failure. There are many different methods for conducting inductive analysis, such as preliminary hazard analysis (PHA), failure mode and effect analysis (FMEA), and failure mode effect and criticality analysis (FMECA) and event tree analysis (Thomas 2006; Czichos 2013).

PHA is an initial study used in the early stages of designing systems to avoid costly redesign if a hazard is discovered later. It is a broad approach, and its main focus consists of the following elements:

1. Identify apparent hazards.
2. Assess the potential accidents (and their severity) that might occur as the consequence of a hazard.
3. Identify preventive measures to be used to reduce the risk.

FMEA first focuses on identifying potential failure modes, based on either PoF or earlier experience with the same or similar products. This information is used in the design and life-cycle phases of equipment, especially in support of diagnostics, maintenance, and prognostics. Effective FMEA is useful in identifying candidate signals and nodes to measure and capture leading indicators of failure that, when conditioned and collected, form signatures that are valuable for producing prognostic information (Hofmeister et al. 2017).

FMECA is an extension of FMEA that adds a criticality analysis to find the probabilities of different failure modes and the consequent severity of those failures. It is usually combined with event tree analysis, which is a forward, causal analytical technique. It gives the failure results (responses) and consequences of a single failure event on related or higher-level system components. Following the path from the initial event, it helps to assess the probability of the outcomes; as a result, overall system analysis can be performed (ETA 2017).

Analytical Modeling: Deductive Approach

The *deductive* approach is based on reasoning, using quantitative data, from general to specific events. For example, if a system failed, we wish to find out which component's behavior was the cause of the problem. A typical example of the deductive approach is the well-known fault tree analysis (FTA). It is similar to an event tree, where the direction of the analysis starts at the highest level and proceeds to lower levels. Any fault tree is based on primary events that are not further developed. A directed graph is constructed, where the primary events are the nodes; then an arc or line is placed from an event to another event, if failure of the initial node might generate failure in the terminal node. Therefore, if a failure occurs in any element of the system, we can move along the graph's tree, backward and forward, and find possible immediate reasons for the failing element and the failure mode. Thus we have the information required to know what to fix and how to fix it, to eliminate the problem (Thomas 2006; Czichos 2013).

Fault trees are closely related to Bayesian networks, if at each node failures are characterized by probability distributions determined by Bayesian rules from those of lower levels (Zhang and Poole 1996). A similar approach is offered by using Markov chains, when the degradation has several levels and the transition probabilities between the different levels are given, or shown from past observations (Ahmed and Wu 2013).

Example 2.1   Consider a machine with two parts, both of which are subject to random failures. The system has four possible states:

1. Both are in working condition.
2. Part 1 is good; part 2 is broken.
3. Part 1 is broken; part 2 is good.
4. Both parts are broken.

The following conditional probabilities are assumed to be known:

$$P\,(\text{part 1 breaks down}|\text{part 2 is good}) = P_{11}$$

$$P\,(\text{part 1 breaks down}|\text{part 2 is down}) = P_{12}$$

$$P\,(\text{part 2 breaks down}|\text{part 1 is good}) = P_{21}$$

$$P\,(\text{part 2 breaks down}|\text{part 1 is down}) = P_{22}$$

Assume that no repair is possible and that at most one part can break down in a single time period. The transition matrix based on the previous probabilities is given as follows:

$$T = \begin{pmatrix} 1 - P_{11} - P_{21} & P_{21} & P_{11} & 0 \\ 0 & 1 - P_{12} & 0 & P_{12} \\ 0 & 0 & 1 - P_{22} & P_{22} \\ 0 & 0 & 0 & 1 \end{pmatrix}$$

The matrix elements give the transition probabilities from each state to all states (including itself):

$$P(1|1) = P(\text{none of the parts breaks down}) = 1 - P_{11} - P_{21}$$

$$P(2|1) = P(\text{part 2 breaks down}|\text{part 1 is good}) = P_{21}$$

$$P(3|1) = P(\text{part 1 breaks down}|\text{part 2 is good}) = P_{11}$$

$$P(2|2) = 1 - P_{12}$$

$$P(4|2) = P_{12}$$

$$P(3|3) = 1 - P_{22}$$

$$P(4|3) = P_{22}$$

$$P(4|4) = 1$$

All other transition probabilities are equal to zero. Notice that the sum of the elements in each row equals unity. In this example, we consider discrete time scales. Let $P_i(t)$ denote the probability that at time period t, the machine is in state i. Then by using the Total Probability Theorem (Papoulis 1984), we see that

$$P_1(t + 1) = (1 - P_{11} - P_{21})P_1(t)$$

$$P_2(t + 1) = P_{21}P_1(t) + (1 - P_{12})P_2(t)$$

$$P_3(t + 1) = P_{11}P_1(t) + (1 - P_{22})P_3(t)$$

$$P_4(t + 1) = P_{12}P_2(t) + P_{22}P_3(t) + P_4(t)$$

If we introduce the vector $P(t) = (P_1(t), P_2(t), P_3(t), P_4(t))$, then these equations can be written in a matrix form:

$$P(t + 1) = P(t) * T \tag{2.1}$$

Example 2.2   Consider the same machine as before, but under different conditions. The time to failure (TTF) distribution of both parts is exponential, with failure rate λ. When any part breaks down, repair starts; its length is also exponential, with parameter μ. We consider a continuous time scale, when there is no "next time period" as

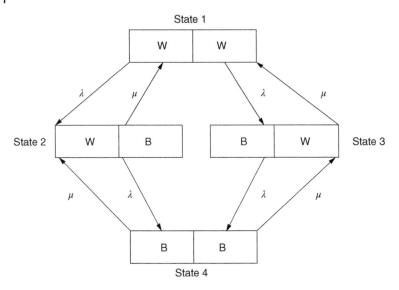

Figure 2.6 Transition diagram.

in the discrete time case; therefore, the rate of change in the probabilities $P_i(t)$ can be given. Figure 2.6 shows the diagram of the possible state transitions, where the rate of transitions is also given next to the arrows. It is assumed that at each time, only one broken part can go through repairs. Four differential equations can describe the process; and, similar to the discrete case, $P_i(t)$ is the probability that the machine is in state i at time t.

In computing the rate of change $P_1'(t)$ for state 1, we see that it can change four different ways: part 2 breaks down, part 1 breaks down, part 2 is repaired from state 2, or part 1 is repaired from state 3. The first two cases decrease the value of $P_1(t)$, and the other two cases increase its value:

$$P_1'(t) = \mu P_2(t) + \mu P_3(t) - 2\lambda P_1(t)$$

Similarly,

$$P_2'(t) = \lambda P_1(t) + \mu P_4(t) - \mu P_2(t) - \lambda P_2(t)$$

$$P_3'(t) = \lambda P_1(t) + \mu P_4(t) - \mu P_3(t) - \lambda P_3(t)$$

$$P_4'(t) = \lambda P_2(t) + \lambda P_3(t) - 2\mu P_4(t)$$

This is a system of linear differential equations that can be solved by simple algorithms using any known software packages, such as Mathematica (Yakowitz and Szidarovszky 1989).

Example 2.3 Another form of fault tree is shown in Figure 2.7. Here, a fuel-error fault has been evaluated –using, for example, FMEA and/or FMECA – as most likely being caused by one of four failure modes: air flow, pressure, temperature, or fuel pump. In this example, a temperature error is determined to be the likely cause of the fuel error. The fault tree indicates a temperature error that is likely to be the result of

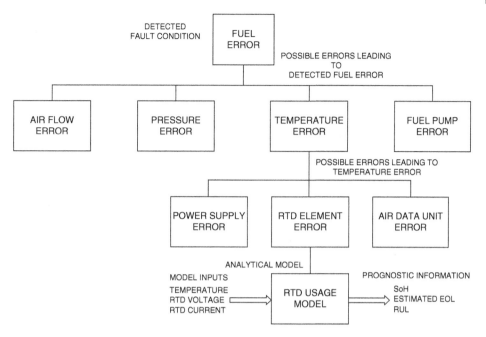

Figure 2.7 Example of a fault tree showing an RTD fault and an RTD-usage model.

three failure modes: power supply, resistive-temperature detector (RTD), or an air-data unit. For an RTD failure, a suitable analytical model is employed to produce prognostic information.

Example 2.4 A simple example of an analytical model is the power model used for degradation in capacitors and resistors. When power is applied, failure results because of increasing temperature (Viswanadham and Singh 1998):

$$t_f = A * P^{-n}, \tag{2.2}$$

where t_f is time to fail, A and n are constants, and P is electrical power.

Statistical models, including usage and distribution models, are predictive expressions that relate dependent variables, such as time to fail (t_f), and independent variables, such as power (P) in Eq. (2.2).

2.2.2 Distribution Modeling

When many physical causes and/or complex reactions result in failure, a distribution model is often used instead of a physical model. Engineering experience can relate special distribution types to given failure types. Examples of distribution models and associated applications include those shown in Table 2.2 (Medjaher and Zerhouni 2013; Viswanadham and Singh 1998; Hofmeister et al. 2006, 2013; Hofmeister and Vohnout 2011; Silverman and Hofmeister 2012).

Table 2.2 Failure distributions and example applications.

Distribution	Example applications
Exponential	Fatigue, wear caused by constant stress: resistors
Gamma	Failures caused by shock and vibration: boards, package connections
Lognormal	Failures caused by failure of insulation resistance, crack growth, and rate-dependent processes: encapsulation failures
Gumbel	Failures caused by corrosion, shear breaks (strength), dielectric breakdown: conductor connections, interconnects
Weibull	Life (use) and breakdown failures: capacitors, cables

The following distributions form characteristic curves that are useful for modeling complex failure mechanisms (a more complete list can be found in Chapter 1):

$$Exponential: f(x, \theta) = \frac{1}{\theta} e^{-\frac{x}{\theta}}$$

$$Gamma: f(x, \theta, k) = \frac{1}{\Gamma(k)\theta^k} x^{k-1} e^{-x/\theta}$$

$$Lognormal: f(x, \mu, m, \sigma) = [e^{-(\ln(x-\mu)-m)^2/(2\sigma^2)}]/[(x - \mu)\sigma\sqrt{2\pi}]$$

$$Gumbel: f(x, \sigma, \mu) = \frac{1}{\sigma} e^{\left(\frac{x-\mu}{\sigma}\right)} \exp\left[-e^{\frac{x-\mu}{\sigma}}\right]$$

$$Weibull: f(x, \beta, \eta, \gamma) = \frac{\beta}{\eta}\left(\frac{x - \gamma}{\eta}\right)^{(\beta-1)} e^{-\left(\frac{x-\gamma}{\eta}\right)^\beta})$$

In engineering applications, the Weibull, an extreme-value type of distribution, and the lognormal distribution are frequently used for modeling because they can be fitted to data from a large number of applications: especially lifetime distributions where failures are bounded below zero (Xu et al. 2015). The versatility of the Weibull distribution is evidenced by the following example life-time applications and by the example Weibull plots in Figure 2.8:

- Maximum flood levels of rivers; maximum wind gusts; steam boiler pressure
- Fatigue life of bearings; blending time of powder coating materials
- Tensile strength of wires; yield strength of beams
- Lifetimes of passive electronic components, such as resistors and capacitors
- Time-dependent dielectric breakdown (TDDB) phenomena; count of detectable high-resistance events in solder joints

2.2.3 Physics of Failure (PoF) and Reliability Modeling

PoF used in reliability modeling and simulation is significantly different from the constant-failure rate (CFR) modeling (based on exponential distribution) used as the basis for the Military Handbook 217 series (MIL-HDBK-217C). The PoF approach has dominated since the 1980s: root causes of failure, such as fatigue, fracture, wear, and corrosion, are studied and corrected to achieve lifetime design requirements by

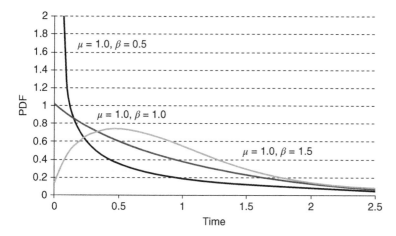

Figure 2.8 Example plots of Weibull distributions.

designing out causes of wear-out failures in components. Such modeling is used to study system performance and reduce failures. The following is a summary of a PoF approach (Weisstein 2015):

1. Identify potential failure modes.
2. Design and perform highly accelerated life tests (HALTs) or highly accelerated stress tests (HASTs) to verify and select dominant failure modes leading to failure.
3. Model the failure mode, and fit data to the model using statistical distributions.
4. Develop an equation for a PoF model to calculate mean time to failure (MTTF) and/or mean time before failure (MTBF).
5. Develop design, materials, and manufacturing improvements to increase MTTF to meet requirements.

The benefits of a PoF approach include the following:

- Reliability is designed in.
- Failures are reduced or eliminated prior to testing.
- Reliability of fielded systems is increased.
- Operational costs are decreased.

The reliability of a system can be obtained from the reliability of its building blocks.

- Assume first that the blocks have a series combination, meaning the system works if all blocks are in working condition. Let n be the number of blocks, and let $R_1(t)$, ..., $R_n(t)$ be the reliability functions of the blocks. Then that of the entire system can be obtained as

$$R(t) = P(X > t) = P((X_1 > t) \cap (X_2 > t) \cap \ldots \cap (X_n > t))$$

where $X_1, \ldots, X_n$ denote the failure times of the blocks. Assuming that the blocks are independent, then

$$R(t) = P(X_1 > t)P(X_2 > t) \ldots P(X_n > t) = R_1(t)R_2(t) \ldots R_n(t) \qquad (2.3)$$

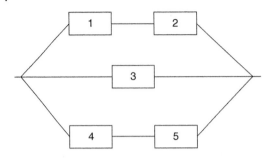

Figure 2.9 A special system structure.

showing that the reliability functions are multiplied. If a new block is added to the system, then a new factor is added to the product, which is less than one, so the system reliability decreases.

- Assume next that the connection is parallel, meaning the system works if at least one of the blocks works. In this case

$$R(t) = 1 - P(X < t) = 1 - P((X_1 < t) \cap (X_2 < t) \cap \ldots \cap (X_n < t))$$
$$= 1 - P(X_1 < t)P(X_2 < t) \ldots P(X_n < t)$$
$$= 1 - (1 - R_1(t))(1 - R_2(t)) \ldots (1 - R_n(t)) \tag{2.4}$$

If a new block is added to the system, then the second term decreases, so $R(t)$ increases.

In many systems, the connection of the blocks is a combination of series and parallel combinations. Then Eqs. (2.3) and (2.4) are used repeatedly, as shown in the following example.

Example 2.5   Consider a system with five blocks that are connected as shown in Figure 2.9. Assume the reliability function of each block is e^{-2t}. Blocks 1 and 2, as well as blocks 4 and 5, have series connections, so the reliability functions of these simple combinations are

$$R_{12}(t) = R_1(t)R_2(t) = e^{-2t} * e^{-2t} = e^{-4t}$$

and

$$R_{45}(t) = R_4(t)R_5(t) = e^{-2t} * e^{-2t} = e^{-4t}$$

Notice next that the combinations 1-2 and 4-5 have a parallel combination with block 3; therefore, by Eq. (2.4), we conclude that the reliability function of the system becomes

$$R(t) = 1 - (1 - R_{12}(t))(1 - R_3(t))(1 - R_{45}(t))$$
$$= 1 - (1 - e^{-4t})(1 - e^{-2t})(1 - e^{-4t})$$
$$= 1 - (1 - e^{-2t})(1 - e^{-4t})^2$$

For example, the probability that the system will operate at least until time $t = 1$ equals

$$R(1) = 1 - (1 - e^{-2})(1 - e^{-4})^2 = 1 - 0.8647 * 0.9817^2 = 0.1167$$

2.2.4   Acceleration Factor (AF)

Chapter 1 discussed how TTF distributions change when an item is subject to extreme stresses. These formulas can be used backward, when observations are made under extreme conditions, to estimate failure times: based on the data, distributions of TTFs can be estimated under normal conditions. This idea is known as accelerated testing (AT), ALT, and so on.

Reliability projections, such as MTTF based on accelerated testing (such as a HALT) when the object is subject to extreme stress, are projected estimates from test results to a future, slower rate of failure at lower levels of stress under normal use conditions. Such projections assume the use of a correct model for life distribution and the use of a correct acceleration model (Tobias 2003; Nelson 2004). In an acceleration model, the TTF (or t_F) at a stress level (s) is given by the following (Kentved and Schmidt 2012):

$$t_F = A\, G(s) \tag{2.5}$$

where $A =$ constant, $G(s) =$ stress function.

The acceleration factor (AF) is the ratio of the TFFs at two different levels of stress:

$$AF = t_{F1}/t_{F2} = G(s1)/G(s2) \tag{2.6}$$

AFs used in reliability estimations (or usage modeling) include the following (White and Bernstein 2008).

Arrhenius Law for Temperature

AFs used in reliability estimations (or usage modeling) include the following (White and Bernstein 2008):

$$AF_T = \exp\left[\frac{E_a}{k}\left(\frac{1}{T_{use}} - \frac{1}{T_{test}}\right)\right] \tag{2.7}$$

where

$T_{use} =$	Product temperature in service use
$T_{test} =$	Product temperature in laboratory test
$E_a =$	Activation energy for damage mechanism and material
$k =$	Bolzman's constant $= 8.617 * 10^{-5} eV/^\circ K$

Kemeny Law for Voltage

$$AF_T = \exp\left[C_0 - \frac{E_a}{kT_j}\right]\exp\left[C_1\left(\frac{V_{cb}}{V_{cbmax}}\right)\right] \tag{2.8}$$

where in addition

$C_0, C_1 =$	Material constants
$T_j =$	Junction temperature
$V_{cb} =$	Collector-base voltage
$V_{cbmax} =$	Maximum collector-base voltage before breakdown

Peck's Law for Temperature and Humidity

$$AF_{TM} = \left(\frac{M_{use}}{M_{test}}\right)^{-n}\exp\left[\frac{E_a}{k}\left(\frac{1}{T_{use}} - \frac{1}{T_{test}}\right)\right] \tag{2.9}$$

where in addition

M_{use} =	Moisture level in service use
M_{test} =	Moisture level in test

This formula can be derived from the temperature-humidity relationship discussed in Chapter 1.

Coffin-Manson Law of Fatigue

$$AF_T = \left(\frac{\Delta T_{test}}{\Delta T_{use}}\right)^n \tag{2.10}$$

where

ΔT =	Difference between the high and low temperatures for the product in service use and in the laboratory test.

Notice that (2.10) is derived from the inverse power law.

Eyring Formula

$$AF_T = \frac{\frac{1}{T_{use}} \exp\left(-\left(A - \frac{B}{T_{use}}\right)\right)}{\frac{1}{T_{test}} \exp\left(-\left(A - \frac{B}{T_{test}}\right)\right)} = \frac{T_{test}}{T_{use}} \exp\left(\frac{B}{T_{use}} - \frac{B}{T_{test}}\right) \tag{2.11}$$

Similar formulas can be derived from any other known rules dealing with extreme stress levels such as the generalized Eyring, temperature non-thermal relations, or the general log-linear law discussed in Chapter 1.

2.2.5 Complexity Related to Reliability Modeling

In addition to requiring the use of a correct distribution model and a correct acceleration model, reliability modeling is further complicated by an almost infinite number of methods, like those in Table 2.3. Each has advantages and disadvantages in comparison to the others for a given application.

Reliability modeling is even more complicated because of the variability in the fitting of values used within the models of each version of a procedure. This variance is evidenced by the modeling examples of the AFs for temperature as shown in Table 2.4.

In addition to differences in acceleration factors and parameter values used in modeling, there are differing versions of distribution models. Examples include three different Coffin-Manson models for calculating a probable test cycle (N_f) that a solder joint fails when subjected to cyclic loading of temperature during accelerated testing (Viswanadham and Singh 1998):

$$N_f = A * (\Delta T)^{1/2} \qquad \text{for } \Delta T < 0.5\, T_{\text{MELT}} \tag{2.12}$$

$$N_f = (f_u)^{1/3} (\Delta T_u)^{1.9} e^{(0.01\, T_t)} \qquad \text{for eutectic PbSn (90/10) solder} \tag{2.13}$$

$$N_f = (f_u)^{1/3} (1/T_u)^{2.0} e^{[\Delta E/k(T_t/T_u)]} \qquad \text{parameter } \Delta E \text{ equals 0.4 eV} \tag{2.14}$$

Table 2.3 Examples of reliability procedures and applications (White and Bernstein 2008).

Reliability procedure/method	Example applications
MIL-HDBK-217	Military
Telecordia SR-332	Telecom
CNET	Ground military
RDF-93 and 2000	Civil equipment
SAE Reliability Prediction Method	Automotive
BT-HRD-5	Telecom
Siemens SN29500	Siemens products
NTT Procedure	Commercial and military
PRISM	Aeronautical and military
FIDES	Aeronautical and military

Table 2.4 Examples of temperature acceleration models (White and Bernstein 2008).

Procedure	Acceleration factor for temperature
MIL-HDBK-217F	$AFT = 0.1 \exp\left[-A(1/T_j - 1/298)\right]$
HRD4	$AFT = 2.6*104 \exp[-3500/T_j)] + 1.8*1013 \exp[-11600/T_j)]$ for $T_j \geq 70°\text{C}$
NTT	$AFT = \exp[3480(1/339 - 1/T_j)] + \exp[8120(1/356 - 1/T_j)]$
CNET	$AFT = A1 \exp[-3500/T_j] + A2 \exp[11600/T_j]$
Siemens	$AFT = A \exp[Ea*11605(1/T_{ji} - 1/T_{j2})] + (1 - A) \exp[-Ea*11605(1/T_{j1} - 1/T_{j2})]$

where subscript u refers to a use value, subscript t refers to a test value, parameter T is in $°\text{K}$, and f_u is a model parameter.

Example 2.6 As an example of testing for reliability modeling, suppose a set of 32 Xilinx FG1156 field programmable gate array (FPGA) devices is subjected to a HALT using the following regime (Xilinx 2003; Hofmeister et al. 2010):

- Test period: temperature cycling until 30 of the 32 FPGAs fail.
- Each temperature cycle of the HALT is a transition from −55 to 125 °C in 30 minutes: 3-minute ramps and 12-minute dwells.
- FPGA daisy-chain type of test package in which the solder balls are series connected such that instances of connectivity failures (high-resistance opens) are captured.
- Each logged failure of an FPGA (diamond symbols in Figure 2.10) represents at least 30 events of high resistance:
  - OPEN is defined as a measured resistance of 500 units or higher having a duration of at least 2 units.
  - A FAIL event is defined as at least two OPENs within a one temperature cycle.
  - FAILURE is defined as 15 FAIL events.

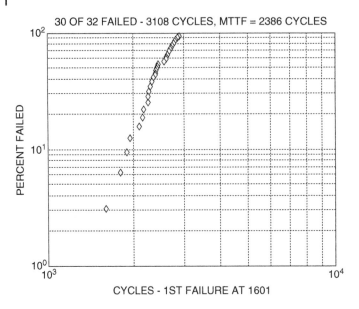

Figure 2.10 HALT result – 30 of 32 FPGA devices failed (Hofmeister et al. 2006).

- A single OPEN in a given temperature cycle is not counted as a FAIL event.
- The HALT results in 30 of the 32 FPGA devices failing, as shown in Figure 2.10.

Now, suppose a prognostic target, such as an FPGA attached to a fiber-resin (FR-4) printed wire board (PWB), which is formerly and sometimes still referred to as printed circuit board (PCB), is operated such that the temperature varies in any given 24-hour period of time from less than −40 °C to over 100 °C, with different temperature ramp-up rates, different dwell times at high temperature, different ramp-down rates, and different dwell times at low temperature. Also suppose that during any given 24-hour period of time, the PWB is subjected to different rates of different magnitudes of vibration and shock (such as might be experienced from being mounted in an engine compartment of a vehicle). Further, suppose that the PWB comprises over a dozen different FPGAs, some of which use standard PbSn solder, some of which use lead-free solder balls, some of which use plastic grid array (PGA) die packages with and without staking, and some of which use ceramic-column grid array (CCGA); all are mounted at different distances from centers of maximum stress-strain. Estimating with a high degree of accuracy when the primary clock-input pin of a specific FPGA will fail and thereby cause the PWB to fail becomes a daunting task (Javed 2014).

2.2.6 Failure Distribution

As the independent variables change in distribution models, the rate of change of a given curve varies, which creates a family of failure curves having a failure distribution with a TTF as illustrated in Figure 2.11. The failure distribution is a probability density function (PDF), and the TTF is the expectation or the 0.50 value of the cumulative distribution function (CDF) of that PDF: note that TTF is not the same entity as MTTF – more on this topic is presented in Chapter 7.

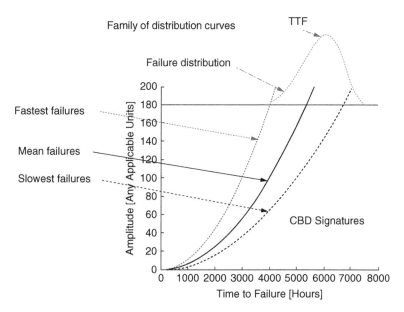

Figure 2.11 Family of failure curves, failure distribution, and TTF.

2.2.7 Multiple Modes of Failure: Failure Rate and FIT

When prognostic targets are prone to more than one dominant failure mode, multiple distribution models and/or multiple-parameter models must be used in a PoF approach. For example, a transistor device could fail because of temperature cycling and also because of failure related to high levels of voltage. Two distribution models might then apply: Arrhenius temperature, Eq. (2.7); and Kemeny distribution, Eq. (2.8). A simplifying approach is to assume that all failures are random and all failure modes are equally dominant. Overall MTTF and failure in time (FIT) values can be calculated by applying a sum-of-failure-rate model and an improved AF to account for two different temperatures, use and test, as shown in Eq. (2.9). One FIT is equal to one failure in 1 billion part hours (White and Bernstein 2008):

$$\text{Failure rate} = \lambda = \frac{(\text{number of failures})}{(\text{number of tested parts})(\text{hours of test}) \, AF} 10^9 \, \text{FIT} \qquad (2.15)$$

2.2.8 Advantages and Disadvantages of Model-Based Prognostics

Advantages of model-based prognostics are many and include the following: (i) such modeling leads to a better understanding of how prognostic targets, especially devices, fail because of defects and weaknesses in manufacturing processes and materials, and how and why they fail because of loading and environmental stresses and strain; (ii) manufacturing processes, materials, electrical and physical designs, and control of operational loading and environmental conditions can be improved to increase reliability; and (iii) simple estimates of state of health (SoH) and remaining useful life (RUL) are possible.

Disadvantages of model-based prognostics are also many and include the following: (i) modeling for other than single-mode failures is complex; (ii) simple models for non-steady state and multiple and variable environment loading generally do not exist; (iii) modeling of large, complex systems of hundreds or thousands of different parts becomes extremely difficult, if not computationally intractable; and, perhaps most important, (iv) model-based approaches are not applicable to a specific prognostic target in a system, as exemplified by Figure 2.11: MTTF, for example, applies to a population of like prognostic targets rather than a fielded, specific prognostic target in a specific system in an operational, non-test environment.

2.3 Data-Driven Prognostics

Data-driven (DD) prognostics (Figure 2.12) comprises two major approaches, statistical and machine learning (ML), that use acquired data to statistically and probabilistically produce prognostic information such as decisions, estimates, and predictions. Statistical approaches include parametric and nonparametric methods; ML approaches include supervised and unsupervised classification and clustering, and regression and ranking (Pecht 2008). This book will not discuss regression and ranking.

2.3.1 Statistical Methods

Statistical methods can be divided into parametric and nonparametric methods (Pecht 2008), including those shown in Table 2.5.

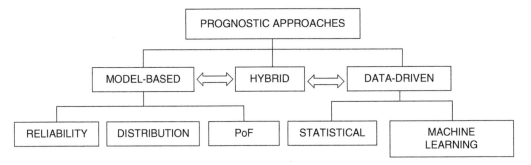

Figure 2.12 Diagram of data-driven approaches.

Table 2.5 Parametric and nonparametric methods.

Parametric technique
Maximum likelihood (MLE)
Likelihood ratio test (LRT)
Minimum mean square error (MSE)
Maximum a posteriori estimation (MAP)
Nonparametric technique
K-nearest neighbor classifier (kNN)
Kernel density estimation (KDE)
Chi square test (CST)

Maximum Likelihood Method

The maximum likelihood method (Ross 1987) is a common procedure to estimate unknown parameters of probability distributions. As in Section 1.4, let $f(t|\theta)$ denote the PDF where θ is unknown. Assume we have a random sample $t_1, t_2, \ldots, t_N$ from this distribution. The likelihood function is defined as

$$L(\theta) = f(t_1|\theta)f(t_2|\theta)\ldots f(t_N|\theta) \tag{2.16}$$

which represents the probability of the sampling event that actually occurred. Since the logarithmic function strictly increases, instead of $L(\theta)$, its logarithm is maximized:

$$\text{Maximize } \ln(L(\theta)) \tag{2.17}$$

and the optional θ value is accepted as the estimate of the unknown parameter.

Example 2.7   Assume that the distribution is Weibull with a known value of β and an unknown value of η. We discussed in Chapter 1 that

$$f(t\,|\beta,\eta) = \frac{\beta}{\eta^\beta}\, t^{\beta-1} e^{\left(\frac{t}{\eta}\right)^\beta}$$

and therefore

$$L(\eta) = \frac{\beta^N}{\eta^{N\beta}}\left(\prod_{i=1}^{N} t_i\right)^{\beta-1} e^{-\sum_{i=1}^{N}(t_i)^\beta \eta^{-\beta}}$$

so

$$l(\eta) = \ln(L(\eta)) = N\ln(\beta) - N\beta\ln(\eta) + (\beta-1)\ln\left(\prod_{i=1}^{N} t_i\right) - \eta^{-\beta}\sum_{i=1}^{N} t_i^\beta$$

By differentiation,

$$\frac{\partial}{\partial\eta}l(\eta) = -\frac{N\beta}{\eta} + \beta\eta^{-\beta-1}\sum_{i=1}^{N} t_i^\beta$$

So, the first-order condition can be written as

$$-\frac{N\beta}{\eta} + \frac{\beta}{\eta^{\beta+1}}\sum_{i=1}^{N} t_i^\beta = 0$$

showing that

$$\eta = \left(\frac{1}{N}\sum_{i=1}^{N} t_i^\beta\right)^{\frac{1}{\beta}} \tag{2.18}$$

Likelihood Ratio Test

The likelihood ratio test (Casella and Berger 2002) is usually used to determine the validity of an estimate. For example, let θ_0 be an estimate of an unknown parameter of a PDF $f(t|\theta)$. The likelihood ratio test is based on the likelihood ratio

$$r = \frac{L(\theta_0)}{\max_\theta L(\theta)} \tag{2.19}$$

where the likelihood function is denoted by $L(\theta)$ and the denominator is the maximal value of $L(\theta)$. In other cases, two estimates are compared. Let θ_1 and θ_2 be two estimates for θ; then

$$r = \frac{L(\theta_1)}{L(\theta_2)} \tag{2.20}$$

If the value of r is small, then in the first case θ_0 is unacceptable, and in the second case θ_2 is a much better estimate.

Example 2.8 Assume an exponential distribution with a parameter λ and sample elements $t_1, t_2, ..., t_N$. Then

$$f(t|\lambda) = \lambda e^{-\lambda t}$$

so

$$L(\lambda) = \lambda e^{-\lambda t_1} \lambda e^{-\lambda t_2} \dots \lambda e^{-\lambda t_N} = \lambda^N e^{-\lambda \sum_{i=1}^{N} t_i}$$

Therefore, the likelihood ratio test in comparing two estimates, λ_1 and λ_2, can be written as

$$r = \frac{\lambda_1^N e^{-\lambda_1 \sum_{i=1}^{N} t_i}}{\lambda_2^N e^{-\lambda_2 \sum_{i=1}^{N} t_i}} = \left(\frac{\lambda_1}{\lambda_2}\right)^N e^{(\lambda_2-\lambda_1)\sum_{i=1}^{N} t_i} \tag{2.21}$$

by Eq. (2.20).

Minimum Mean Square Error

Minimum mean square error is mainly used in fitting function forms with unknown parameters. For example, a density histogram is obtained from a sample with points $(t_1, f_1), (t_2, f_2), ..., (t_N, f_N)$, and it is known that the corresponding PDF is $f(t|\theta)$. The least square estimate of θ is obtained by minimizing the overall squared error:

$$Q = \sum_{i=1}^{N} (f(t_i|\theta) - f_i)^2 \tag{2.22}$$

Example 2.9 Assume that $f(t|\theta) = \theta e^{-\theta t}$; then

$$Q = \sum_{i=1}^{N} (\theta e^{-\theta t_i} - f_i)^2 \tag{2.23}$$

The first-order condition can be written as

$$\frac{\partial Q}{\partial \theta} = 2 \sum_{i=1}^{N} (\theta e^{-\theta t_i} - f_i)(e^{-\theta t_i} - \theta^2 e^{-\theta t_i}) = 0$$

which is a nonlinear equation and can be solved by standard methods.

Maximum a Posteriori Estimation

The maximum a posteriori estimation (Stein et al. 2002) is based on Bayesian principles. Consider again a PDF $f(t|\theta)$ depending on the unknown parameter θ. The likelihood function is given as $L(\theta)$, which is maximized in order to get the best estimate for θ. In practical cases, usually its logarithm is maximized. Assume now that a prior distribution is known for θ, with PDF $g(\theta)$. By the Theorem of Bayes

$$L(\theta|t_1, \ldots, t_N) = \frac{L(\theta)g(\theta)}{\int L(\bar{\theta})g(\bar{\theta})d\bar{\theta}} \tag{2.24}$$

where the integration domain is the domain of all possible values of θ. Since the denominator is independent of θ, we need to optimize only $L(\theta)g(\theta)$.

Example 2.10 Assume a normal distribution with parameters μ and σ^2, where σ^2 is known and only μ is unknown. Then

$$L(\mu) = \prod_{i=1}^{N} \frac{1}{\sigma\sqrt{2\pi}} e^{-\frac{(t_i - \mu)^2}{2\sigma^2}}$$

From statistics, it is known that the normal distribution is its conjugate prior, so we can select

$$g(\mu) = \frac{1}{\bar{\sigma}\sqrt{2\pi}} e^{-\frac{(\mu - \bar{\mu})^2}{2\bar{\sigma}^2}}$$

And therefore

$$L(\mu)g(\mu) = \frac{1}{\bar{\sigma}\sqrt{2\pi}} e^{-\frac{(\mu - \bar{\mu})^2}{2\bar{\sigma}^2}} \prod_{i=1}^{N} \frac{1}{\sigma\sqrt{2\pi}} e^{-\frac{(t_i - \mu)^2}{2\sigma^2}}$$

which can be maximized. Its logarithm equals

$$-\ln(\bar{\sigma}\sqrt{2\pi}) - \frac{(\mu - \bar{\mu})^2}{2\bar{\sigma}^2} - N\ln(\sigma) - N\ln(\sqrt{2\pi}) - \sum_{i=1}^{N} \frac{(t_i - \mu)^2}{2\sigma^2}$$

This is a concave function in μ, and its derivative has the special form

$$\frac{-2(\mu - \bar{\mu})}{2\bar{\sigma}^2} + \sum_{i=1}^{N} \frac{2(t_i - \mu)}{2\sigma^2} = \frac{1}{\bar{\sigma}^2\sigma^2} \left\{ -\sigma^2(\mu - \bar{\mu}) + \bar{\sigma}^2 \left(\sum_{i=1}^{N} t_i - N\mu \right) \right\}$$

giving the solution

$$\mu = \frac{\sigma^2\bar{\mu} + \bar{\sigma}^2 \sum_{i=1}^{N} t_i}{\sigma^2 + N\bar{\sigma}^2} \tag{2.25}$$

K-Nearest Neighbor Classifier

The *KNN* classifier (Cover and Hart 1967) is based on the following simple procedure. Assume we have N vectors; each of them is attached with a class label. We need to put a given vector into the most appropriate class. We select a positive integer $k \geq 1$ and determine the k closest vectors from the given N vectors by using any distance measure, such as the Euclidean distance. Then the selected k closest vectors "vote" about the class by selecting the class that appears the most times among the k closest vectors. In this way, we can order any set of vectors into given classes based on a given sample.

In a two-dimensional case, the Euclidean distance of vectors (x_c, y_c) and (x_i, y_i) is the following:

$$D_i = \sqrt{(x_c - x_i)^2 + (y_c - y_i)^2} \tag{2.26}$$

The Euclidean distance is based on the Pythagorean theorem. If the dimension is larger than two, say n, then similarly

$$D_i = \sqrt{\sum_{j=1}^{n} (x_c^{(j)} - x_i^{(j)})^2} \tag{2.27}$$

where the components of vector x_c are $x_c^{(j)}$ and those of vector x_i are $x_i^{(j)}$.

Kernel Density Estimation

This method is based on the following equation:

$$p(x) = \frac{1}{Nh} \sum_{i=1}^{N} K\left(\frac{x - x_i}{h}\right) \tag{2.28}$$

which can be interpreted as follows (Wand and Jones 1995). Let $x_1, x_2, \ldots, x_N$ be the sample elements. A kernel function $K(x)$ is selected that is nonnegative and that has an integral of one (like the properties of any PDF). A parameter $h > 0$ is also chosen, which is called the *bandwith*. Then Eq. (2.28) gives an estimate of the PDF, from which the sample is generated, at point x. Table 2.6 gives a collection of commonly used kernel functions.

Chi-Square Test

The chi-square test (Ross 1987) is used to test whether a given sample comes from a population with a specific distribution, and therefore it does not provide the distribution; it can only be used to check whether a user-selected distribution is appropriate.

Table 2.6 Kernel functions.

Kernel	$K(x)$	Domain
Uniform	$\dfrac{1}{2}$	$(-1, 1)$
Triangle	$1 - |x|$	$(-1, 1)$
Epanechnikov	$\dfrac{3}{4}(1 - x^2)$	$(-1, 1)$
Quartic	$\dfrac{15}{16}(1 - x^2)^2$	$(-1, 1)$
Twiweight	$\dfrac{35}{32}(1 - x^2)^3$	$(-1, 1)$
Gaussian	$\dfrac{1}{\sqrt{2\pi}}e^{-\frac{x^2}{2}}$	$(-\infty, \infty)$
Cosinus	$\dfrac{\pi}{4}\cos\left(\dfrac{\pi}{2}x\right)$	$(-1, 1)$

The data is divided into K bins, with y_k and y_{k+1} being the lower and upper limits of class k. It is assumed that the bins are defined by subintervals between the consecutive nodes $y_0 < y_1 < \ldots < y_K$. Let O_k be the observed frequency for bin k, and E_k; then the expected frequency is defined as

$$E_k = N(F(y_{k+1}) - F(y_k))$$

where N is the total number of sample elements and $F(y)$ is a CDF. The chi-square test computes the value of χ^2 as

$$\chi^2 = \sum_{i=1}^{K} \frac{(O_i - E_i)^2}{E_i} \tag{2.29}$$

and the distribution $F(y)$ is rejected if

$$\chi^2 > \chi^2_{1-\alpha, K-c}$$

where α is a user-selected significance level, and c is the number of the unknown parameters. The threshold $\chi^2_{1-\alpha, K-c}$ can be found in the chi-square test tables.

2.3.2 Machine Learning (ML): Classification and Clustering

Machine learning (ML), a form of artificial intelligence, predicts future behavior by learning from the past: classification and clustering are forms of ML divided into supervised and unsupervised techniques, which are further divided into discriminative and generative approaches. Certain ML approaches, such as regression and ranking, are less useful compared to classification and clustering, which use computational and statistical methods to extract information from data (Pecht 2008). Table 2.7 is a summary list of some of the ML techniques.

These techniques are well presented in the literature; therefore we do not discuss them in detail. Instead, we will select one method from each category. The other methods are briefly described as examples.

Discriminant Analysis

The objective of discriminant analysis (Fukunaga 1990) is to classify objects (usually given as multidimensional vectors) into two or more groups based on certain features that describe the objects by minimizing the total error of classification. This is done by assigning each object to the group with the highest conditional probability. The mathematical solution requires sophisticated techniques of matrix analysis.

Neural Networks

Neural networks are database input-output relations (Faussett 1994). They are "connectionist" computer systems. Let the input vector be denoted as $x = (x_1, x_2, \ldots, x_m)$ and the output vector as $y = (y_1, y_2, \ldots, x_n)$. The transformation $x \to y$ is performed in several stages. The initial nodes of the network are the input variables, the final (terminal) nodes are the output variables, and the different stages of the transformation are represented by hidden layers including the hidden nodes. Figure 2.13 shows a neural network structure with three input, three output, and two hidden nodes.

Table 2.7 Supervised and unsupervised classification and clustering.

	Technique
Supervised	
Discriminative	Linear discriminant analysis (LDA)
	Neural networks (NNs)
	Support vector machine (SVM)
	Decision tree classifier
Generative	Naive Bayesian classifier (NBC)
	Hidden Markov model (HMM)
Unsupervised	
Discriminative	Principal component analysis (PCA)
	Independent component analysis (ICA)
	HMM-based approach
	SVM-based approach
	Particle filtering (PF)
Generative	Hierarchical classifier
	k nearest neighbor classifier (kNN)
	Fuzzy C-means classifier

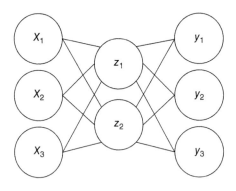

Figure 2.13 A special neural network.

The first step is to transform the input and output variables into the same order of magnitude. The hidden variables are linear combinations of the transformed inputs as

$$z_1 = w_{11}f(x_1) + w_{21}f(x_2) + w_{31}f(x_3)$$

$$z_2 = w_{12}f(x_1) + w_{22}f(x_2) + w_{32}f(x_3) \tag{2.30}$$

where $f()$ represents the transform function of the variables x. In many applications

$$f(x) = \frac{e^x}{1 + e^x} = 1 - \frac{1}{1 + e^x} \tag{2.31}$$

is selected, since it is strictly increasing, $f(-\infty) = 0$, and $f(\infty) = 1$; that is, the input values are transformed into the unit interval $(0, 1)$. The transformed output variables are also linear combinations of the hidden variables:

$$g(y_1) = \overline{w}_{11}z_1 + \overline{w}_{21}z_2$$

$$g(y_2) = \overline{w}_{12}z_1 + \overline{w}_{22}z_2 \tag{2.32}$$

$$g(y_3) = \overline{w}_{13}z_1 + \overline{w}_{23}z_3$$

In Eqs. (2.30) and (2.32) the coefficients w_{ij} and $\overline{w}_{ij}$ are the unknowns, and their values are determined so that the resulting input/output relation

$$g(y_j) = \overline{w}_{1j} \sum_{i=1}^{3} w_{i1}f(x_i) + \overline{w}_{2j} \sum_{i=1}^{3} w_{i2}f(x_i)$$

has the best fit to the measured input and output data.

Assume that we have N input-output data sets:

$$(x_1^{(k)}, x_2^{(k)}, x_3^{(k)}; y_1^{(k)}, y_2^{(k)}, y_3^{(k)}) \qquad\qquad (k = 1, 2, \dots, N)$$

then similar to the least squares method, the overall fit is minimal as measured by

$$Q = \sum_{k=1}^{N} \left[g(y_j^{(k)}) - \overline{w}_{1j} \sum_{i=1}^{3} w_{i1}f(x_i^{(k)}) - \overline{w}_{2j} \sum_{i=1}^{3} w_{i2}f(x_i^{(k)}) \right]^2 \tag{2.33}$$

where the unknowns are the w_{ij} and $\overline{w}_{ij}$ values. The optimization can be done by using software packages or by using special neural network algorithms like back propagation.

In practical cases, the structure (number of hidden layers and number of nodes on them) of the neural network is selected, and the weights w_{ij} and $\overline{w}_{ij}$ are determined.

For the optimization, usually only half of the data set is used; the other half is then used for validation, when Q is computed based on data that was not used in determining the weights. If Q is sufficiently small, then the structure and weights of the network are accepted; otherwise, a new structure (with new added hidden layers and/or added nodes) is chosen, and the procedure is repeated. The optimal choice of the weights is usually called the *training* of the network.

Support Vector Machines

Support vector machines (Cortes and Vapnik 1995) can be described for the case of two groups of vectors. Assume there are N training data points (x_1, y_1), ..., (x_N, y_N), where $y_i = +1$ or -1, indicating the class the vectors x_i belong to. The objective is to find a *maximum-margin hyperplane* that divides the set of data points (x_i, y_i) into two groups. In one, $y_i = +1$; and in the other, $y_i = -1$. The hyperplane is selected so the distance between the hyperplane and the closest point from either group is maximized. If the training vectors are heavily separable, then there is a vector w such that the hyperplane $w^T x + b = 0$ satisfies the following property:

$$w^T x_i + b \geq 1 \ \text{ if } y_i = 1$$

$$w^T x_i + b \leq -1 \ \text{ if } y_i = -1$$

These relations can be summarized as

$$y_i(w^T x_i + b) \geq 1 \text{ for all } i = 1, 2, \dots, N \tag{2.34}$$

In order to maximize the distance between the hyperplanes

$$w^T x_i + b = 1 \text{ and } w^T x_i + b = -1$$

we have to minimize the length of $w = (w_i)$:

$$D = \sqrt{\sum w_i^2} \tag{2.35}$$

subject to the constraints $y_i(w^T x_i + b) \geq 1$. The vector w and scale b define the classifier as

$$x \rightarrow \text{sign}\{w^T x_i + b\}$$

The vectors closest to the separating hyperplane are called *support vectors*.

Decision Tree Classifier

The decision tree classifier technique (Rokach and Maimon 2008) is based on a logically based tree containing the test questions and conditions. The process uses a series of carefully selected questions about the test records of the attributes. Depending on the answer to a question, a well-selected follow-up question is asked; based on the answer, either a new question follows, or the process terminates with a decision about the category the object belongs to. A typical everyday example is the series of questions a doctor asks a patient to make the right diagnosis.

Naive Bayesian Classifier

The naive Bayesian classifier technique (Webb et al. 2005) is also based on m classes $C_1, \ldots, C_m$. An attribute vector x belongs to class C_i if and only if for the conditional probabilities,

$$P(C_i|x) > P(C_j|x) \tag{2.36}$$

for all $j \neq i$. By the Bayesian theorem

$$P(C_i|x) = P(x|C_i)P(C_i)/P(x)$$

And since $P(x)$ is independent of the classes, optimal class C_i is selected by maximizing

$$P(x|C_i)P(C_i) \tag{2.37}$$

The value of $P(C_i)$ is usually taken as the relative frequency of the sample vectors belonging to class C_i. The naive Bayesian classifier assumes the class conditional independence as

$$P(x|C_i) = \prod_k P(x_k|C_i)$$

where the components of vector x are the x_k values.

Hidden Markov Chains

Hidden Markov chains (Ghahramani 2001) are probabilistic extensions of finite Markov chains illustrated earlier in Example 2.1. The states are not known, but certain probability values are assigned to them in addition to the state transition matrix. The states cannot be directly observed, but observations are made for outputs each state can produce with certain probabilities. This type of Markovian model is called *hidden*: since the states are hidden, only their outputs can be observed.

Principal Component Analysis

PCA (Jolliffe 2002) is a statistical method. Assume there are N vectors, which are usually closely related to each other. This technique uses a linear transformation to convert the observation set into a collection of linearly uncorrected variables called the *principal components*. The method is mathematically based on ideas of matrix analysis.

Independent Component Analysis

Independent component analysis (Stone 2004) is a procedure that finds underlying factors or components from multivariate or multidimensional data. Let the observation of random variables be denoted by $x_1(t)$, $x_2(t)$, ..., $x_N(t)$. The method finds a matrix M and variables $y_j(t)$ such that

$$y = Mx \tag{2.38}$$

where the components of y and x are y_j and x_i, respectively.

The objective is to find the minimal number of independent components y_j.

HMM-Based and SVN-Based Approaches

The HMM-based approach (Ghahramani 2001) and the support-vector network SVN based approach (Cortes and Vapnik 1995) are both often used in unsupervised ML.

Particle Filtering

Particle filtering (Andrieu and Doucet 2002) is a sequential Monte Carlo method based on a large sample. The estimate of the PDF converges to the true value as the number of sample elements tends to infinity.

Hierarchical Classifier

The KNN process is very close to a well-known clustering algorithm sometimes called the *hierarchical classifier* (Alpaydin 2004). Assume that we have N vectors and want to organize them into k clusters where the distances between the vectors of the same cluster need to be as small as possible. At the initial step, each vector is a one-element cluster. At each subsequent step, the number of clusters decreases by one until the required number of clusters is reached. Each cluster is represented by the algebraic average of its elements, and then the distances of these average vectors are determined. The two closest averages are selected, and their clusters are merged. In this way, we will have one less cluster in each step.

Fuzzy C-Means Classifier

The fuzzy C-means classifier approach (Bezdek et al. 1999) allows each piece of data to belong to two or more clusters. Let x_1, ..., x_N be the data set, and let u_{ij} denote the degree of membership of data vector x_i in cluster j ($j = 1, 2, ..., K$). The membership values are determined by using an iterative procedure as follows:

Step 1: Select an initial set of u_{ij} values, $u_{ij(0)}$.
Step 2: At each later step k, compute the center vectors for each cluster as

$$c_j = \frac{\sum_{i=1}^{N} u_{ij(k)}^m x_i}{\sum_{i=1}^{N} u_{ij(k)}^m}$$

where $m > 1$ is a real number selected by the user and $u_{ij(k)}$ is the current degree of membership of x.

Step 3: Update the u_{ij} values as

$$u_{ij(k+1)} = \frac{1}{\sum_{l=1}^{K} \left(\frac{\|x_i - c_j\|}{\|x_i - c_l\|} \right)^{\frac{2}{m-1}}}$$

Step 4: Stop if for all i and j,

$$|u_{ij(k+1)} - u_{ij(k)}| < \varepsilon$$

where ε is a threshold.

The final u_{ij} values are accepted as degrees of membership of the data vectors in the clusters. In this approach, the number of clusters is assumed to be given. If some of the resulting cluster centers become close to each other, then we can reduce the number of clusters by merging and repeating the process.

2.4 Hybrid-Driven Prognostics

A model-based approach, especially PoF, is generally chosen as a prognostic health monitoring (PHM) approach when highly accurate prognostics are desirable. However, this approach is often difficult to design and develop and not very accurate for applying to a specific prognostic target. A data-driven approach is much easier to design and develop, compared to a model-based approach, but is often evaluated as producing less accurate prognostic information, as illustrated by Figure 2.14. Some advantages and disadvantages of the two approaches are listed in Table 2.8 (Medjaher and Zerhouni 2013; Javed 2014; Lebold and Thurston 2001).

One hybrid approach combines model-based and data-driven prognostics in two phases: offline and online. The first phase comprises the construction of the nominal and degradation models, and the definition of the faults and performance thresholds

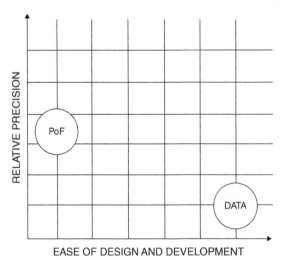

Figure 2.14 Comparison of model-based (PoF) and data-driven prognostic approaches.

Table 2.8 Some advantages and disadvantages of model-based and data-driven prognostics.

Model-based prognostics	Data-driven prognostics
Advantages	
High precision compared to data-driven	Less dependence on material and electrical properties
Deterministic	
Thresholds can be defined and related to performance measures such as stability	Low cost of design and development
	Easier to apply to complex systems
Useful for evaluating performance of materials and electrical properties	
Disadvantages	
Difficult to apply to complex systems	Lower precision compared to model-based non-deterministic
High cost of design and development	
Complexity and variability of model parameters related to material and electrical properties of materials	Not useful for evaluating performance of material and electrical properties of prognostic targets

needed to calculate the RUL of the system. The second phase comprises the use of models and thresholds to detect the onset of faults, assess the state of SoH of the system, and predict future SoH and RUL. The models are verified and fitted to data from life-based and stress-based experiments and tests intended to mimic real-use conditions. Sensors are then developed and used to collect data from fielded systems to monitor and manage the health of those systems (Medjaher and Zerhouni 2013).

An advantage of the hybrid approach is a relative precision that is higher than that achieved by using only a model-based approach and higher than that achieved by using only a data-driven approach (Figure 2.15). This is especially true when a PoF-based model is adapted to sensor data and the adapted model is used to produce prognostic information. A disadvantage is the added complexity of adapting the model to sensor data.

Figure 2.15 Relative comparison of PHM approaches – PoF, data-driven, and hybrid.

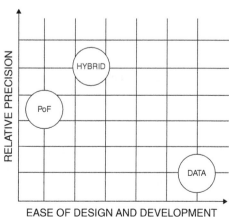

2.5 An Approach to Condition-Based Maintenance (CBM)

Modern prognostic-enabled systems comprise (i) prognostic methods to support prognosis and (ii) health management. In such systems, prognostic-enabling services sense, collect, and process condition-based data (CBD) to provide prognostic information; and health management services use that prognostic information for prognosis to make decisions and issue imperatives related to maintenance and service: CBM. The major capabilities are the following: advanced diagnostics to detect leading indicators of failure, advanced prognostics to predict RUL, and advanced maintenance and logistics to manage the health of the system (Hofmeister et al. 2013; IEEE 2017).

One approach to CBM is to use CBD as input to traditional models, such as PoF and reliability, to produce prognostic information more closely related to a specific prognostic target. Difficulties with such an approach remain, especially the following: the complexity of the modeling; the time and cost required to develop, verify, and qualify a model; and the tendency of the model to be sensitive to a specific set of environment and use conditions.

An alternative approach to CBM is to use modeling of CBD signatures instead of, for example, PoF or reliability modeling of a prognostic target. It should be noted that traditional modeling, such as PoF, is still an important tool for analyzing CBD: understanding and selecting which features and leading indicators to use for prognostic enabling.

2.5.1 Modeling of Condition-Based Data (CBD) Signatures

An example of an alternative approach to CBM based on CBD signatures is shown in Figure 2.16. A sensor framework senses, collects, and transmits sensor output data to a feature-vector framework that performs data processing such as data conditioning, data fusing, and data transforming to transform CBD into failure-progression signatures: fault-to-failure progression (FFP) signature data, degradation progression signature (DPS) data, and functional failure signature (FFS) data.

Any of these signatures can be used as input to a prediction framework to produce prognostic information such as estimates of RUL, prognostic horizon (PH), and SoH for processing by a health-management framework to make intelligent decisions about the health of the system and initiate, manage, and complete service, maintenance, and logistics activities to maintain health and ensure the system operates within functional specifications.

2.5.2 Comparison of Methodologies: Life Consumption and CBD Signature

A common model-based approach to PHM is a life-consumption methodology defined by the Center for Advanced Life Cycle Engineering (CALCE), University of Maryland at College Park, Maryland. As seen in the simplified diagrams in Figure 2.17, that model-based approach (Pecht 2008) is similar to, but different from, using CBD signature models (Hofmeister et al. 2013). The primary differences (see Table 2.9) are related to the difference in modeling and data: modeling using (i) physical, reliability, and/or statistical modeling or (ii) modeling based on empirical data – CBD signatures; and data using (i) environmental, usage, and operational data such as voltage, current,

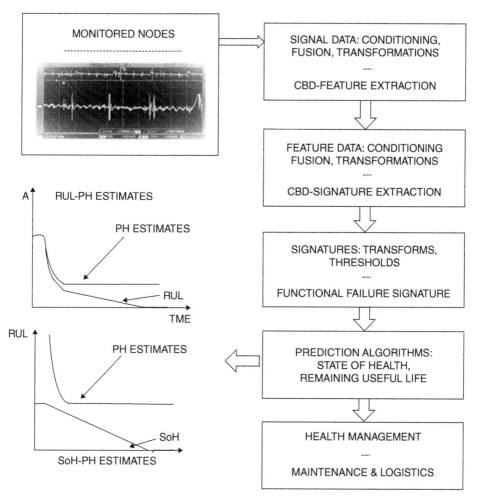

Figure 2.16 Example diagram of a heuristic-based CBM system using CBD-based modeling.

and temperature or (ii) CBD signatures at nodes with environmental, usage, and operational data used to condition signature data.

2.5.3 CBD-Signature Modeling: An Illustration

A switch mode power supply (SMPS) such as that shown in Figure 2.18 is used to illustrate modeling of CBD signatures with an analysis of a circuit or assembly, which in this case is the output filter of the SMPS. The output filter has also been simplified: for example, to exclude such components and subcircuits as a feedback loop, diodes, and high-frequency noise filters. Additionally, the filter has been further simplified to lump inductance, capacitance, and resistance into three passive components (L1, C1, and RL).

Example Backup and Setup
Suppose you are asked to prognostic enable a SMPS, and it is known that the supply has a high failure rate caused by failure of tantalum oxide capacitors used in the

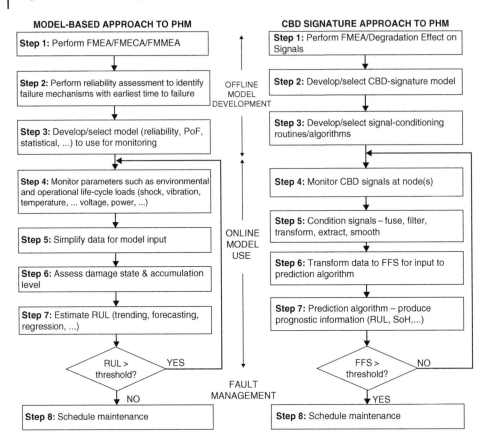

Figure 2.17 Diagram comparison of model-based and CBD-signature approaches to PHM.

output filter. Failure and repair information reveals that the capacitors fail short and then burn open. The fail-short condition causes high current, resulting in an overload condition that turns off the SMPS – but, at the same time, damage continues and the capacitor burns open. Subsequently, maintenance personnel declare a no fault found (NFF)/cannot duplicate (CND) repair condition. The SMPS tests okay – the DC output voltage, ripple voltage, noise levels, and regulation of the supply under loading conditions are all within specifications. Upon further inspection after disassembling the circuit card assembly (CCA) – an enclosed metal frame – physical evidence in the form of color and smell reveals 1 of 22 parallel-connected capacitors has failed. Removal (desoldering) and testing of the capacitor confirms the capacitor has burned open, which reduces the effective total capacitance in the output filter.

Also suppose the customer specifications for prognostic enabling of the SMPS includes the following requirements: (i) the prognostic system shall detect degradation at least 200 operational days prior to functional failure; and (ii) the prognostic system shall provide estimates of RUL that shall converge to within 5% accuracy at least 72 operational days, ±2 days, before functional failure occurs. *Functional failure* is defined as a level of damage at which an object, such as a power supply, no longer operates within specifications.

Table 2.9 Differences in focus of model-based and heuristic-based approaches to PHM.

Step	Model-based focus	Heuristic-based focus
1	Identify failure mode, effects analysis.	Identify failure mode, effects analysis; identify nodes and signatures comprising leading indicators of failure.
2	Identify failure modes having the earliest time-to-failures.	Characterize the basic curve of the signature(s) related to a failure mode.
3	Develop the model to use for predicting time of failure.	Develop the algorithms to transform CBD signatures into fault to failure (FFP) degradation progression signature (DPS) data and then further transform into functional failure signature (FFS) data.
4	Monitor environmental, usage, and operational loads: the model inputs.	Monitor signals: the model inputs; monitor selected environmental, usage, and operational loads as required for conditioning signals.
5	Simplify and condition data for model input.	Condition and transform data. Use environmental, usage, and operational data to condition data rather than as model inputs.
6	Assess the state and level of accumulated damage.	Use FFS to detect damage and as input to prediction algorithms.
7	Produce prognostic information.	Same
8	Perform fault management.	Same

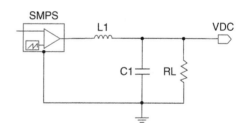

Figure 2.18 Simplified diagram of a switch-mode power supply (SMPS) with an output filter.

Evaluation of a Reliability-Modeling Approach

You happen to know that tantalum oxide capacitors are a high-failure-rate component in a SMPS. PoF analysis reveals the following as dominant failure modes caused by the aging effects of voltage and temperature: changes in (i) capacitance (C), (ii) dissipation factor (DF), and (iii) insulation resistance (IR) (Alan et al. 2011). Loss of IR includes abrupt, avalanche failures that, for example, occur when IR drops sufficiently low to cause a high enough increase in leakage current to generate sufficient heat to melt the solder connecting the capacitor to the external leads of a package.

Example 2.11 For the IR failure mode, the following statistical distribution model might be used for MTTF as a function of the operating voltage and temperature (Alan et al. 2011; Prokpowicz and Vaskas 1969).

$$\text{MTTF} = (C/V^n)\exp(E_a/kT) \qquad (2.39)$$

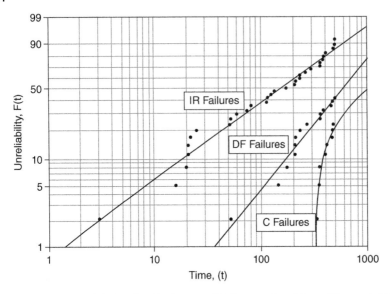

Figure 2.19 Unreliability plots for three models for capacitor failures (Alan et al. 2011).

This formula is a special case of the temperature-nonthermal relationship introduced in Chapter 1 when we select voltage as U, temperature as V, and E_a/k as β in the general equation. The values for the voltage exponent (n) and activation energy (E_a) have differing values and are highly dependent on experimental setup, materials, and physical structure of the manufactured capacitors. This is most likely why a major difficulty with this model, and others for change in C and DF, is a large uncertainty (unreliability in modeling), as illustrated by Figure 2.19.

The unreliability of using the model from Eq. (2.39) might be due to the limited number of parameters in the model. Uncertainty might be improved by using a model that takes into account the switching rate, operating temperature, humidity, load current, switching-induced spikes in the output voltage, and so on. To be more exact (and presumably more accurate), similar treatment might be applied for L and R; and because R includes the load, and so on: *it becomes clear that an exact, highly accurate PoF and/or statistical model to predict time of failure of the capacitance of the output filter of a particular, fielded, in-use power supply is not practical because it is too complex and too costly to develop and verify.*

Perhaps more important, the complexity of traditional modeling for complex systems becomes computationally intractable.

Mahalanobis Distance Modeling of Failure of Capacitors

One data-driven approach to modeling the failure of capacitors uses the Mahalanobis distance (MD) method, which reduces multiple parameters to a single parameter using correlation techniques after normalizing data. MD values are calculated between incoming data and a baseline representative of a healthy state to detect anomalies in the following way (Alan et al. 2011).

Assume we have n vectors with m components, and each component represents a parameter. Let $(x_{i1}, x_{i2}, \ldots, x_{im})$ denote the ith vector. The mean of the kth parameter is given as

$$m_k = \frac{1}{n} \sum_{i=1}^{n} x_{ik} \tag{2.40}$$

The standard deviation of the kth parameter is similarly

$$s_k^2 = \frac{1}{n-1} \sum_{i=1}^{n} (x_{ik} - m_k)^2 \tag{2.41}$$

And so the standardized value of x_{ik} becomes

$$z_{ik} = \frac{x_{ik} - m_k}{s_k} \tag{2.42}$$

Define

$$Z_i = (z_{i1}, z_{i2}, \ldots, z_{im})$$

Then the correlation matrix of the standardized values is given as

$$C = \frac{1}{n-1} \sum_{i=1}^{n} Z_i Z_i^T$$

And finally, the MD value can be obtained as

$$MD = \frac{1}{m} Z_i C^{-1} Z_i^T \tag{2.43}$$

where T denotes the transpose.

MD values are not normally distributed and need to be further transformed by using a Box Cox transformation and maximizing the logarithm of the likelihood function. However, in a highly accelerated test experiment, the MD method only predicted failure in 14 out of 26 samples (Alan et al. 2011). Even if the degree of accuracy were acceptable, that still leaves, for example, the problem of predicting when the power supply will fail because of the loss of sufficient capacitance for any specific failure mode. To address those kinds of problems, heuristic-based modeling of CBD can be used rather than traditional models. We conclude that neither model-based nor data-driven approaches is likely to meet the requirements for resolution, precision, and prognostic accuracy.

There are other techniques as well, to reduce multiple measurements into a single health indicator. One example, for instance, is the multivariate state estimation technique (Cheng and Pecht 2007), which is based on the best healthy estimates of the sample vectors obtained by the use of least squares. However, other metrics can be also selected for the distance of the sample vectors and their estimates. These distances are indicative of SoH, since smaller distances indicate better health. Other examples include autoassociative kernel regression (Guo and Bai 2011), a method based on kernel smoothing (Wand and Jones 1995) discussed in Section 2.3.1. After the sequence of single health indicators is determined, the resulting time series can be used as input data for all procedures being discussed in this book.

CBD: Noise and Features

The output of an SMPS, as seen at the top of Figure 2.20, is very noisy. It consists of DC voltage (the desired output) plus all manner of noise in the form of signal variations

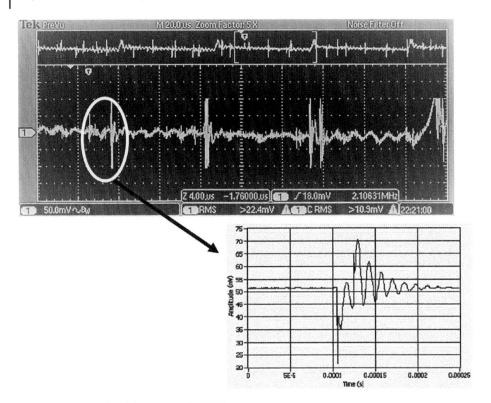

Figure 2.20 Example of the output of a SMPS.

that include the following: thermal noise; ripple-voltage variations; switching noise; harmonic distortion; and responses to load variations such as damped-ringing responses, an example of which is shown at the bottom of Figure 2.20 (Hofmeister et al. 2017).

Noise contains information, and when noise in CBD is appropriately isolated and conditioned, leading indicators of failure can be extracted as feature data (FD). Useful FD at the output node of an SMPS includes, but is not limited to, ripple voltage (amplitude and frequency), switching amplitude and frequency, and other features found in a damped-ringing response. An important first step in prognostic-enabling an object such as a component, circuit, or assembly in an electrical system is to perform analyses, such as PoF and FMEA, to identify FD of interest associated with failure modes of interest.

Example 2.12   Circuit, PoF, and FMEA analyses indicate that the identified failure mode should result in measurable changes in several features. Of those, you decide to focus on ripple voltage (increase in amplitude) and the damped-ringing response, increase in resonance frequency (f_R), instead of a change in decay time $(1/\tau)$ or a change in amplitude (A_R); see Figure 2.21.

CBD: Prognostic Requirements
It is not cost-justifiable to select all applicable features of CBD for use as FD in signatures: you need to select one or more FD that lets you meet prognostic requirements. Important requirements include the following:

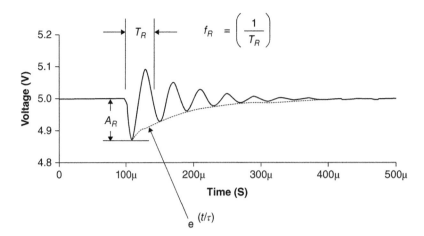

Figure 2.21 Example of a ringing response from an electrical circuit to an abrupt stimulus.

- Sensor cost to isolate, condition, and extract one or more features. Cost includes weight, footprint, power, and reliability of the sensor.
- Cost of signal conditioning by the sensor and by the PHM framework.
- Prognostic distance (PD): the time between detection of damage leading to failure and the time of functional failure. Functional failure is defined as a level of damage at which an object, such as an SMPS, no longer operates within specifications.
- Prognostic resolution and precision of estimates of a future point in time at which functional failure is likely to occur.
- Prognostic accuracy (α) and α-distance (PD_α) within which estimates of RUL and SoH must converge with respect to functional end of life (EoL).

Example 2.13 Suppose the customer specifications for prognostic enabling of the SMPS include the following:

- The prognostic system shall detect degradation at least 200 operational days prior to functional failure.
  - You interpret this as a requirement that the minimum PH must be equal to 200 days. Factors related to this are your ability to design or choose a sensor and data-conditioning algorithms that provide sufficient noise mitigation and selectivity to isolate FD with sufficient lead time before failure.
  - You also interpret this as being a required resolution in terms of days rather than, for example, weeks or hours.
  - You obtain customer confirmation.
- The prognostic system shall provide estimates of RUL that shall converge to within 5% accuracy at least 72 operational days, ±2 days, before functional failure occurs.
  - You interpret this as a requirement that sampling must occur at least once every day because of the precision specification of ±2 days.
  - You also interpret the specifications as a requirement for a minimum PD_α of 72 days with all subsequent estimates within 5% accuracy.
  - You obtain customer confirmation.

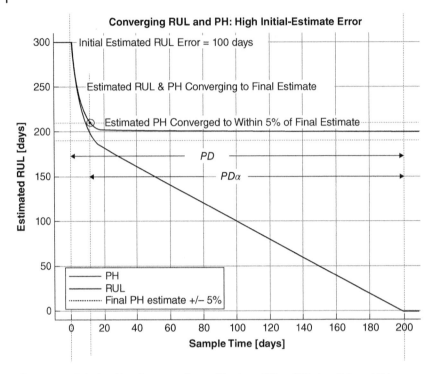

Figure 2.22 Relationship of prognostic specifications (PD and PDα) to RUL and PH.

- You create a plot that illustrates the relationship of PD, PD$_\alpha$, RUL, and PH (see Figure 2.22) and obtain customer confirmation of your interpretation of the prognostic specifications with respect to prognostic information.

CBD: Test Bed Experimentation and Effects Analysis

When you are pretty sure you understand the failure mode (the problem) and the effects of that failure on measurable signals, in all likelihood you will conclude that you need to experiment and perform further effects analysis. So, you design and build a test bed that allows you to inject faults (loss of filtering capacitance) in an exemplary SMPS. Such a test bed lets you perform experiments to verify your understanding of the problem, measurable effects of that problem, and how well (or not) various solutions are likely to meet prognostic requirements.

Example 2.14 In all likelihood, you will quickly conclude that using ripple-voltage amplitude as FD is not satisfactory: there are too many other sources of high-amplitude noise that require filtering and mitigation. This leads you to design and develop a prototype method of injecting a low-power, abrupt change in load that causes the SMPS to produce a damped-ringing response: one that your sensor can isolate, digitize, and use digital-signal processing (DPS) to extract FD.

You use PoF and FMEA to arrive at the signal modeling of Eq. (2.44): there is a DC component (V_{DC}); there is noise (A_{NOISE}); and there is a sinusoidal component having a frequency (ω), phase (ϕ), and amplitude (A_R) that decays exponentially with a time constant (τ):

$$V_o = V_{DC} + A_R\{\exp(-t/\tau)\}\{\cos(\omega t + \phi)\} + A_{NOISE} \tag{2.44}$$

You also develop other modeling expressions that show frequency (ω) in the signal model of Eq. (2.44) is a function of the resistance R, the capacitance C, the inductance L, and the feedback gain A of the output filter:

$$\text{Natural resonant frequency}: \omega_0 = \sqrt{A+1}\ (1/\sqrt{L_0 C_0}) \tag{2.45}$$

$$\text{Circuit quality}: Q = \sqrt{A+1}\ (1/R)(\sqrt{L/C})\ \text{series} \tag{2.46}$$

$$Q = \sqrt{A+1}\ (R)(\sqrt{C/L})\ \text{parallel} \tag{2.47}$$

$$\text{Measurable frequency}: \omega = \omega_0 \sqrt{1 - 1/(4\,Q^2)} \tag{2.48}$$

For the case when C degrades and letting $C = C_0 - \Delta C$, Eq. (2.45) becomes

$$\omega = \sqrt{A+1}\ (1/\sqrt{L_0 C}) = \sqrt{A+1}\ (1/\sqrt{L_0(C_0 - \Delta C)})$$
$$= \sqrt{A+1}\ (1/\sqrt{L_0 C_0(1 - \Delta C/C_0)}) = \sqrt{A+1}\ (1/\sqrt{L_0 C_0})(1/\sqrt{(1 - \Delta C/C_0)})$$
$$\omega = \omega_0\ (1/\sqrt{(1 - \Delta C/C_0)}) \tag{2.49}$$

where ω_0 is the nominal value of the resonant frequency of the damped-ringing response at the nominal capacitance C_0 of the output filter and ΔC is zero (no change in C).

You simulate Eq. (2.49), plot the simulation results (Figure 2.23), and compare that to your plots of experimental data (Figure 2.24). The experimental results are similar to the simulation results – the primary differences are that the experimental data is noisier and is more linear at lower values of frequency change. Since you have knowledge of a

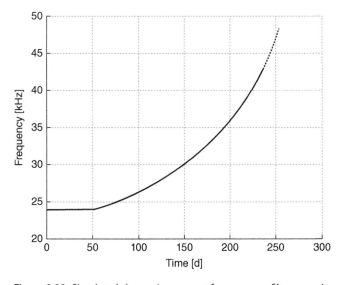

Figure 2.23 Simulated change in resonant frequency as filter capacitance degrades.

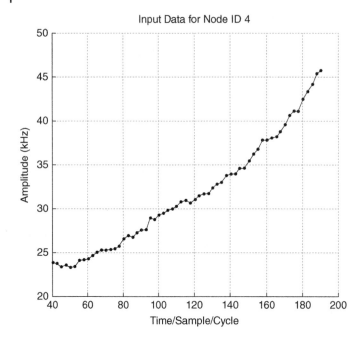

Input Data for Node ID 4

Figure 2.24 Experimental change in resonant frequency as filter capacitance degrades.

prediction algorithm that uses graphical methods, a Kalman-like filtering approach, and random-walk solution approaches, you are happy with your results.

Figure 2.24 is the CBD signature: the change in frequency is highly correlated to degradation (loss of filtering capacitance) and can be used as input to any number of prediction algorithms to produce prognostic information such as RUL.

CBD: Feature Data and Degradation Function

From the results of deriving Eq. (2.49), we use the inductive approach to analytical modeling to state the following:

- Given a feature, FD, that has a nominal value FD_0 in the absence of degradation, then
- Given a degradation of a parameter that has a value of P_0 in the absence of degradation, and
- Let dP represent the amount of change in the value of P_0 caused by degradation,
- Then the value of a given feature is the non-degraded value of the feature times a degradation function of a parameter that changes.

In Eq. (2.49), if we let $FD = \omega$, $FD_0 = \omega_0$, $P_0 = C_0$ and $dP = \Delta C$ we have

$$FD = FD_0 \left(1/\sqrt{(1 - dP/P_0)}\right)$$

and, in general,

$$FD = FD_0 f(dP, P_0) \tag{2.50}$$

The model for a given degradation function, $f(dP, P_0)$, depends on, for example, the component that is degrading, the failure mode, and the PoF. This is an important result, and more information is presented in following chapters of this book.

2.6 Approaches to PHM: Summary

This chapter presented an overview of traditional approaches to PHM including model-based prognostics, data-driven prognostics, and hybrid-driven prognostics. Model-based prognostics is potentially the most accurate, but it is the most difficult to apply to complex systems; data-driven prognostics is the least difficult to apply, but also is the least accurate; and hybrid-driven prognostics provides the greatest accuracy and is the most difficult to apply, especially to complex systems. A brief presentation of a CBD approach to CBM concluded the chapter: the output of an SMPS; identification and selection of a useful FD (resonant frequency) extracted from a damped-ringing response from CBD; the effect of prognostic requirements on operational requirements and accuracy of prognostic information; and a test bed and simulation to assist in and verify modeling of a CBD signature. An important result is Eq. (2.50), which relates changes in signatures to a degradation function:

$$FD = FD_0 f(dP, P_0)$$

The next chapter develops and describes in more detail the approach and introduces the methods used to transform CBD signatures into signatures that are more amenable to be processed by prediction algorithms.

References

Ahmed, W. and Wu, Y.W. (2013). Reliability prediction model of SOA using hidden Markov model. 8th China Grid Annual Conference, Changchun, China, 22–23 August.

Alan, M., Azarian, M., Osterman, M., and Pecht, M. (2011). Prognostics of failures in embedded planar capacitors using model-based and data-driven approaches. *Journal of Intelligent Material Systems and Structures* 22: 1293–1304.

Alpaydin, E. (2004). *Introduction to Machine Learning*. Cambridge, MA: MIT Press.

Andrieu, C. and Doucet, A. (2002). Particle filtering for partially observed Gaussian state space models. *Journal of the Royal Statistical Society B* 64 (4): 827–836.

Bezdek, J., Keller, J., Krishnapuram, R., and Pal, N. (1999). *Fuzzy Models and Algorithms for Pattern Recognition and Image Processing*. New York: Springer.

Casella, G. and Berger, R.L. (2002). *Statistical Inference*, 2e. Pacific Grove, CA: Duxbury.

Cheng, S. and Pecht, M. (2007). Multivariate state estimation technique for remaining useful life prediction of electronic products. Association for the Advancement of Artificial Intelligence. www.aaai.org.

Cortes, C. and Vapnik, V. (1995). Support-vector networks. *Machine Learning* 20 (3): 273–297.

Cover, T. and Hart, P. (1967). Nearest neighbor pattern classification. *IEEE Transactions on Information Theory* 13 (1): 21–27.

Czichos, H. (ed.) (2013). *Handbook of Technical Diagnostics: Fundamentals and Application to Structures and Systems*. Heidelberg/New York: Springer.

Faussett, L.V. (1994). *Fundamentals of Neural Networks: Architecture, Algorithms and Applications*. Upper Saddle River, NJ: Prentice Hall.

Fukunaga, K. (1990). *Introduction to Statistical Pattern Recognition*. San Diego, CA: Academic Press.

Ghahramani, Z. (2001). An introduction to hidden Markov models and Bayesian networks. *Journal of Pattern Recognition and Artificial Intelligence* 15 (1): 9–42.

Guo, P. and Bai, N. (2011). Wind turbine gearbox condition monitoring with AAKR and moving window statistic methods. *Energies* 4: 2077–2093.

Hofmeister, J., Goodman, D., and Wagoner, R. (2016). Advanced anomaly detection method for condition monitoring of complex equipment and systems. 2016 Machine Failure Prevention Technology, Dayton, Ohio, US, 24–26 May.

Hofmeister, J., Lall, P. and Graves, R. (2006). In-situ, real-time detector for faults in solder joint networks belonging to operational, fully programmed field programmable gate arrays (FPGAs). 2006 IEEE Autotest Conference, Anaheim, California, US, 18–21 September.

Hofmeister, J., Szidarovszky, F., and Goodman, D. (2017). An approach to processing condition-based data for use in prognostic algorithms. 2017 Machine Failure Prevention Technology, Virginia Beach, Virginia, US, 15–18 May.

Hofmeister, J. and Vohnout, S. (2011). Innovative cable failure and PHM toolset. MFPT: Applied Systems Health Management Conference, Virginia Beach, Virginia, US, 10–12 May.

Hofmeister, J., Vohnout, S., Mitchell, C. et al. (2010). HALT evaluation of SJ BIST technology for electronic prognostics. 2010 IEEE Autotest Conference, Orlando, Florida, US, 13–16 September.

Hofmeister, J., Wagoner, R., and Goodman, D. (2013). Prognostic health management (PHM) of electrical systems using conditioned-based data for anomaly and prognostic reasoning. *Chemical Engineering Transactions* 33: 992–996.

Hollander, M., Wolfe, D.A., and Chicken, E. (2014). *Nonparametric Statistical Methods*, 3e. Hoboken, NJ: Wiley.

IEEE. (2017). Draft standard framework for prognosis and health management (PHM) of electronic systems. IEEE 1856/D33.

Javed, K. (2014). A robust & reliable data-driven prognostics approach based on extreme learning machine and fuzzy clustering. Université de Franche-Comté, English, 21 July.

Jolliffe, I.T. (2002). *Principal Component Analysis*. New York: Springer.

Kentved, A. and Schmidt, K. (2012). Reliability – acceleration factors and accelerated life testing. Input to SEES meeting 2012-05-14. www.sees.se/$-1/file/delta-alt-sees-may-2012-ver-0.PDF.

Kumar, S. and Pecht, M. (2010). Modeling approaches for prognostics and health management of electronics. *International Journal of Performability Engineering* 6 (5): 467–476.

Lebold, M. and Thurston, M. (2001) Open standards for condition-based maintenance and prognostic systems. 5th Annual Maintenance and Reliability Conference (MARCON 2001), Gatlinburg, Tennessee, US, 6–9 May.

ManagementMania. (2017). ETA (event tree analysis) https://managementmania.com/en/eta-event-tree-analysis.

Medjaher, K. and Zerhouni, N. (2013). Framework for a hybrid prognostics. *Chemical Engineering Transactions* 33: 91–96. https://doi.org/10.3303/CET1333016.

National Science Foundation Center for Advanced Vehicle and Extreme Environment Electronics at Auburn University (CAVE3). (2015). Prognostics health management for electronics. http://cave.auburn.edu/rsrch-thrusts/prognostic-health-management-for-electronics.html (accessed November 2015).

Nelson, W. (2004). *Accelerated Testing: Statistical Models, Test Plans, and Data Analysis.* New York: Wiley.

O'Connor, P. and Kleyner, A. (2012). *Practical Reliability Engineering.* Chichester, UK: Wiley.

Papoulis, A. (1984). *Probability, Random Variables, and Stochastic Processes,* 2e. New York: McGraw-Hill.

Pecht, M. (2008). *Prognostics and Health Management of Electronics.* Hoboken, NJ: Wiley.

Prokpowicz, T.I. and Vaskas, A.R. (1969). Research and development, intrinsic reliability subminiature ceramic capacitors. Final Report, ECOM-90705-F, 1969 NTIS AD-864068.

Rokach, L. and Maimon, O. (2008). *Data Mining with Decision Trees: Theory and Applications.* Singapore: World Scientific.

Ross, M.S. (1987). *Introduction to Probability and Statistics for Engineers and Scientists.* New York: Wiley.

Sheppard, J.W. and Wilmering, T.J. (2009). IEEE standards for prognostics and health management. *IEEE A&E Systems Magazine* (September 2009): 34-41.

Silverman, M. and Hofmeister, J. (2012). The useful synergies between prognostics and HALT and HASS. IEEE Reliability and Maintainability Symposium (RAMS), Reno, Nevada, US, 23–26 January.

Speaks, S. (2005). Reliability and MTBF overview. Vicor Reliability Engineering. http://www.vicorpower.com/documents/quality/Rel_MTBF.pdf (accessed August 2015).

Stein, D., Beaver, S., Hoff, L. et al. (2002). Anomaly detection from hyperspectral imagery. *IEEE Signal Processing Magazine* 19 (1): 58–69.

Stone, J.V. (2004). *Independent Component Analysis: A Tutorial Introduction.* Cambridge, MA: MIT Press.

Thomas, D.R. (2006). A general inductive approach for analyzing qualitative evaluation data. *American Journal of Evaluation* 27 (2): 237–246.

Tobias, P. (2003). How do you project reliability at use conditions? In: *Engineering Statistics Handbook.* National Institute of Standards and Technology https://www.itl.nist.gov/div898/handbook/apr/section4/apr43.htm.

Vichare, N. (2006). Prognostics and health management of electrics by utilization of environmental and usage loads. Doctoral thesis. Department of Mechanical Engineering, University of Maryland.

Vichare, N., Rodgers, P., Eveloy, V., and Pecht, M. (2007). Environment and usage monitoring of electronic products for health assessment and product design. *Quality Technology and Quantitative Management* 4 (2): 235–250.

Viswanadham, P. and Singh, P. (1998). *Failure Modes and Mechanisms in Electronic Packages.* New York: Chapman and Hall.

Wand, M.P. and Jones, M.C. (1995). *Kernel Smoothing.* London: Chapman & Hall/CRC.

Webb, G.I., Boughton, J., and Wang, Z. (2005). Not so Naive Bayes: aggregating one-dependence estimations. *Machine Learning* 58 (1): 5–24.

Weisstein, E. (2015). Extreme value distribution. MathWorld, a Wolfram web resource. http://mathworld.wolfram.com/ExtremeValueDistribution.html.

White, M. and Bernstein, J. (2008). Microelectronics reliability: physics-of-failure based modeling and lifetime evaluation. NASA Electronic Parts and Packaging (NEPP) Program, Office of Safety and Mission Assurance, NASA WBS: 939904.01.11.10, JPL Project Number: 102197, Task Number: 1.18.15, JPL Publication 08-5.

Xilinx. (2003). The reliability report. xgoogle.xilinx.com, 225–229.

Xu, Z., Hong, Y., and Meeker, W. (2015). Assessing risk of a serious failure mode based on limited field data. *IEEE Transactions on Reliability* 64 (1): 51–62.

Yakowitz, S. and Szidarovszky, F. (1989). *An Introduction to Numerical Computations*. New York: Macmillan.

Zhang, N.L. and Poole, D. (1996). Exploiting causal independence in Bayesian network inference. *Journal of Artificial Intelligence Research* 5: 301–328.

Further Reading

Filliben, J. and Heckert, A. (2003). Probability distributions. In: *Engineering Statistics Handbook*. National Institute of Standards and Technology http://www.itl.nist.gov/div898/handbook/eda/section3/eda36.htm.

Saxena, A., Celaya, J., Saha, B. et al. (2009). On applying the prognostic performance metrics. Annual Conference of the Prognostics and Health Management Society (PHM09), San Diego, California, US, 27 Sep.–1 Oct.

Tobias, P. (2003). Extreme value distributions. In: *Engineering Statistics Handbook*. National Institute of Standards and Technology https://www.itl.nist.gov/div898/handbook/apr/section1/apr163.htm.

3

Failure Progression Signatures

3.1 Introduction to Failure Signatures

Chapter 2 introduced three classical prognostic approaches for prognostics and health management/monitoring (PHM): model driven, data driven, and hybrid driven. You were also introduced to usage-based and condition-based approaches. You learned the primary disadvantages of classical and usage-based approaches for prognostics: they are not applicable to a specific prognostic target in a system, and/or they are nondeterministic and not suitable for application to prognostic targets, and/or it is complex to adapt them to sensor data. You also learned that leading indicators of failure can be extracted from sensor data and collected to form condition-based data (CBD) signatures; the modeling and processing of such signatures is a condition-based approach to condition-based maintenance (CBM). Figure 3.1 shows the relationship of an approach using CBD signatures to classical PHM approaches; although the block diagram indicates the approaches are different, a conditioned-based approach often employs analysis and modeling techniques such as reliability modeling, physics of failure (PoF) analysis, and failure mode and effect analysis (FMEA) (Hofmeister et al. 2013, 2016, 2017; Medjaher and Zerhouni 2013; Pecht 2008).

3.1.1 Chapter Objectives

The primary objective of this chapter is to take a more in-depth look at CBD signatures and the desirability of transforming those signatures into other signatures that are more amenable as input to prediction algorithms that produce prognostic information. Those other signatures are the fault-to-failure progression (FFP) signature, degradation progression signature (DPS), and functional failure signature (FFS).

A secondary objective is to describe and show how those signatures are used for reliable condition-based monitoring: detecting the onset of degradation, monitoring the increasing progression of damage due to degradation, making prognostic estimates for when damage is likely to reach a level defined as functional failure, and detecting functional failure.

Another objective is to show an important advantage of DPS data and DPS-based FFS compared to CBD signature or FFP signature curves: the former is more linear. A method for quantifying the nonlinearity of FFP-based and DPS-based FFS curves shall be developed and presented in this chapter.

Prognostics and Health Management: A Practical Approach to Improving System Reliability Using Condition-Based Data, First Edition. Douglas Goodman, James P. Hofmeister and Ferenc Szidarovszky.
© 2019 John Wiley & Sons Ltd. Published 2019 by John Wiley & Sons Ltd.

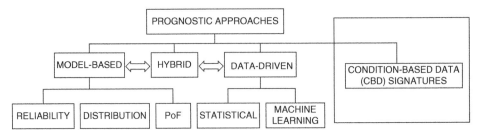

Figure 3.1 Diagram of classical and CBD prognostic approaches for PHM systems. Source: based on Pecht (2008).

3.1.2 Chapter Organization

The remainder of this chapter is organized to present and discuss a heuristic-based approach to modeling CBD signatures as follows:

3.2 Basic Types of Signatures
  This section presents information on basic types of signatures: CBD signatures, FFP signatures, FFP-based FFS, transforming FFP signatures into DPS, and DPS-based FFS.
3.3 Model Verification
  This section discusses model verification along with signal classification and verifying CBD, FFP, DPS, and DPS-based models.
3.4 Evaluation of FFS Curves: Nonlinearity
  This section discusses the evaluation of FFS curves in the context of a transfer curve for a sensing system and how to calculate metrics that quantify the nonlinearity of that transfer curve.
3.5 Summary of Data Transforms
  This section summarizes the material presented in this chapter.

3.2 Basic Types of Signatures

A signature characterizes a feature of interest, such as feature data (FD) amplitude with respect to time: signature $=$ **FD** $=$ a set of FD_i data points over a period of time $T = \{t_m, t_{m+1}, \ldots t_{m+n}\}$, where amplitude refers to a characteristic value such as voltage, current, resistance, force, energy, and so on that changes as the magnitude of damage/degradation increases over time:

$$\text{signature} = \mathbf{FD} = \{FD_m, FD_{m+1}, \ldots FD_{m+n-1}, FD_{m+n}\} \tag{3.1}$$

There are three signatures of interest related to signals and prognostic processing: CBD, FFP, and DPS. An FFP signature is a transform of a CBD signature to reduce modeling complexity, and a DPS signature is a transform of an FFP signature that further reduces modeling complexity and lends itself to increased accuracy of prognostic information. Another signature is particularly amenable to processing by prediction algorithms: an FFS that is a transform of either an FFP or a DPS signature. We will next show how the different signatures can be obtained.

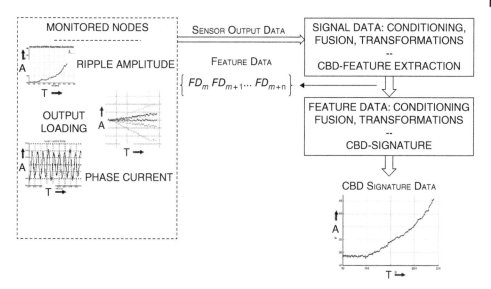

Figure 3.2 Functional block diagram for CBD signature data and processing flow.

3.2.1 CBD Signature

As illustrated in Figure 3.2, a sensor is located at a node of prognostic interest, such as the output of a switch mode power supply (SMPS). The sensor is designed to capture and process CBD, such as that shown in Figure 3.3, to support the isolation and extraction of FD: a leading indicator of failure. Isolation and extraction of a particular value of FD, such as the resonant frequency of a damped-ringing response of an SMPS to an abrupt change in output load, almost always requires signal conditioning. Signal conditioning includes, but is not limited to, the following: background noise and harmonic filtering and mitigation; gating of the signal to specific, timed events to sufficiently isolate a feature of interest; fusing of two or more signals to transform two types of data to another data type – such as using voltage and current signals to calculate impedance; domain transforms such as time-to-frequency and frequency-to-time; and fusing of signals to cancel or mitigate environmental variations such as resistive dependence on temperature. Chapter 5 provides more information on signal conditioning.

A CBD data point in a leading indicator of failure comprises a feature of interest, FD, in a signal that changes as degradation progresses, plus all other variations, referred to as noise (N), that are not related to degradation and the feature of interest:

$$\text{CBD data} = FD + N \tag{3.2}$$

Signals at nodes within a system typically exhibit more than just a particular CBD of interest. An example of signals is shown in Figure 3.4; the node is the output of a SMPS, and a particular CBD of interest is a damped-ringing response such as that shown in Figure 3.5, for which an idealized plot is shown in Figure 3.6 (Erickson 1999; Judkins et al. 2007; Judkins and Hofmeister 2007; Hofmeister et al. 2013, 2016, 2017).

The signals shown in Figure 3.4 consist of multiple types of CBD and noise, including damped-ringing responses, switching noise from power transistors, ripple voltage, pulse-width modulator effects, effects of voltage and current regulation and feedback,

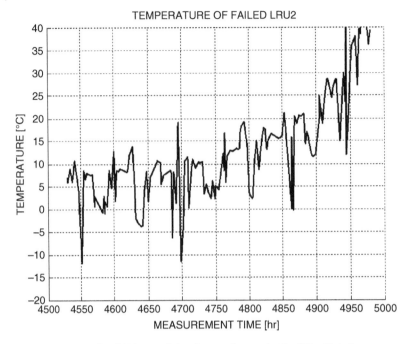

Figure 3.3 Example of CBD containing feature data and noise (FD + Noise).

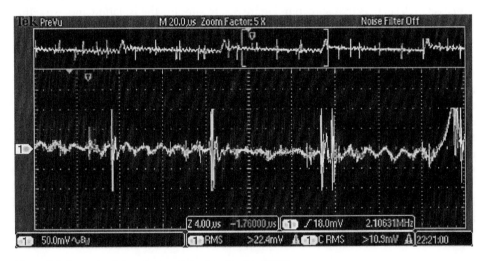

Figure 3.4 Example of signals at an output node of an SMPS.

and background noise:

$$\text{SMPS output} = CBD_1 + CBD_2 + \ldots CBD_n + N_1 + N_2 + \ldots N_m \tag{3.3}$$

where CBD_i are signal terms and N_j are noise terms. An example of a CBD term is a damped-ringing response, such as that shown in Figure 3.5. An idealized

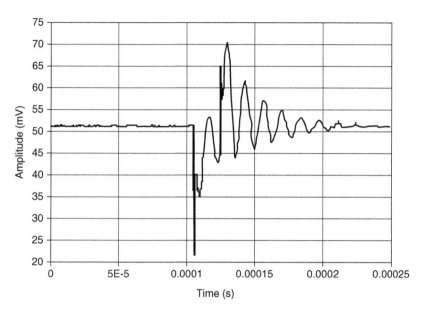

Figure 3.5 Example of a damped-ringing response (Judkins and Hofmeister 2007).

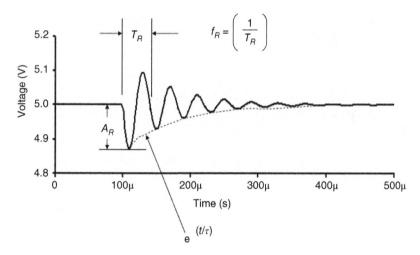

Figure 3.6 Modeling a damped-ringing response. Source: based on Judkins and Hofmeister (2007).

damped-ringing response is shown in Figure 3.6 and is modeled as

$$V_0 = V_{DC} + A_R\{\exp(-t/\tau)\}\{\cos(\omega t + \phi)\}$$

which includes five possible features of interest: a direct current (DC) voltage V_{DC}, a response amplitude A_R, a dampening time constant τ, a frequency ω, and a phase shift ϕ.

Even noise, with appropriate signal conditioning and processing, can be used as CBD – for example, to prognostic-enable a noise-filtering subcircuit in an SMPS. The primary objective of sensors used in prognostic-enabling applications is to isolate CBD that forms leading indicators of failure, which, after suitable signal conditioning, are transformed into FD in signatures for use as input to prediction algorithms.

Example 3.1 Suppose you have been asked to prognostic-enable an SMPS used by a customer for two different applications: a high-end application with conservative failure requirements and a low-end application with less stringent failure requirements. You decide to select the resonant frequency of a damped-ringing response (example presented in Chapter 2) as your leading indicator of failure: (i) loss of filtering capacitor is the predominant failure mode; (ii) therefore as the power supply degrades due to that loss, the resonant frequency increases; and (iii) changes in resonant frequency are less dependent on load changes compared to, for example, changes in amplitude of ripple voltage.

Starting with Eq. (2.49) and letting C_0 represent the value of an undamaged capacitor,

$$\omega \approx \omega_0 \sqrt{C_0/(C_0 - \Delta C)}$$

where frequency in Hz is given by

$$f = \omega/2\pi = FD \tag{3.4}$$

then the signal model becomes

$$FD = FD_0 \sqrt{C_0/(C_0 - \Delta C)} \tag{3.5}$$

where a scalar value of frequency, FD, is returned by an intelligent sensor attached to the output node of the SMPS. Then from Eqs. (3.2) and (3.5),

$$FD = FD_0 \sqrt{C_0/(C_0 - \Delta C)} = CBD - N \tag{3.6}$$

Figure 3.6 shows an idealized plot of a damped-ringing response and features of interest such as the resonant frequency $f_R = 1/T_R$, which is related back to Eq. (3.4).

Example 3.2 You design and develop or obtain a sensor that presents an abrupt, low-value change in load to the SMPS. That abrupt change in load elicits a damped-ringing response that is captured and digitized by the sensor, and digital signal processing (DSP) methods and techniques are used to calculate the resonant frequency. The sensor does this a large number of times: for example, 100 times in a sampling period of one second at a sampling rate of once per hour. The resultant 100 values of frequency are averaged and output from the sensor as a single value for FD.

The sensor output is $FD + N$ and is plotted in Figure 3.7.

Example 3.3 Now that you have selected the feature of interest (frequency) and have a model, Eq. (3.1), you need to design a solution, develop a test bed, and complete a design of experiment(s) (DOE) to be used to verify your chosen solution. Referring back to Figure 3.1, you determine that the PHM solution needs to handle the following variability in the data:

- Noise, in the form of frequency variability (refer back to Figure 3.1).
- The nominal value of frequency (FD_0) in the absence of degradation: $\Delta C = 0$.
- The frequency amplitude at which the SMPS is defined as functionally failed for the high-end application and for the low-end application.

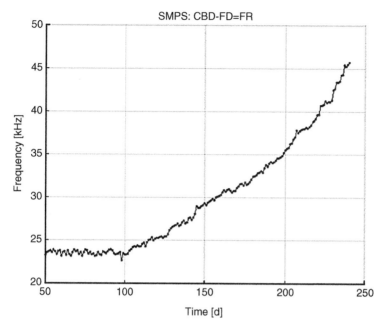

Figure 3.7 Example of a CBD signature: FD is the resonant frequency of a damped-ringing response.

As a result of interaction with customer representatives and engineers, you determine that your design and the technology you are using let you specify 0.6 kHz as a noise margin (NM) to mitigate unfiltered noise: 2.5% of the nominal value of 24 kHz as the resonant frequency of an average, undamaged SMPS. You also define the failure level (FL) for the high-end application as a loss of 60% filtering capacitance and the FL for the low-end application as a loss of 70% filtering capacitance.

Evaluation of the SMPS specifications indicates that manufacturing variations result in a resonant frequency that varies between 21 and 27 kHz. When you apply this information into a test-bed design, build the test bed, execute your DOE, and collect experimental data, you end up with plotted results as shown in Figure 3.8 and the tabulated results shown in Table 3.1.

CBD Signature Evaluation
Evaluation of the CBD signature as input to an intended prediction algorithm leads to the following conclusions:

- The prediction algorithm is likely to produce estimates of prognostic information – remaining useful life (RUL), state of health (SoH), and prognostic horizon (PH) – that meet the accuracy requirements of your customer, especially with regard to a required minimum prognostic distance (PD) within a specified accuracy (PD_α).
- A large number of models are required to correctly process CBD signatures, especially in a large, complex system.
- Using CBD signatures as an approach to PHM is perhaps as complex as classic model-driven approaches.

Table 3.1 Range of failure thresholds: Lot 1 and Lot 2.

Boundary condition	High-end failure level	Low-end failure level
Lot 1: Nominal frequency 27 kHz	43.2	45.9
Lot 2: Nominal frequency 21 kHz	33.6	35.7

Example 3.4   You evaluate the results of your CBD modeling and decide that you need to define at least two signature models for every SMPS lot in which the nominal resonant frequency varies from a nominal value by more than, for example, 0.3 kHz (1.25% of 24 kHz); the exact range is application dependent. Additionally, manufacturing variations cause the output filtering capacitance to vary, and you calculate the nominal resonant frequency of a given SMPS as ranging from 21 to 27 kHz (refer back to Table 3.1): the spread of 6 kHz and a 0.3 kHz increment mean you need to define and use 20 CBD-based models.

3.2.2   FFP Signature

One way to simplify modeling is to normalize CBD signature data to create FFP signature data, and at the same time apply a noise margin to the data (see Figure 3.9). The procedure is to subtract from CBD an NM to create a FD_i value, then subtract a nominal FD_0 value and divide the result by the nominal FD_0 value:

$$FD_i = CBD_i - NM \tag{3.7}$$

$$FFP_i = (FD_i - FD_0)/FD_0 \tag{3.8}$$

$$\text{FFP signature} = \textbf{FFP} = \{FFP_m, FFP_{m+1}, \dots FFP_{m+n-1}, FFP_{m+n}\} \tag{3.9}$$

FFP signatures are preferable to CBD signatures because the effect of noise on prediction accuracy is mitigated by the use of a noise margin and because normalization results in relative units of measure in amplitude instead of units of measure such as kHz. Instead of a model for a nominal frequency of 24 kHz, a model for a nominal frequency of 24.3 kHz, a model for a nominal frequency of 24.6 kHz, and so on, you need only one model, which reduces both modeling and processing complexity when producing prognostic information.

Example 3.5   You decide to transform your CBD signature data into FFP signature data by using Eq. (3.8) and the following parameter values: $FD_0 = 24.0$ kHz and $NM = 0.6$ kHz. The plot of the resulting FFP signature is the left-side plot in Figure 3.10. You examine the plot and conclude the following:

- The amplitude values are no longer Hz or kHz units of measure: they are relative units of measure having no dimension.
- Instead of labeling the y-axis as [AU] for arbitrary units or [ratio], you choose to use a more informative designation, [kHz/kHz].
- The signature starts at a value of about 0.1 instead of the less-than-0 value you expected.

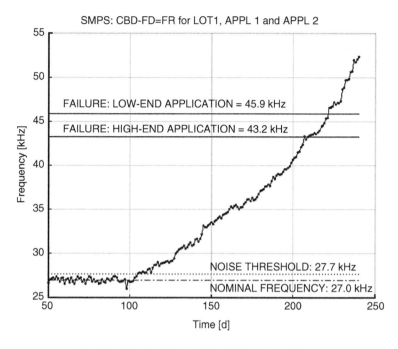

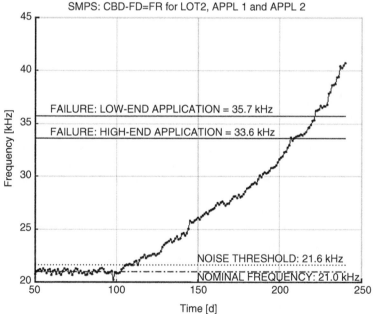

Figure 3.8 CBD signature and levels for SMPS lot 1 (top) and SMPS lot 2 (bottom).

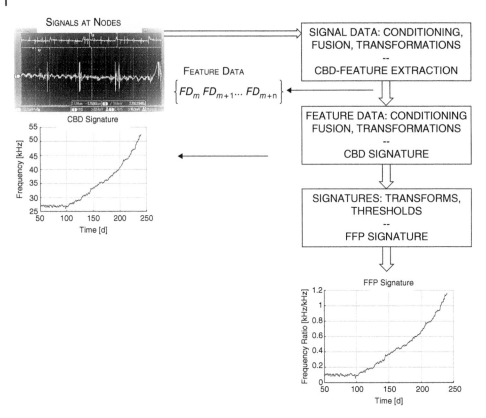

Figure 3.9 Functional block diagram for FFP signature and processing flow.

Example 3.6   You realize that using a fixed value of 24 kHz as the nominal frequency introduces an offset type of error in the FFP signature data. From Table 3.1, you realize that either you need to use 27 kHz or you need to use calibration. You evaluate your prognostic requirements and choose to employ the following in your design: (i) leave the design of the sensor as is, and (ii) employ a node-based architecture that includes support for specifying a calibration phase in which a number of samples from the sensor are averaged, saved, and used as a nominal frequency. The bottom plot in Figure 3.10 shows the result of using a calibrated value of nominal frequency.

An FFP signature point at or below 0 in amplitude indicates no damage. A functional failure level is application-dependent as illustrated by two functional failure thresholds: 0.6 for high-end applications (top plot in Figure 3.11) and 0.7 for low-end applications (bottom plot in Figure 3.11).

FFP Signature: Nominal Value and Calibration for FD

As we have discussed, a prevailing problem with modeling data is variations in the data: signal-induced noise, thermal-induced noise, operational-induced noise, environmental-induced noise, manufacturing-induced noise, and so on. A significant challenge in designing and developing PHM systems to achieve reliable conditioning monitoring is reducing and/or mitigating such variation to reduce the complexity in data processing to obtain affordable, reliable results. The previous example introduced

Figure 3.10 FFP signatures using a fixed value and a calibrated value for nominal frequency.

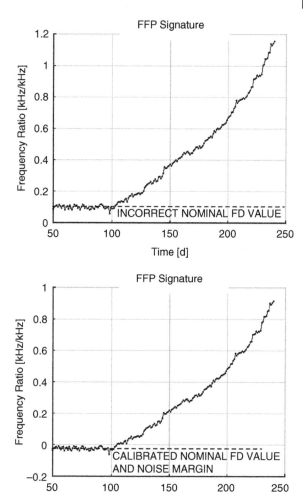

complexity associated with the nominal value used to transform CBD signature data into FFP signature data: the nominal value for FD plus a noise margin must be equal to or greater than the maximum variation in data that results from manufacturing and all other variations in the data not associated with degradation.

There are two basic methods for determining the nominal value to use: (i) perform sufficient experiments, measurements, and analysis to make an accurate determination of the maximum value to use for all instantiations of a prognostic target; and/or (ii) perform a calibration step that sufficiently measures, calculates, and saves a nominal value to use for each instantiation. For applications requiring high precision and high accuracy in prognostic estimates, calibration may be best – otherwise, use high values.

FFP Signature Data: Failure Threshold

To use FFP signatures as input to a prediction algorithm, either models with a defined failure threshold need to be used, or the prediction algorithm must be told the value of the failure threshold.

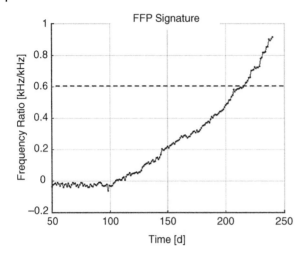

Figure 3.11 FFP signature, calibrated value for nominal frequency, failure threshold at 0.6 and 0.7.

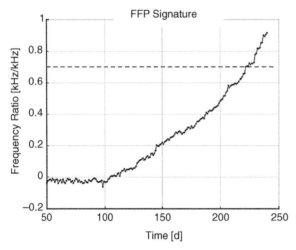

FFP Signature Data: Benefits Compared to CBD Signature Data

FFP signatures require less modeling compared to CBD signatures: one model for each failure level compared to 20 models (refer back to 3.4). In addition, prediction-algorithm processing of FFP signature data compared to that for CBD signature is greatly simplified:

- An FFP point less than or equal to 0 represents a state of zero degradation: RUL = maximum (specified or system-determined value) and SoH = 100%.
- An FFP point between 0 and less than a specified failure threshold represents a degraded state: $0 < \text{RUL} < \text{maximum}$ and $0\% < \text{SoH} < 100\%$.
- An FFP point at or above a failure threshold represents a functionally failed state: RUL = 0 and SoH = 0%.

3.2.3 Transforming FFP into FFS

One technique for handling different levels/thresholds of functional failure is to divide FFP signature data by the value of the failure level (FL) and, at the same time, multiply the result by 100 to create FFS data related to percent of failure:

$$FFS_i = 100\,FFP_i/FL \tag{3.10}$$

$$\text{FFS signature} = \boldsymbol{FFS} = \{FFS_m,\,FFS_{m+1},\,\ldots\,FFS_{m+n-1},\,FFS_{m+n}\} \tag{3.11}$$

FFS Signature Data: Benefits Compared to FFP Signature Data

FFS signatures require the least modeling compared to CBD and FFP signatures: one model for all failure levels. In addition, prediction-algorithm processing of FFS signature data is greatly simplified. Referring to Figure 3.12:

- An FFS point less than or equal to 0 represents 100% health: RUL = maximum and SoH = 100%.
- An FFS point greater than zero and less than 100 represents a degraded state: 0 < RUL < maximum and 0% < SoH < 100%.
- An FFS point at or above 100 represents 0% health – that is, a functionally failed state: RUL = 0 and SoH = 0.

Figure 3.12 FFS signatures for FL = 0.6 (top) and FL = 0.7 (bottom).

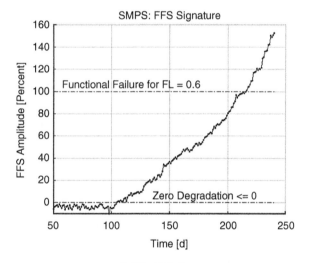

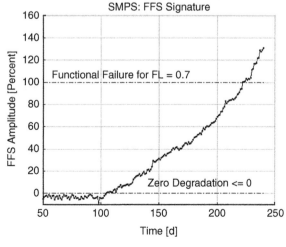

Disadvantage of FFP-Based FFS Data
A disadvantage of FFP-based FFS data is that the signature is generally curvilinear: it has the same characteristic curve as the original CBD signature.

Example 3.7   Referring to Figure 3.12, when you transform FFP signatures using $FL = 0.6$ and $FL = 0.7$, you obtain the plotted results shown. The plots demonstrate that the signatures have been correctly transformed: although the failure amplitude is the same (100%), the time of failure for the left-hand plot occurs at time ~ 215 days, and the time of failure for the right-hand plot occurs at time ~ 230 days.

3.2.4   Transforming FFP into a Degradation Progression Signature (DPS)

The existence of a DPS that is a transform of an FFP signature has been asserted: now we present a general definition and a derivation of a particular FFP. We will also explain and demonstrate why a DPS approach to prognostic enabling for PHM is extremely advantageous compared to other CBD-based approaches, including FFP and FFS for FFP signatures. Further, we have already explained and demonstrated that CBD-based approaches are advantageous compared to classical model-driven, data-driven, and hybrid-driven approaches to prognostic enabling for PHM.

Definition of a DPS
A DPS is a function of a change in value of a parameter of interest, dP:

- When there is no degradation, the value of the parameter is unchanged and $dP = 0$.
- As degradation progresses, the magnitude of dP increases.

$$DPS_i = dP_i/P_0 \tag{3.12}$$

$$\mathbf{DPS} = \{DPS_m, DPS_{m+1}, \ldots DPS_{m+n-1}, DPS_{m+n}\} \tag{3.13}$$

Then defining dP_i to be an absolute value, $dP_i = \text{abs}(dP_i)$, a particular DPS point becomes

$$DPS_i = \text{abs}(dP_i)/P_0 \tag{3.14}$$

Derivation of a DPS Model for an Exemplary FFP Signature
A derivation of a DPS model begins with the general model for a point in an FFP signature, Eq. (3.8),

$$FFP_i = (FD_i - FD_0)/FD_0$$

to which we apply a model for an FD of interest, such as that for resonant frequency of a damped-ringing response of an SMPS, Eq. (3.5):

$$FD_i = FD_0 \left(\sqrt{C_0/(C_0 - \Delta C_i)} \right)$$

Let $P_0 = C_0$ and $dP_i = \Delta C$. Then

$$FD_i = FD_0 \left(\sqrt{P_0/(P_0 - dP_i)} \right)$$

so that

$$\sqrt{P_0/(P_0 - dP_i)} = FD_i/FD_0 \rightarrow P_0/(P_0 - dP_i) = (FD_i/FD_0)^2$$

Therefore,

$$P_0/(FD_i/FD_0)^2 = P_0 - dP_i$$

and, rearranging terms,

$$P_0(FD_0)^2/(FD_i)^2 = P_0 - dP_i$$

or

$$(FD_0/FD_i)^2 = (P_0 - dP_i)/P_0 = 1 - dP_i/P_0$$

And finally,

$$DPS_i = dP_i/P_0 = 1 - (FD_0/FD_i)^2 \qquad (3.15)$$

Comparison of DPS and FFP

A DPS is more linear compared to the FFP signature from which it is derived. When Eq. (3.15) is applied to an experimental-based FFP signature, such as that shown in Figure 3.11, the result is the DPS in the top plot of Figure 3.13. The comparison of both the DPS and its FFP is shown in the bottom plot.

Figure 3.13 DPS and FFP signatures for data shown in Figure 3.11.

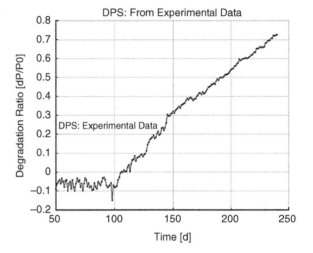

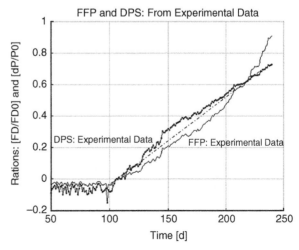

Comparison of FFP-Based and DPS-Based Failure Thresholds

Comparing the DPS and FFP right-side plots of Figure 3.13 indicates the amplitudes are different for a given point in time. This means an FFP signature at an amplitude ratio of 0.4 is not at the same level of degradation as for a DPS at that same amplitude ratio. This is because a failure level for an FFP signature is defined as follows:

$$FL_{FFP} = FFP_i \tag{3.16}$$

Then from Eq. (3.8),

$$FL_{FFP} = (FD_i - FD_0)/FD_0$$

then

$$FD_i - FD_0 = FD_0 FL_{FFP} \rightarrow FD_i = FD_0 FL_{FFP} + FD_0$$

Then we have

$$FD_0/FD_i = 1/(FL_{FFP} + 1)$$

And from Eq. (3.15),

$$FL_{DPS} = 1 - (1/(1 + FL_{FFP}))^2 \tag{3.17}$$

Using a DPS-Based Failure Level

Since an FFP signature is a FFP, it is logical to define a failure level using Eq. (3.16) and then calculate the equivalent level for dP/P_0 using Eq. (3.17).

Example 3.8 Suppose you wish to use $FL_{FFP} = FD_i = 0.4$ as the failure level. To calculate the equivalent FL_{DPS}, you would use Eq. (3.17):

$$FL_{DPS} = 1 - (1/(1 + FL_{FFP}))^2 = 1 - (1/(1 + 0.4))^2$$

$$= 1 - (1/1.4)^2 = 0.49 = 0.5$$

Figure 3.14 shows an $FL = 0.5$ superimposed on DPS and an $FL = 0.4$ superimposed on FFS: functional failure occurs at the same time for both, which is the expected, desirable result.

3.2.5 Transforming DPS into DPS-Based FFS

A DPS, similar to an FFP, is more useful when transformed into a DPS-based FFS: divide DPS data signature data by the value of the failure level (FL) and, at the same time, multiply the result by 100 to create FFS data related to percent of failure:

$$FFS_{DPSi} = 100 \, DPS_i/FL \tag{3.18}$$

$$DPS - based \, FFS = FFS_{DPS} = \{FFS_{DPSm,} \, FFS_{DPSm+1}, \ldots, FFS_{DPSm+n}\} \tag{3.19}$$

Similar to an FFP-based FFS, there is one model for all failure levels, and prediction-algorithm processing of FFS signature data is greatly simplified.

Example 3.9 Use the results of 3.8 and transform the FFP data into an FFS using an FL value of 0.7; transform the DPS data into an FFS using an FL value of 0.65; and then

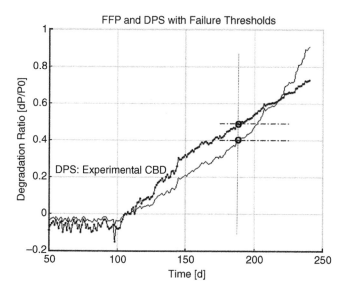

Figure 3.14 FFP and DPS Showing FL = 0.4 and FL = 0.5.

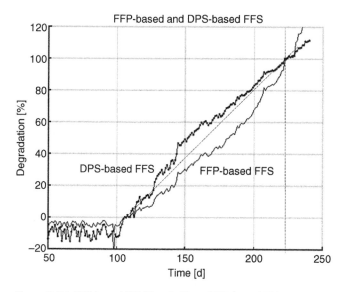

Figure 3.15 DPS-based FFS (FL = 0.65) and FFP-based FFS (FL = 0.7).

plot the results. Functional failure using an FFP-based FFS occurs at the same time as when using a DPS-based FFS, as shown in Figure 3.15 which has an ideal FFS transfer curve superimposed between the 0% and 100% points. Visual analysis of the plots leads to a qualitative assessment that the DPS-based FFS is less curvilinear compared to the FFP-based FFS.

3.3 Model Verification

It is important that all models used in a PHM system are verified: simulate each model, and compare the model against experimental and/or actual fielded data. There needs to be high correlation between the simulated data results and the results you obtain using experimental and/or field data. Most likely there will be discrepancies, but you should be able to explain those discrepancies to make an informed decision as to whether the discrepancies are acceptable and/or modeling and processing changes are necessary and/or desirable.

The verification focus of CBD modeling is comparing the characteristic curve exhibited by the CBD signatures against an ideal curve obtained from simulating your PoF-based and/or FMEA-based model. The objective of the modeling is not to exactly replicate CBD signatures with model-simulated curves; rather, it is to ensure the totality of model-simulated curves: the models for CBD signatures, FFP signatures, DPS data, and FFS data are sufficiently correlated and accurate. When FFS data is input to prediction algorithms in your PHM system, the resultant prognostic information needs to meet the accuracy, convergence, and reliability requirements for reliable condition-based monitoring.

3.3.1 Signature Classification

There is little use for prognostic-enabling a component that typically degrades from zero degradation (100% health) to functional failure (0% health) in less than a second. Instead, diagnostic detection, reporting, and remedial actions and/or disaster avoidance and/or remedial measures are really all that can be done for such "avalanche" failure modes. We are interested in degradation-induced signatures: those signatures having changes that are highly correlated to the levels of degradation. Further, the time between the onset of degradation and a level of degradation that results in functional failure must be sufficiently long to enable the prognostic system to detect degradation, produce prognostic estimates that converge to a level of required accuracy, and do so on or before a required minimum length of time (a prognostic distance [PD]). We model signatures using the following general form we previously developed (Eq. (2.50) in Chapter 2):

$$FD_i = FD_0 f(dP_i, P_0) \tag{3.20}$$

where $f(dP, P_0)$ is a degradation model, FD_i is a signature data point, FD_0 is the nominal value of a measurable feature absent any degradation, dP_i is a change in value of a parameter, and P_0 is the nominal value of that parameter absent any degradation.

As we shall see in this and the next chapter, the signature identification, conditioning, and transform methods used means it is not necessary to determine or use exact parameter values, but we need to identify the class of each signature to select a correct degradation model. The degradation-related signatures result from failure modes that can be modeled as power functions or exponential functions.

Although degradation signatures are modeled as power functions or exponential functions that have either increasing or decreasing amplitudes, constant signatures are only applicable in the sense that they indicate the absence of degradation. We shall only

Figure 3.16 Examples of signatures: decreasing (top) and increasing (bottom) slope angles.

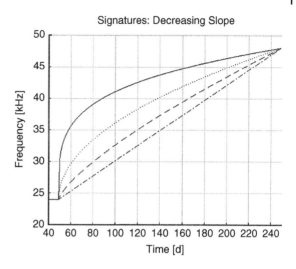

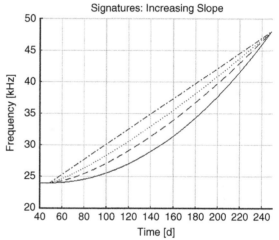

use models for increasing amplitudes; signatures having decreasing amplitudes are transformed into a complementary, increasing-amplitude signature and degradation model – more information on this is provided in the next chapter. Referring to Figures 3.16 and 3.17, degradation signatures are further classified by whether the signature has a decreasing or increasing slope angle as degradation progresses: a linear, straight-line degradation signature is a special case that is classified as having a constant slope angle. The slope angle is the arc between a tangent to the curve and the horizontal axis.

3.3.2 Verifying CBD Modeling

Verify that the model you have selected and/or developed is sufficiently correlated to the FD you have selected for processing. For example, when the system produces a signature

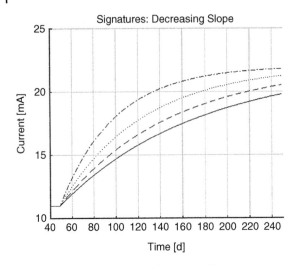

Signatures: Decreasing Slope

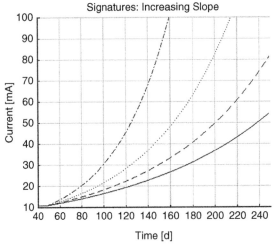

Signatures: Increasing Slope

Figure 3.17 Other examples of signatures: decreasing (top) and increasing (bottom) slope angles.

that has a decreasing slope angle, your CBD modeling should produce a signature that has a decreasing slope angle, and both should have the same rate of change.

Example 3.10 For example, use Eq. (3.2), CBD data $= FD + N$, and Eq. (3.6), $FD_i = CBD_i - NM$, and do the following:

- Set NM to 0.7 kHz, which you previously determined and used.
- Use the CBD previously shown back in Figure 3.8, left side, for which FD_0 was calculated to be 27.0 kHz.
- Examine the data to determine a constant rate of degradation, dP, per day:

$$dP \sim (0.75/140) \text{ per day.}$$

- Calculate and plot simulated FD points (Figure 3.18).
- Compare the simulated plot against the plot of the experimental CBD signature minus a 0.7 kHz noise margin (Figure 3.19).

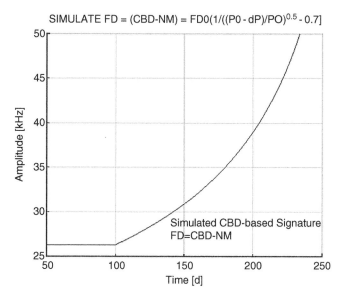

SIMULATE FD = (CBD-NM) = FD0(1/((P0 - dP)/PO)$^{0.5}$ - 0.7]

Figure 3.18 Simulated CBD-based signature: FD = CBD − NM.

Example 3.11 You compare the plots and conclude they are very closely correlated. To complete your verification that Eqs. (3.2) and (3.6) are correct for your application, you examine the design of your experimental test bed and the design of the SMPS you are prognostic enabling and determine the following:

• The SMPS uses pulse-width modulation with voltage feedback, which linearizes the change in frequency: the circled 1 in Figure 3.19, bottom plot.
• Capacitance degradation is an injected fault mode wherein capacitance is discretely removed, which causes a step-like change in the output: see (2) in Figure 3.19, bottom plot.

3.3.3 Verifying FFP Modeling

You need to verify your FFP modeling even when you do not intend to use FFP signatures. Your modeling should produce a characteristic FFP curve that is the characteristic CBD curve, except that the curve is shifted downward; it is at or below 0 in the absence of degradation; and the units of measure have changed from physical units such as voltage, amperes, and resistance to a ratio of values such as [voltage/voltage], and so on.

Example 3.12 An example of verifying FFP modeling is to verify that Eqs. (3.7) and (3.8) are correct for your application:

$$FFP_i = (FD_i - FD_0)/FD_0$$

$$\text{FFP signature} = \textbf{FFP} = \{FFP_m, FFP_{m+1}, \dots FFP_{m+n-1}, FFP_{m+n}\}$$

• Use your simulated FD data to create and plot a simulated FFP signature such as that shown in the top plot in Figure 3.20.

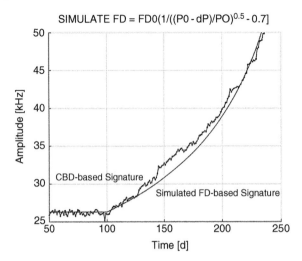

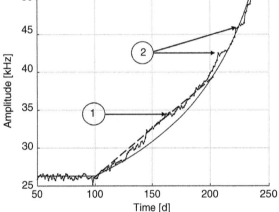

Figure 3.19 Differences: experimental and simulated signatures.

- Compare the plots of the simulated and experimental-based FFP signatures, such as that shown in the bottom plot in Figure 3.20.

3.3.4 Verifying DPS Modeling

Similar to verifying FFP modeling, you need to verify your DPS modeling.

Example 3.13 Use Eq. (3.16) and plot the DPS data using both the simulated data and experimental data:

- The DPS using simulated data from a square root function is ideal (see Figure 3.21, top plot).

Figure 3.20 Simulated (top) and comparison of experiment (bottom) FFP signatures.

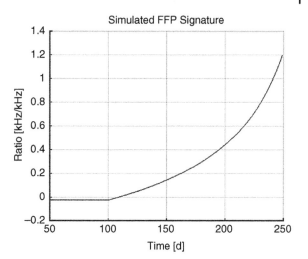

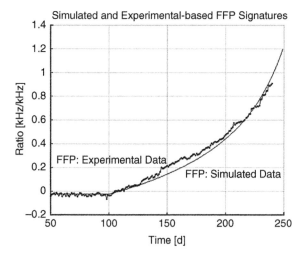

- The DPS using experimental data, while not ideal, is less curvilinear compared to the CBD-based and FFP-based signatures (see Figure 3.21, bottom plot).

3.3.5 Verifying DPS-Based FFS Modeling

A final step is to verify the model for a DPS-based FFS.

Example 3.14   Transform your simulated DPS data into a DPS-based FFS (see the top plot in Figure 3.22, and verify that it is ideal (for a constant rate of degradation). Then compare your simulated result to your experimental result (right-side plot in Figure 3.22).

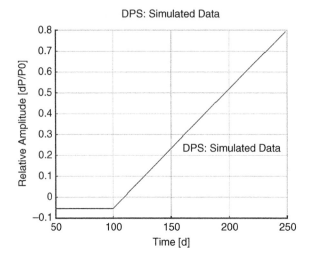

DPS: Simulated Data

Figure 3.21 Simulated DPS (top) and experimental DPS (bottom).

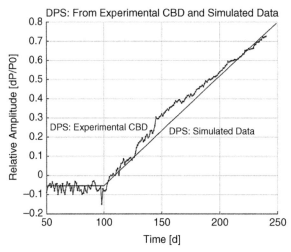

3.4 Evaluation of FFS Curves: Nonlinearity

So far in this chapter, we have successfully developed and verified modeling to produce signatures:

$FD_i = CBD_i - NM$	Feature data point – Eq. (3.7)
$FFP_i = (FD_i - FD_0)/FD_0$	FFP signature point – Eq. (3.8)
$DPS_i = 1 - (FD_0/FD_i)^2$	DPS point – Eq. (3.15)
$FFS_i = 100\, DPS_i/FL$	FFS point (using DPS data) – Eq. (3.18)

The final transform uses Eq. (3.18) to produce a DPS-based FFS to achieve an objective of transforming CBD signature data to FFS data: an ideal DPS-based FFS is a straight line

Figure 3.22 Simulated FFS from simulated (top) and from experimental DPS (bottom): FL = 0.65.

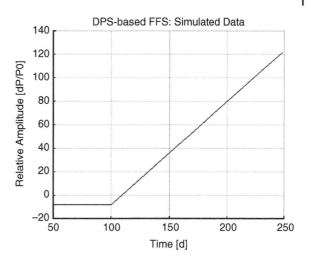

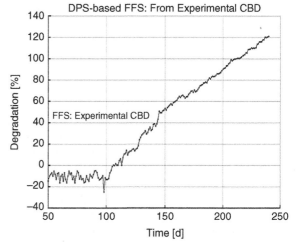

between the 0% and 100% data points. Prior to that, we developed a model, Eq. (3.10), for transforming FFP signature data into FFP-based FFS data, and we made the following qualitative assessment:

> A disadvantage of FFP-based FFS data is that the signature is generally curvilinear: it has the same characteristic curve as the original CBD signature.

Figure 3.15 contains three plots of example FFS data: (i) a DPS-based FFS, (ii) an ideal FFS transfer curve, and (iii) an FFP-based FFS. A qualitative assessment of those curves might include the following:

- The example DPS-based FFS is more curvilinear compared to the ideal FFS.
- The example FFP-based FFS is also more curvilinear compared to the ideal FFS.
- The example DPS-based FFS is less curvilinear compared to the FFP-based FFS.

But a visual inspection of the plots in Figure 3.15 indicates the latter qualitative assessment is not true for every data-point-by-data-point comparison of the DPS-based FFS to the FFP-based FFS. What we need is quantitative assessment – and we advocate a quantitative method for assessing the nonlinearity of FFS curves that is based on integral nonlinearity (INL) measurements that are commonly used for assessing the linearity of output-transfer curves of data converters (Texas Instruments 1995; Carr and Brown 2000; Jenq and Qiong 2002).

3.4.1 Sensing System

A sensing system comprises everything between a monitored node and the input port of a predication system: the node, the sensor, and all the intervening frameworks. The prediction system acts upon the input from the sensing system to produce prognostic information used to provide a prognosis about a future failure: sensing hardware, data/domain converters, signal conditioning, data fusion, and so on.

In a PHM system, the sensing system provides input to a prediction system that processes the input from the sensing system to produce prognostic information that includes (i) RUL, (ii) SoH, and (iii) PH. An ideal RUL curve is a decreasing straight line between the point in time of the onset of degradation and the point in time of functional failure: the amplitude of that line is the time between functional failure and the onset of degradation. Similarly, an ideal SoH curve is a decreasing straight line between the point in time of the onset of degradation and the point in time of functional failure: an ideal SoH curve decreases linearly from 100% to 0% in amplitude. An ideal PH curve is a horizontal straight line comprising the sum of the current time of a sample and the RUL at that time.

The linearity of a sensing system is an assessment of how the measured curve produced by the sensing system deviates from the ideal curve (TI 1995; Carr and Brown 2000). The output of the sensing system is an FFS, and an ideal FFS input curve to the predication system is a straight line having a value of 0 just before the instant in time that degradation begins and a value of 100 at the instant when the monitored prognostic target is deemed to have functionally failed. For example, the left side of Figure 3.22 is a plot of an ideal FFS curve, and the right side of that figure is a plot of our example DPS-based FFS; as shown, that curve deviates from an ideal FFS curve.

3.4.2 FFS Nonlinearity

We define functional-failure signature nonlinearity (FNL) as a measure of the deviation between the actual FFS and an ideal FFS, very much like an INL assessment of an analog-to-digital data converter (ADC) or transducer output transfer curve (Texas Instruments 1995; Carr and Brown 2000; Jenq and Qiong 2002). A point-by-point FNL is a measurement of the error between an actual FFS value and the ideal FFS value at that point in time (see Figure 3.23). A total error (FNL_E) provides an assessment of the nonlinearity of the FFS curve (see Figure 3.24).

Calculating FFS Nonlinearity (FNL)
Calculating FNL is a post-processing type of operation using FFS data, because we need to first calculate the total time between the point in time of the onset of degradation and the point in time when functional failure occurs:

$$TTF = t_{FAILURE} - t_{ONSET} \tag{3.21}$$

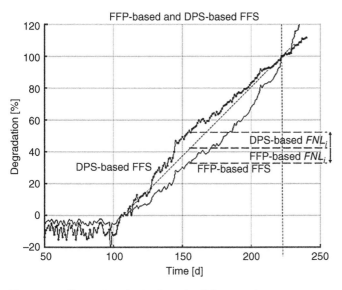

Figure 3.23 Illustration of point-by-point FNL comparison.

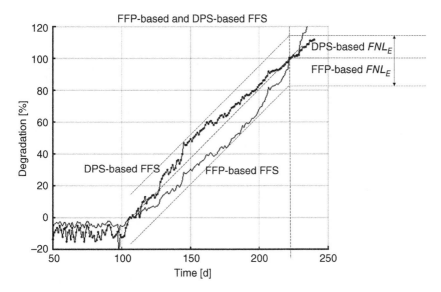

Figure 3.24 Illustration of total FNL$_E$ comparison.

Then for each point in the set $\{FFS_i\}$, we create an ideal point,

$$IDEAL\_FFS_i = 100(t_i - t_{ONSET})/TFF \qquad (3.22)$$

and calculate point-by-point nonlinearity,

$$FNL_i = FFS_i - IDEAL\_FFS_i \qquad (3.23)$$

and calculate total nonlinearity,

$$FNL_E = \max(\text{abs}(\{FNL_i\})) \qquad (3.24)$$

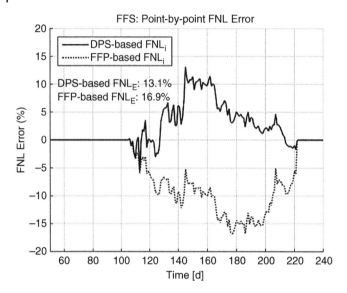

Figure 3.25 Example plot of FFP-based and DPS-based FNL_i.

where max is a maximum function and abs is an absolute-value function. Example plots on nonlinear errors are shown in Figure 3.25.

Example 3.15 Previously, you used an FL value of 0.7 and created an FFP-based FFS (in 3.7); you transformed the FFP-based FL value of 0.7 to a DPS-based value of 0.65; then you created a DPS-based FFS (3.9); and you compared the results (Figure 3.15) and qualitatively concluded that the DPS-based FFS was less curvilinear compared to the FFP-based FFS. Now, verify the qualitative conclusion using Eqs. (3.21–3.24) for both sets of FFS data and make a quantitative assessment.

3.5 Summary of Data Transforms

Using an SMPS as an example, we described and showed the following signals and the transforms that were performed:

- **SMPS output** (left side of Figure 3.26). Analog voltage of direct current (DC) and alternating current (AC) component features and noise. One of the features is damped-ringing responses to abrupt changes in load.
- **Ringing response** (right side of Figure 3.26). An intelligent sensor might do the following: present a series of abrupt, low-power changes in load to the SMPS at timed intervals (sampling periods); capture the ringing responses; convert analog voltage to digital data; calculate an average frequency for each of several sinusoids in the response; calculate an average value of the average resonant frequency of each of the series of ringing responses; calculate the average frequency as a sample value; and output that average as a scalar CBD_i value.
- **CBD signature** (left side of Figure 3.27). Sensor data is transformed into a CBD signature by subtracting the value of a noise margin from each received CBD_i to

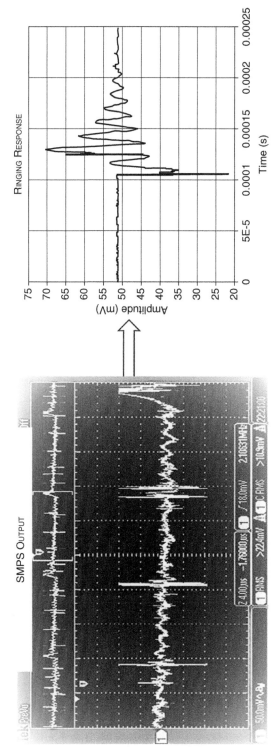

Figure 3.26 SMPS output and extracted damped-ringing response.

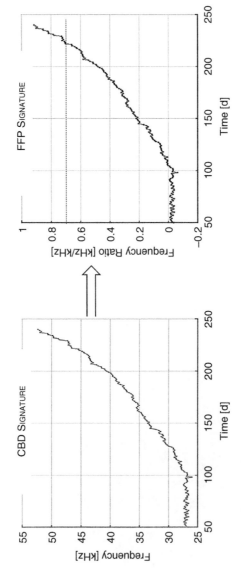

Figure 3.27 CBD signature and FFP signature.

create a feature point, FD_i. The set of FD_i values, $\{FD_i\}$, is a CBD signature, as in Eq. (3.7):

$$FD_i = CBD_i - NM$$

- **FFP signature** (right side of Figure 3.27). A set of FD_i values is transformed into a set of FFP_i values by subtracting from each FD_i value a nominal value of FD and dividing the result by that nominal value. The set of FFP_i values, $\{FFP_i\}$, is an FFP signature, as in Eq. (3.9):

$$FFP_i = (FD_i - FD_0)/FD_0$$

- **FFP-based FFS** (left side of Figure 3.28). An FFP set, $\{FFP_i\}$, is transformed into $\{FFS_i\}$ by dividing each value in the set by a functional-failure level, FL, and multiplying the result by 100, as in Eq. (3.10):

$$FFS_i = 100 \, FFP_i/FL$$

- **DPS** (right side of Figure 3.28). An FFP set, $\{FFP_i\}$, is transformed into $\{DPS_i\}$ by transforming the FFP model into a new model that is a function of a parameter, P_0, and a change in that parameter, dP, and then solving for dP_i/P_0 in terms of the feature parameters FD_i and FD_0, as in Eq. (3.15):

$$DPS_i = 1 - (FD_0/FD_i)^2 \text{ for this particular FFP model}$$

- **DPS-based FFS** (Figure 3.29). A set of DPS values, $\{DPS_i\}$, is transformed into a DPS-based FFS, $\{FFS_i\}$ by dividing each DPS_i value in the set by an FL value and multiplying the result by 100, as in Eq. (3.18),

$$FFS_i = 100 \, DPS_i/FL$$

- where FL is calculated using Eq. (3.17):

$$FL_{DPS} = 1 - (1/(1 + FL_{FFP}))^2$$

High-Level Procedure for Producing a DPS-Based FFP

A high-level procedure for producing a DPS-based FFP is the following (refer to Figure 3.30): (i) select one or more output nodes to monitor for prognostic purposes; (ii) develop or acquire sensors to isolate CBD leading indicators of interest and to extract selected FD; (iii) signal condition and transform the collected FD to produce a CBD-based signature – Eq. (3.6); (iv) further condition as necessary and transform CBD signature data into FFP signature data – Eqs. (3.7) and (3.8); (v) transform FFP signature data into DPS data using Eqs. (3.8) and (3.15); and (vi) transform DPS data into an FFS using Eqs. (3.17) and (3.18). You use Eqs. (3.21)–(3.24) to quantitatively assess the nonlinearity of FFS data.

Qualitative Assessment

The biggest advantage of the DPS approach is that DPS-based FFS data is more linear compared to that based on FFP data. Additional advantages are the following:

- Similar to an FFP signature, the amplitude of $\{DPS_i\}$ data is dimensionless.
- The number of required models is reduced.

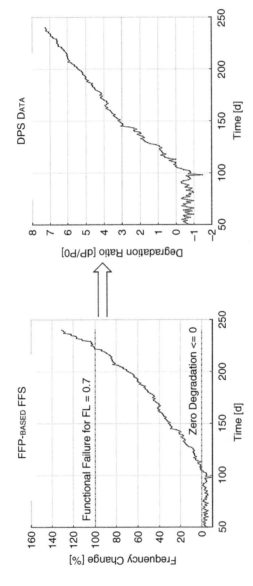

Figure 3.28 FFP-based FFS and DPS.

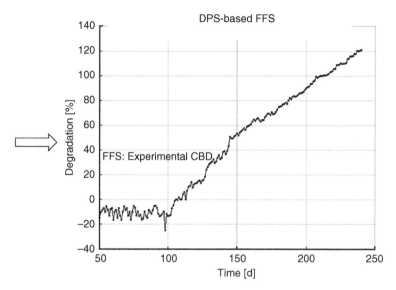

Figure 3.29 DPS-based FFS.

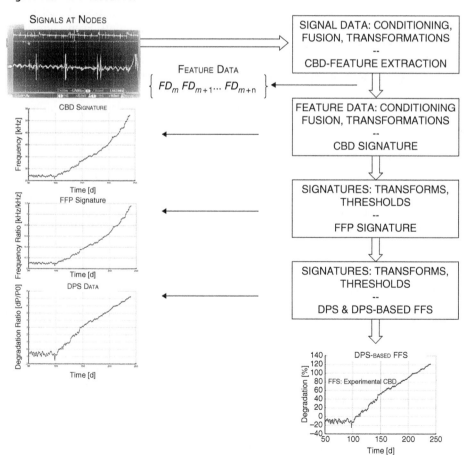

Figure 3.30 Procedural diagram for producing a DPS-based FFS.

Using Eq. (3.14) and the results of the examples used to verify modeling, we make the following inductive observations regarding the set $\{DPS_i\}$, regardless of the type of an underlying FD model (such as a linear or power function) and regardless of how FD changes with respect to time:

- When a parameter dPi changes as a linear function with respect to time, $\{DPS_i\}$ changes linearly with respect to time.
- When a parameter dPi changes as a power function with respect to time, $\{DPS_i\}$ changes as power function with respect to time.
- When a parameter dPi changes exponentially with respect to time, $\{DPS_i\}$ changes exponentially with respect to time.

The biggest advantage is the following: $\{DPS_i\}$ is less curvilinear than its underlying $\{FFP_i\}$ signature data, and so $\{DPS_i\}$ data is particularly useful as input to prediction algorithms for processing to predict a future time of failure for a prognostic target.

3.6 Degradation Rate

In the previous section of this chapter, we saw that an ideal DPS-based FFS is a linearly increasing line from 0 or less (no degradation) to 100 or more (functionally failed): an ideal DPS-based FFS is a straight line between the 0% and 100% points. Even a non-ideal DPS-based FFS was shown to be less curvilinear than its underlying FFP signature (refer back to Figure 3.29). The linearity of a DPS-based FFS is dependent on the linearity of the rate of change in a degrading parameter, dP: if the degradation rate with respect to time is constant, the DPS and the DPS-based FFS will be linear; otherwise the degree of non-linearity of the DPS and the DPS-based FFS is dependent on the degree of nonlinearity in the degradation rate.

3.6.1 Constant Degradation Rate: Linear DPS-Based FFS

As we have seen, an idealized DPS for a constant degradation rate results in an idealized linear DPS-based FFS. And even when a DPS is not ideal, the DPS-based FFS is more linear (less curvilinear) compared to the characteristic curve for its DPS. The examples and exemplary data we have been using are a result of assuming a constant degradation rate such as that modeled as follows:

$$dP_i = D_0(t_i)^\delta \tag{3.25}$$

in which $\delta = 1.0$ for a constant degradation rate and t_i equals

$$t_i = (t_S - t_D) \tag{3.26}$$

where t_S is the time of a sample and t_D is the time of the onset of degradation: $t_S \geq t_D$.

Example 3.16 Show that the DPS-based signature for our example is always linear given a constant degradation rate and nonlinear for a nonconstant degradation rate. Using Eqs. (3.25) and (3.12),

$$DPS_i = dP_i/P_0 = D_0/P_0(t_i)^\delta$$

so that for $\delta = 1$, DPS_i is linear and nonlinear otherwise.

Example 3.17 Show that the FFP-based signature for our example is always nonlinear, even when $\delta = 1$. Using Eq. (3.15),

$$DPS_i = 1 - (FD_0/FD_i)^2 \rightarrow FD_i/FD_0 = \sqrt{1 - (D_0/P_0)(t_i)^\delta}$$

and using Eq. (3.8),

$$FFP_i = (FD_i - FD_0)/FD_0 = FD_i/FD_0 - 1$$

$$= (1/\sqrt{1 - (D_0/P_0)(t_i)^\delta}) - 1$$

which is nonlinear for all nonzero values of δ.

3.6.2   Nonlinear Degradation Rate

A non-idealized DPS occurs when the degradation rate is not constant: it is nonlinear with respect to time. However, we will see that even when a DPS is not ideal, a DPS-based curve is still more linear (less curvilinear) compared to its FFP signature.

Example 3.18 Plot a DPS, dP/P_0, using Eq. (3.25), where $D_0 = 0.90/200$ is valid after the onset of degradation, degradation begins at $t = 51$, and $\delta = 1.0$ for a time period of 1 to 250 days. Compare the dotted-line plot in Figure 3.31 to the solid-line plot, in which $dP = (0.75/200)t^{1.05}$.

Notice that the differential nonlinearity (DNL) between the ideal, linear DPS and the non-ideal, curvilinear DPS transfer curves is very small. The nonconstant degradation rate, in this case, increases the nonlinearity of the CBD and FFP signature curves from that seen on the left side of Figure 3.32. The nonlinearity increases even further, as shown

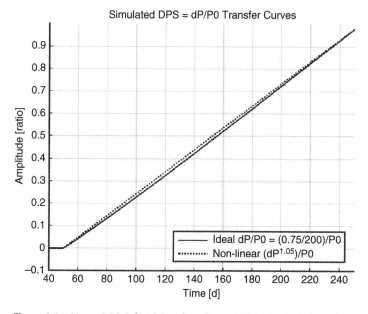

Figure 3.31 Linear DPS (left side) and nonlinear DPS (right side) data plots.

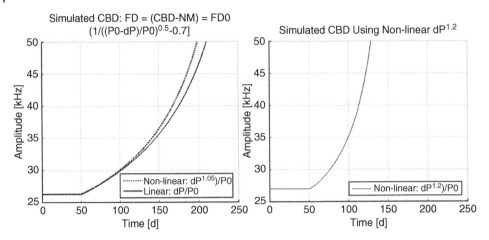

Figure 3.32 Simulated CBD plots using nonlinear and linear degradation rates.

on the right side of Figure 3.32, when the nonlinearity in the degradation rate is increased even further: for example, to $dP = (0.75/200)t^{2.0}$.

The DNL term introduced in 3.18 and a related term, integral nonlinearity (INL), are measurements of the deviation between two analog values: in this book, the two analog values are an ideal (straight line) signature and the signature of the data, such as an FFS, that is input to prediction algorithms (Texas Instruments 1995). In this book, we introduce a new term, FFS nonlinearity (FNL), which is a measurement of the deviation between ideal an RUL and an ideal SoH transfer curve and actual curves of the estimates of RUL and SoH from such prediction algorithms (Hofmeister et al. 2018).

Example 3.19 Transform the simulation data you created for the right-hand plot shown in Figure 3.32 to DPS data, assuming a constant degradation rate. When you do so, you create a nonlinear DPS, a transfer curve, as shown in Figure 3.33; an ideal DPS is also shown for comparison. Even though the DPS is nonlinear, it is more linear than its FFP signature data (refer back to the right-hand plot in Figure 3.32).

3.7 Failure Progression Signatures and System Nodes

In Chapter 1, we used an example illustration of a framework for a PHM system for CBM. Referring to Figure 3.34, the framework comprises (i) a sensor framework, (ii) a feature-vector framework, (iii) a prediction framework, (iv) a health-management framework, (v) a performance-validation framework, and (vi) a control- and data-flow framework. The sensor framework is attached to a node, and in a system there might be hundreds or thousands of nodes, each of which is a point in a system that is a source of CBD: acoustic, electrical, magnetic, mechanical stresses and strains, vibration-based, radiation, optical, and so on. It makes sense, then, to use a node-based architecture as an anchor for a PHM system that collects CBD for reliable condition monitoring to support CBM.

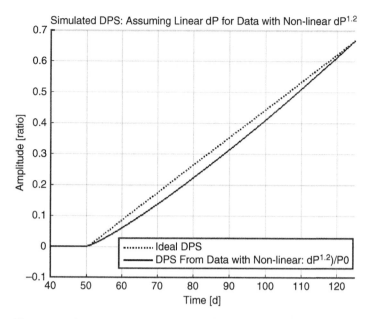

Figure 3.33 Comparison of ideal DPS to DPS from data for a nonconstant degradation rate.

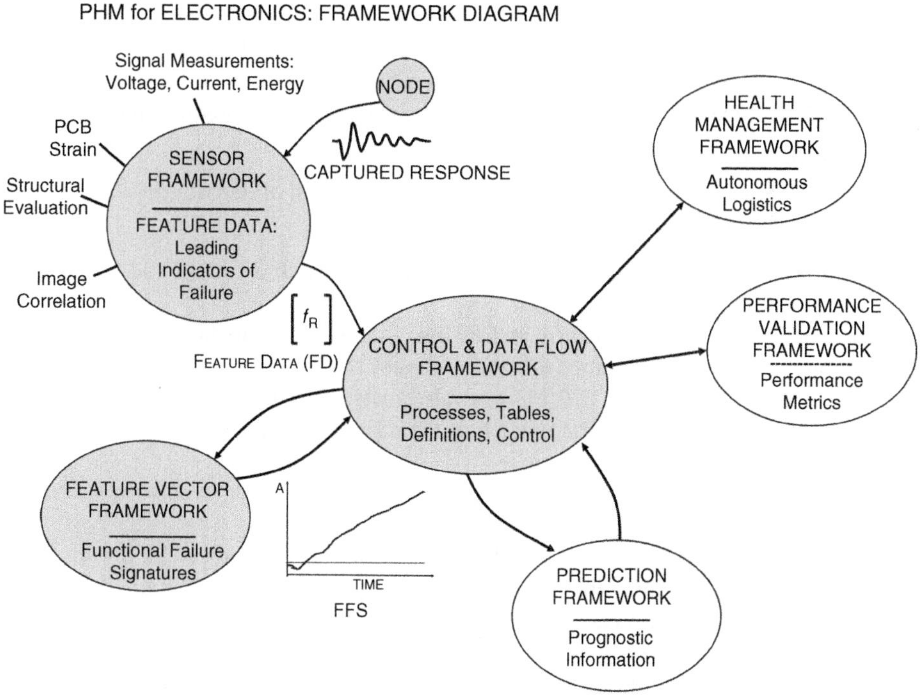

PHM for ELECTRONICS: FRAMEWORK DIAGRAM

Figure 3.34 Node-based framework for supporting failure progression signatures.

In this chapter, we presented and discussed sensors attached to nodes to monitor signals, to isolate CBD and FD using noise margins; calibration to use as nominal values for FD; and various transforms and data processing associated with CBD signatures, FFP signatures, DPS data, FFS data, and predications (prediction framework). It makes sense to use a node-based architecture for a framework for a PHM system as opposed to, for example, designing and developing the details for each node in a control- and data-flow framework. A sensor framework of sensors and drivers captures signals at nodes, performs data conditioning, and produces FD. A feature-vector framework of processors performs additional conditioning and transforms to create FFSs used as input data to a prediction framework that produces prognostic information. A health-management framework processes prognostic information to create diagnostic, prognostic, and logistics directives and actions to maintain the health and reliability of the system. A performance-validation framework provides means and methods to produce prognostic-performance metrics for evaluation of accuracy of prognostic information. A control- and data-flow framework manages PHM functions and actions (CAVE3 2015; Hofmeister et al. 2017; Kumar and Pecht 2010; Pecht 2008).

3.8 Failure Progression Signatures: Summary

This chapter presented a rationale for transforming CBD signatures into other signatures: FFP, DPS, and FFS. You learned that both an FFP and a DPS can be transformed into an FFS. You saw that an FFP is a normalized version of a CBD and that an FFP greatly reduces the number of signature models; and we have shown that transforming an FFP into an FFS produces a very amenable signature for inputting to prediction algorithms:

1. Amplitude values are percentages.
2. Amplitudes of 0 or less indicate the absence of detection of degradation.
3. Amplitudes of 100 or greater indicate functional failure.

Functional failure is defined as a SoH condition in which a prognostic target is no longer operating within specifications.

You learned that you can transform an FFP into a DPS by solving the signature in terms of a change in a parameter value (dP) with respect to a nominal, non-degraded value of the parameter (P_0): $DPS = (dP/P_0)$; you also learned that a DPS is less curvilinear compared to an FFP and that there are two forms of DPS: constant and nonconstant degradation rate with respect to time:

$$f(DPS, t) = (d/dt)DPS$$

You learned that a DPS, like an FFP, can be transformed into an FFS and that a DPS-based FFS is most amenable to processing by prediction algorithms – even when the rate of degradation with respect to time is not constant.

You also learned that you need to verify your modeling choices against the signatures you are modeling and classify whether a signature has a decreasing or increasing slope angle as degradation progresses, to help determine the degradation function for a signature:

$$FD_i = FD_0 f(dP_i, P_0)$$

This chapter presented two quantitative measures for assessing the nonlinearity of an FFS curve: point-by-point nonlinearity and total nonlinearity, using Eqs. (3.23) and (3.24):

$$FNL_i = FFS_i - IDEAL\_FFS_i$$

$$FNL_E = \max(\text{abs}(\{FNL_i\}))$$

In the next chapter, you will learn that signatures can be effectively modeled by a limited number of characteristic curves: 10 power functions and 4 exponential functions. You will also learn that a heuristic approach to modeling signatures is sufficiently accurate for prognostics: formal modeling based on traditional methods such as FMEA and PoF is not necessary.

References

Carr, J.J. and Brown, J.M. (2000). *Introduction to Biomedical Equipment Technology*, 4e. Upper Saddle River, New Jersey: Prentice Hall.

Erickson, R. (1999). *Fundamentals of Power Electronics*. Norwell, MA: Kluwer Academic Publishers.

Hofmeister, J., Wagoner, R., and Goodman, D. (2013). Prognostic health management (PHM) of electrical systems using conditioned-based data for anomaly and prognostic reasoning. *Chemical Engineering Transactions* 33: 992–996.

Hofmeister, J., Goodman, D. and Wagoner, R. (2016). Advanced anomaly detection method for condition monitoring of complex equipment and systems. 2016 Machine Failure Prevention Technology, Dayton, Ohio, US, 24–26 May.

Hofmeister, J., Szidarovszky, F., and Goodman, D. (2017). An approach to processing condition-based data for use in prognostic algorithms. 2017 Machine Failure Prevention Technology, Virginia Beach, Virginia, US, 15–18 May.

Hofmeister, J.P, Goodman, D.L., and Szidarovszky, F. (2018). Transforming condition-based data signatures into functional failure signatures, IEEE 2018 Aerospace Conference, Big Sky, Montana, US, 3–9 March.

Jenq, Y.C. and Li, Q. (2002). Differential non-linearity, integral non-linearity, and signal to noise ratio of an analog to digital converter. Portland, Oregon: Department of Electrical and Computer Engineering, Portland State University.

Judkins, J.B. and Hofmeister, J.P. (2007). Non-invasive prognostication of switch mode power supplies with feedback loop having gain, IEEE Aerospace Conference 2007, Big Sky, Montana, US, 4–9 Mar.

Judkins, J.B., Hofmeister, J., and Vohnout, S. (2007). A prognostic sensor for voltage regulated switch-mode power supplies. IEEE Aerospace Conference 2007, Big Sky, Montana, US, 4–9 Mar, Track 11–0804, 1–8.

Kumar, S. and Pecht, M. (2010). Modeling approaches for prognostics and health management of electronics. *International Journal of Performability Engineering* 6 (5): 467–476.

Medjaher, K. and Zerhouni, N. (2013). Framework for a hybrid prognostics. *Chemical Engineering Transactions* 33: 91–96. https://doi.org/10.3303/CET1333016.

National Science Foundation Center for Advanced Vehicle and Extreme Environment Electronics at Auburn University (CAVE3). (2015). Prognostics health management for electronics. http://cave.auburn.edu/rsrch-thrusts/prognostic-health-management-for-electronics.html (accessed November 2015).

Pecht, M. (2008). *Prognostics and Health Management of Electronics*. Hoboken, New Jersey: Wiley.

Texas Instruments. (1995). Understanding data converters. Application Report SLAA013.

Further Reading

Filliben, J. and Heckert, A. (2003). Probability distributions. In: *Engineering Statistics Handbook*. National Institute of Standards and Technology. http://www.itl.nist.gov/div898/handbook/eda/section3/eda36.htm.

Tobias, P. (2003). Extreme value distributions. In: *Engineering Statistics Handbook*. National Institute of Standards and Technology. https://www.itl.nist.gov/div898/handbook/apr/section1/apr163.htm.

Tobias, P. (2003). How do you project reliability at use conditions? In: *Engineering Statistics Handbook*. National Institute of Standards and Technology. https://www.itl.nist.gov/div898/handbook/apr/section4/apr43.htm.

4

Heuristic-Based Approach to Modeling CBD Signatures

4.1 Introduction to Heuristic-Based Modeling of Signatures

Stresses and strains in systems, cyclic or otherwise, often cause accumulating fatigue damage in prognostic targets, resulting in changes in one or more measurable signals called *condition-based data* (CBD). Those changes are leading indicators of failure that can be extracted, conditioned, and transformed into dimensioned feature data (FD) and then into dimensionless fault-to-failure progression (FFP) signature data. Such data is highly correlated to the progression of a prognostic target from a state of 100% health (zero or insignificant degradation) to a state of zero health (degraded to a level defined as functionally failed). FFP signature data can be further conditioned and transformed into an FFP-based functional-failure signature (FFS) and/or into a degradation-progression signature (DPS) and then into a DPS-based FFS (Viswanadham and Singh 1998; Hofmeister et al. 2013, 2016, 2017; Kwon et al. 2010).

Referring to Figures 4.1 and 4.2, the development or selection of a set of models for CBD signatures of interest is a first step in transforming CBD signature data into an FFS for processing by a predication system to produce prognostic information in a prognostics and health management/monitoring (PHM) system. While it is certainly possible to use, for example, failure-mode and effects analysis (FMEA) and physics of failure (PoF) methods and to develop a degradation-signature model for each CBD signature of interest – such as we did in Chapter 3 – it is asserted that such rigorous, case-by-case modeling is not necessary. Instead, it is generally sufficient to match a CBD signature of interest to one produced by a known degradation-signature model and use that for transforming data into an FFS: a heuristic-based approach to CBD modeling. Differences in signatures due to modeling inexactitude shall be shown to be insignificant compared to differences in signatures due to noise, given a sufficiently robust prediction system, such as one that employs special techniques (similar to Kalman filtering) to model the signatures of data (filtering) and one that employs the concept of inertia and momentum wherein signals do not rapidly change in response to degradation (Hofmeister et al. 2017).

4.1.1 Review of Chapter 3

In Chapter 3, a switched-mode power supply (SMPS) was used as an exemplary prognostic target with an output filter that exhibited a failure mode in the form of loss of filtering capacitance. That loss of capacitance was modeled as a change (dP) in the value

Prognostics and Health Management: A Practical Approach to Improving System Reliability Using Condition-Based Data, First Edition. Douglas Goodman, James P. Hofmeister and Ferenc Szidarovszky.

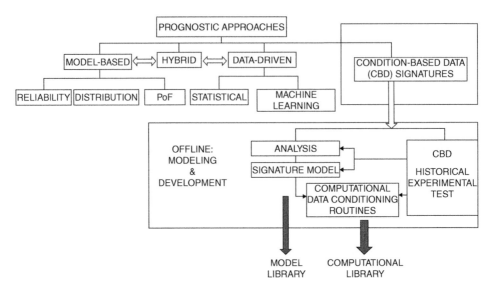

Figure 4.1 Block diagram for offline modeling of CBD signatures.

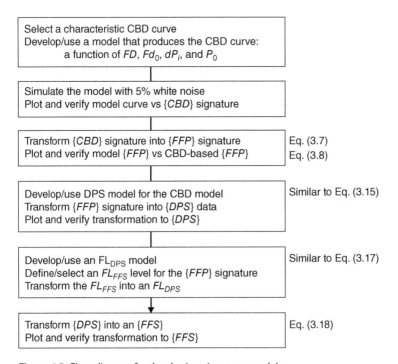

Figure 4.2 Flow diagram for developing signature models.

of a parameter having a nominal value (P_0) to create a model to transform FFP signature data into DPS data (Erickson 1999; Judkins et al. 2007; Judkins and Hofmeister 2007).

Chapter 3 introduced the concept of a sensing system comprising everything between a monitored node and the input port of a system that provides diagnosis and/or prognosis for the monitored system, to include the remaining useful life (RUL), state of health (SoH), and prognostic horizon (PH).

An important function of a sensing system is transforming CBD into a linearized transfer curve. The linearity of a sensing system is an assessment of how the measured curve produced by the sensing system deviates from the ideal curve (Texas Instruments 1995; Carr and Brown 2000).

A set of models was developed in Chapter 3, including the following:

$FD_i = CBD_i - NM$	FD point	Eq. (3.7)
$FFP_i = (FD_i - FD_0)/FD_0$	FFP signature point	Eq. (3.8)
$FFS_i = 100\ FFP_i/FL_{FFP}$	FFP-based FFS point	Eq. (3.10)
$DPS_i = dP_i/P_0 = 1 - (FD_0/FD_i)^2$	DPS point	Eq. (3.15)
$FL_{DPS} = 1 - (1/(1 + FL_{FFP}))^2$	FL relationship	Eq. (3.17)
$FFS_i = 100\ DPS_i/FL_{DPS}$	DPS-based FFS point	Eq. (3.18)

FMEA and PoF methods were used to create models for an example CBD signature resulting from a square root function type of degradation that is inversely proportional to dP. Those models were used to illustrate signature transforms to create DPS-based FFS data particularly amenable to processing by a prediction system to support prognosis in a PHM system:

- *FFP.* Transforms unit-of-measure-dependent data into normalized data in dimensionless ratio values.
- FL_{FFP}. Specifies an FFP value at which functional failure is declared to occur. This value is used to calculate the equivalent DPS value at which functional failure occurs.
- *DPS.* Transforms curvilinear FFP data into linearized data.
- FL_{DPS}. Transforms FFP-based failure-threshold values to DPS-based failure-threshold values.
- *FFS.* Transforms failure-threshold-dependent DPS data into a transfer curve in percent values.

4.1.2 Theory: Heuristic Modeling of CBD Signatures

The theory for heuristic modeling of CBD signatures presented in this chapter is based on the following:z

- We are given a failure mode in which the value of one or more parameters changes in value within a system.
- PoF analysis shows that a measurable signal based on how that parameter changes in response to degradation will produce a family of signatures.
- The difference between each signature in a family is related to the rate of change in a degradation parameter.
- In the presence of, for example, phase changes, the characteristics of a signature might change; however, all signatures within a family are likely to exhibit the same change at approximately the same level of degradation.

- Further, variations related to phase changes are likely to produce insignificant changes in estimated values.
- It is not necessary to develop a 100% accurate model of exactly how a dominant parameter changes in response to a failure mode for each and every target for prognostic enabling.

The theory predicts the following results, which apply to the soundness of a heuristic approach to modeling signatures:

- Different failure modes that result in similar changes in a parameter produce similar changes in measurable signals.
- Given a failure mode that causes a signature that is reasonably close to a signature from a known model, it is sufficient to use that known model to produce an FFS as input to prediction algorithms.
- A caveat is that the use of a known model that significantly differs from an exact PoF model might result in FFS data that produces prognostic information that fails to sufficiently meet accuracy requirements.

4.1.3 Chapter Objectives

In this chapter, the primary objective is to present and discuss a heuristic approach to modeling CBD signatures, as follows: (i) we first develop degradation-signature models that generate signatures with characteristic curves; and (ii) we then develop additional models that are used to transform signatures and define failure levels.

Subsequently, instead of developing specific models for specific CBD signatures, CBD signatures are matched to known characteristic curves produced by existing degradation-signature models. The corresponding transform models are used.

This approach is simpler compared to an approach using FMEA and PoF methods, such as the following: (i) develop a PoF-based degradation-signal model for every prognostic target; (ii) use that model to generate a characteristic curve; (iii) compare the generated curve to the curve of the CBD signature; (iv) modify, as necessary, the PoF-based model; (v) develop the other models for that CBD signature; and then (vi) use the developed models. Those models are for the following functions that produce characteristic signatures for matching to CBD signatures:

- Power functions for $n = 0.25$, $n = 0.5$, $n = 0.75$, $n = 1.0$, $n = 1.25$, $n = 1.50$, and $n = 2$:
  - Increasing and decreasing signatures,
  - Increasing and decreasing slope angles.[1]
- Exponential functions for $\lambda = P_0$ and $P_0 = 50$, $P_0 = 100$, $P_0 = 150$, and $P_0 = 200$:
  - Increasing and decreasing signatures,
  - Increasing and decreasing slope angles.

The developed degradation-signature models are simulated and the results examined to verify that each model produces data having an expected characteristic signature. The sets of signatures are the following: CBD, FFP, degradation progression signature

---

1 A special case occurs when a curve is a straight line: the slope angle is constant – neither increasing nor decreasing.

(DPS), and FFS. Models to transform an FFP-based level of functional failure, FL_{FFP}, to a DPS-based level of functional failure, FL_{DPS}, are verified by comparing the times of functional failure to the time related to the defined value for FL_{FFP}. We use arbitary units of measure when creating simulated CBD as a partial means of verifying correct transformation and also to demonstrate the usefulness of transforms.

4.1.4 Chapter Organization

The remainder of this chapter is organized to present and discuss a heuristic-based approach to modeling CBD signatures as follows:

4.2 General Modeling Considerations: CBD Signatures
   This section presents some general considerations related to the modeling of CBD signatures, including noise margin (NM) and nominal value of features.
4.3 CBD Modeling: Degradation-Signature Models
   This section presents the development of degradation-signature models from which the models to transform FFP signature data into DPS data are developed.
4.4 DPS Modeling: FFP to DPS Transform Models
   This section presents the development of the models to transform FFP signature data into DPS data.
4.5 FFS Modeling: Failure Level and Signature Modeling
   This section presents considerations for selecting a failure level, the development of the models for transforming FFP-based failure levels into DPS-based failure levels, and the models for transforming DPS data into FFS data.
4.6 Heuristic-Based Approach to Modeling Signatures: Summary
   This section summarizes the material presented in this chapter.

4.2 General Modeling Considerations: CBD Signatures

A goal of a PHM system is to collect and prepare CBD for processing, in order to provide a prognosis of the health of a system; such collection and preparation for processing is performed in the sensing system. To do so in our system, models are used to characterize signatures; to transform signatures; to define signature threshold values at which functional failure is defined to occur; and to create a transfer curve as input from the sensing system to a diagnosis/prognosis system in a PHM system. A first consideration is the need to filter out as much noise as possible to meet accuracy requirements, where *noise* is defined as any variation in signature data that is not 100% correlated to degradation. For example, we have a problem if noise introduces a 20% variation in amplitude of the CBD-based data and we are asked to produce RUL estimates within 10% accuracy. Other considerations include the following: definition of a degradation-signature model that produces a characteristic curve; the nominal FD value; matching the characteristic curve of an FFP signature to a degradation-signal model; and selecting and using a set of models to transform data. The set of models related to a degradation-signal model is used to do the following: transform FFP signature data to DPS data, transform FFP-based failure levels to DPS-based failure levels, and transform DPS data to FFS data.

4.2.1 Noise Margin

To filter out all noise is, for practical purposes, neither technically feasible nor cost effective. A next step is to mitigate the effects of unfiltered noise by using a NM having a value sufficiently large to unambiguously declare that a data value is caused by degradation:

$$FD_i = CBD_i - NM$$

which is Eq. (3.7), and where

$$CBD_i = CBD_{FD} + CBD_m + N_n \tag{4.1}$$

where CBD_m and N_n represent any unfiltered variations in the data not related to the degradation of interest and CBD_{FD} represents data related to the degradation of interest, and where the magnitude of NM is defined as the following:

$$NM \geq \max\{CBD_m, NM_n, m, n = 1, 2, \ldots\} \tag{4.2}$$

Of course, any NM you use introduces an offset type of error in the prognostic information; more detail on that is presented in later chapters.

4.2.2 Definition of a Degradation-Signature Model

The NM in Eq. (4.2) is used to mitigate unfiltered noise. Ideally, all noise is filtered and there is no need for a noise margin ($NM = 0$). Then we can consider the more general relation:

$$FD_i = CBD_i - NM = CBD_i = g(FD_0)f(dP_i, P_0) + C_0$$

where, in general, $g(FD_0)$ equals a constant value, and C_0 is a constant including noise. We then define $f(dP_i, P_0)$ to represent a degradation-signature model. From this definition,

$$f(dP_i, P_0) = FD_i/FD_0 - C_1 \quad (C_1 = C_0/FD_0) \tag{4.3}$$

which is a dimensionless, univariate ratio. Later in this chapter, we shall define a set of degradation-signature models and their characteristic signatures.

4.2.3 Feature Data: Nominal Value

An important step is to determine a nominal value, FD_0, to use for FD_i in the absence of degradation. One method of determination is to measure and average a number of data samples for a sample population of the prognostic target: that method introduces sampling variation that, as we saw in Chapter 1, is characterized by metrics such as variance and standard deviation. Sampling variation occurs because it is not possible to manufacture prognostic targets that have exactly the same nondegraded value.

Yet another method is to use a nominal value specified by the manufacturer/supplier of the prognostic target – of course, we then have to account for the specified or presumed manufacturing/measurement tolerances. A more dynamic method is to use data samples taken from each instance of the prognostic target, save the applicable sample average or sample mean, and use that value for FD_0.

These are just some of the methods available: we need to evaluate, select, and implement whatever methods are necessary to filter, cancel, or otherwise mitigate noise and thus enable us to meet the accuracy and precision requirements for our PHM system.

4.2.4   Feature Data, Fault-to-Failure Progression Signature, and Degradation-Signature Model

From Eqs. (3.8) and (4.3),

$$FFP_i = (FD_i - FD_0)/FD_0 = FD_i/FD_0 - 1$$

$$f(dP_i, P_0) = FD_i/FD_0 - C_1$$

we relate FFP-signature data to a degradation function as follows:

$$FFP_i = f(dP_i, P_0) \tag{4.4}$$

We define a degradation-signature model as a function of dP_i and P_0 that generates a characteristic curve representative of an FFP signature. We shall develop a set of such models; and from those models, we shall develop another set of models to transform FFP signature data into DPS data.

We transform a CBD signature of interest to an FFP signature and match that signature to one produced by a known degradation-signature model. Then we use FFP-to-DPS transform models to transform FFP signature data into DPS data, which we further transform into FFS data that is used as input for diagnosis/prognosis of a PHM system.

4.2.5   Approach to Transforming CBD Signatures into FFS Data

The approach to transforming CBD signatures into FFS data comprises a series of transforms and computations that are performed in the sensing system:

- Transform CBD signatures into FFP signatures.
- Transform FFP signatures into DPS data.
- Define an FFP-based threshold level at which functional failure occurs: FL_{FFP}.
- Calculate the DPS-based threshold level, FL_{DPS}, using the FL_{FFP} value.
- Transform DPS data into FFS data: $FFS_i = 100 \, DPS_i/FL_{DPS}$.

Transforming CBD Signatures into FFP Signatures
We use Eq. (3.8) to isolate a feature data of interest (a leading indicator of failure) and to determine what level of NM we need to unambiguously detect the presence of damage:

$$FD_i = CBD_i - NM$$

We use a nominal value of feature data, FD_0, and the methodology of Eqs. (4.3) and (4.4), and complete the transformation of CBD signatures into FFP signatures:

$$FFP_i = FD_i/FD_0 - C_1$$

Transforming FFP Signature Data into DPS Data
Match sample data, FFP_i, to degradation signatures, $f(dP_i, P_0)$, a set of which we shall present. For each degradation signature in that set, we shall develop a model to transform FFP signature data into DPS data:

$$dP_i/P_0 = g(FFP_i) \tag{4.5}$$

which is the result of using Eq. (4.4), a model for $f(dP_i, P_0)$, and solving for dP_i/P_0.

In addition to developing the FFP-to-DPS transform models, $g(FFP_i)$, we shall create plots to verify results.

Calculating a DPS-Based Failure Threshold

An FFS, by definition, defines functional failure as occurring when the value of a FFS_i data point reaches or exceeds 100%. Therefore, to transform DPS data into FFS data, we need to specify a value of DPS data at which functional failure occurs. Since an FFP signature is a function of degradation, it is better to use an FFP signature to define a failure threshold and then transform that threshold value into a value related to DPS data

$$FL_{FFP} = FFP_i \qquad (4.6)$$

at a defined level of functional failure. Then from Eq. (4.5), a DPS-based failure threshold is the following:

$$FL_{DPS} = g(FL_{FFP}) \qquad (4.7)$$

We shall use Eqs. (4.5) and (4.6) to develop an FL_{FFP} model for each set of characteristic degradation signatures and then develop corresponding models to transform each FL_{FFP} model into a DPS-based model: FL_{DPS}.

Transforming DPS Data into FFS Data

From Eq. (3.18), to transform DPS data into FFS data,

$$FFS_{DPSi} = 100 \, DPS_i / FL_{DPS} \qquad (4.8)$$

This is used as input to the prediction system of the PHM system, one data point at a time.

4.3 CBD Modeling: Degradation-Signature Models

In this section, we shall create degradation-signature models that approximate CBD signatures of interest: those caused by actual degradation. Eq. (4.4) defined FFP data to be functionally related to the degradation of a parameter:

$$FFP_i = f(dP_i, P_0)$$

Each unique $f(dP_i, P_0)$ is a degradation-signature model, which has dimensionless values. Examination of CBD signatures of interest reveals the following significant characteristics as degradation progresses:

- Degradation signatures either increase or decrease in amplitude as degradation progresses.
- Degradation signatures exhibit a decreasing or increasing slope angle[2] or, as appropriate, as vertically asymptotic, or as horizontally asymptotic.
- Degradation signatures can be modeled as power or exponential functions.
- Decreasing degradation signatures can be converted to increasing signatures to reduce the number of degradation-signature models required to support prognosis.
- A special case of degradation signature is a straight line: the slope angle is constant. In plots, such straight lines are included in plot examples of increasing or decreasing slope angles.[3]

---

2 For brevity, especially in plots, "decreasing or increasing angle" is used instead of "decreasing or increasing slope angle."

3 It is understood that a straight-line signature is a special case.

4.3.1 Representative Examples: Degradation-Signature Models

In this section, we present 14 non-inclusive, representative examples of degradation-signature models: 10 as power functions and 4 as exponential functions. These are representative of commonly encountered degradation signatures. The examples presented assume that degradation increases at a constant rate with respect to time d: for example, $dP_i = (P_0/150)(d - 50)$. In the following 14 models, this expression of dP_i will be assumed. Later, we shall show that even when dP_i does not increase at a constant rate, the practical effect on the shape of the signature is negligible.

Power Function #1: Increasing Signatures, Decreasing or Increasing Slope Angles

The following model is for a power function that produces characteristic signatures that increase and have either decreasing or increasing slope angles depending on the value of the power n (see Figures 4.3 and 4.4):

$$f(dP_i, P_0) = (dP_i/P_0)^n \tag{4.9}$$

where dP_i is an amount of change in the value of a parameter (degradation) at the time the data was sampled and P_0 is the unchanged (no degradation) value of a parameter: $P = P_0 + dP_i$.

When n is less than 1, $0 < n < 1.0$, the signature has a decreasing slope angle (Figure 4.3); when n is greater than 1, $n > 1.0$, the signature has an increasing slope angle (Figure 4.4); and when $n = 1$, the signature has a constant slope angle. *Slope angle* is the absolute value of the angle between the tangent line of the curve and the positive half of the horizontal axis.

Note that when $dP_i = P_0$, then regardless of the value of n,

$$(dP_i/P_0)^n = (P_0/P_0)^n = 1^n = 1$$

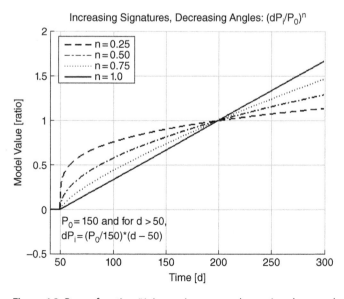

Figure 4.3 Power function #1: increasing curves, decreasing slope angles.

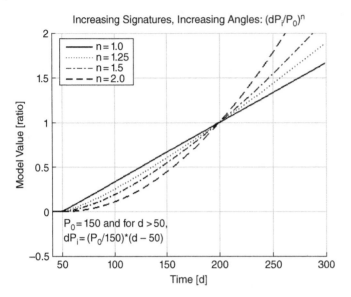

Figure 4.4 Power function #1: increasing curves, increasing slope angles.

Power Function #2: Decreasing Signatures, Decreasing or Increasing Slope Angles
The following model is for a power function that produces characteristic signatures that decrease and have either decreasing or increasing slope angles, depending on the value of the power n (see Figures 4.5 and 4.6):

$$f(dP_i, P_0) = -(dP_i/P_0)^n \tag{4.10}$$

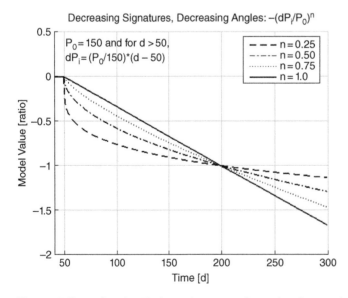

Figure 4.5 Power function #2: decreasing curves, decreasing slope angles.

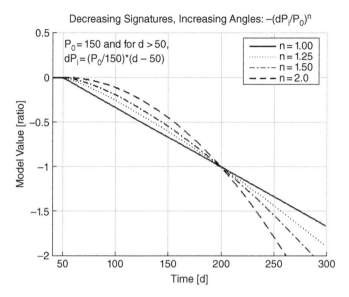

Figure 4.6 Power function #2: decreasing curves, increasing slope angles.

where dP_i is an amount of change in the value of a parameter (degradation) at the time the data was sampled, P_0 is the unchanged (no degradation) value of a parameter, and FD_0 is the nominal value of the signature in the absence of degradation when $dP_i = 0$.

When n is less than 1, $n < 1.0$, the degradation signature has a decreasing slope angle (Figure 4.5); when n is greater than 1, $n > 1.0$, the signature has an increasing slope angle (Figure 4.6); and when $n = 1$, the slope angle of the signature is constant. Examination of Eq. (4.3) and the plots reveals that the plots might not be applicable when $dP_i > P_0$ unless the parameter is an active component that has negative values: passive components have positive values.

Note that when $dP_i = P_0$, then regardless of the value of n,

$$-(dP_i/P_0)^n = -(P_0/P_0)^n = -1^n = -1$$

Notice that Eq. (4.10) is the mirror image of Eq. (4.9) with respect to the horizontal axis.

Example 4.1 When asked to calculate the slope angle for Eq. (4.9) when $n = 2.0$ for $dP_i = 0$ and for $dP_i = P_0$, the slope angles you calculate are $0°$ and $-45°$. The slope angles increase as expected and as shown in the plots in Figure 4.6.

Power Function #3: Increasing Signatures, Vertically Asymptotic
The following model is for a power function that produces characteristic signatures that increase and become vertically asymptotic (see Figure 4.7):

$$f(dP_i, P_0) = [1/(1 - dP_i/P_0)]^n - 1 \tag{4.11}$$

The model implies the following:

- When $dP_i < P_0$, the signatures are increasing for all values of n.

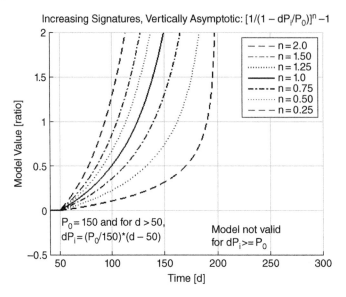

Figure 4.7 Power function #3: increasing curves, vertically asymptotic, $dP_i < P_0$.

- As $dP_i \rightarrow P_0$, the signatures become vertically asymptotic – an unstable system in which minor changes in parameter values result in large changes in signature amplitude.
- When $dP_i > P_0$, the model is declared "not valid."

Power Function #4: Decreasing Signatures, Vertically Asymptotic
The following model is for a power function that produces characteristic signatures that decrease and become vertically asymptotic (see Figure 4.8):

$$f(dP_i, P_0) = 1 - [1/(1 - dP_i/P_0)]^n, \tag{4.12}$$

This is the mirror image of Eq. (4.11) with respect to the horizontal axis. The model implies the following:

- When $dP_i < P_0$, the signatures are decreasing for all values of n.
- As $dP_i \rightarrow P_0$, the signatures become vertically asymptotic – an unstable system in which minor changes in parameter values result in large changes in signature amplitude.
- When $dP_i > P_0$, the model is declared "not valid."

Power Function #5: Increasing Signatures, Horizontally Asymptotic
The following model is for a power function that produces characteristic signatures that increase and become horizontally asymptotic (see Figure 4.9):

$$f(dP_i, P_0) = 1 - [1/(1 + dP_i/P_0)]^n \tag{4.13}$$

As the magnitude of this type of signature becomes increasingly horizontal, changes in magnitude of noise begins to dominate changes in signal magnitude caused by degradation. The level of degradation for which functional failure is defined should be at or below

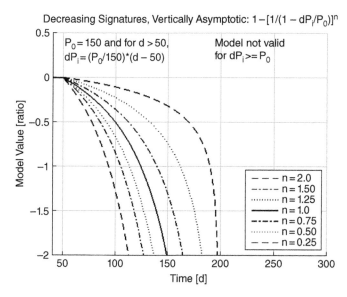

Figure 4.8 Power function #4: decreasing curves, vertically asymptotic, $dP_i < P_0$.

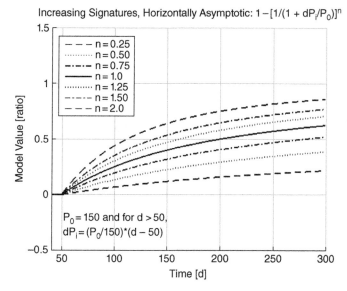

Figure 4.9 Power function #5: increasing curves, horizontally asymptotic.

the level where an incremental change in signal magnitude between samples equals the maximum magnitude of noise.

Power Function #6: Decreasing Signatures, Horizontally Asymptotic

The following model is for a power function that produces characteristic signatures that decrease and become horizontally asymptotic (see Figure 4.10):

$$f(dP_i, P_0) = [1/(1 + dP_i/P_0)]^n - 1, \tag{4.14}$$

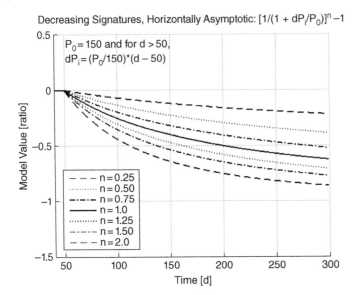

Decreasing Signatures, Horizontally Asymptotic: $[1/(1 + dP_i/P_0)]^n - 1$

Figure 4.10 Power function #6: decreasing curves, horizontally asymptotic.

This is the mirror image of Eq. (4.13) with respect to the horizontal axis. As the magnitude of this type of signature becomes increasingly horizontal, changes in magnitude of noise begin to dominate changes in signal magnitude caused by degradation. The level of degradation for which functional failure is defined should be at or below the level where an incremental change in signal magnitude between samples equals the maximum magnitude of noise.

Power Function #7: Increasing Signatures, Slightly Curvilinear, Decreasing or Increasing Slope Angles

The following model is for a power function that produces characteristic signatures that increase, are slightly curvilinear, and have either a decreasing angle (see Figure 4.11) or an increasing angle (see Figure 4.12) depending on the value of the power n:

$$f(dP_i, P_0) = (1 + dP_i/P_0)^n - 1 \qquad (4.15)$$

At first glance, the signatures shown in the plots might appear to be straight lines; however, except for $n = 1$ (a straight-line curve), the signatures are slightly curvilinear.

Power Function #8: Decreasing Signatures, Slightly Curvilinear, Decreasing or Increasing Slope Angles

The following model is for a power function that produces characteristic signatures that decrease, are slightly curvilinear; and have either a decreasing (see Figure 4.13) or increasing (see Figure 4.14) slope angle depending on the value of the power n:

$$f(dP_i, P_0) = 1 - (1 + dP_i/P_0)^n, \qquad (4.16)$$

This is the mirror image of Eq. (4.15) with respect to the horizontal axis.

At first glance, the signatures might appear to be straight lines; however, except for $n = 1$ (a straight-line curve), the signatures are slightly curvilinear.

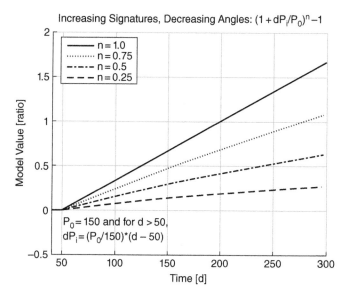

Figure 4.11 Power function #7: increasing curves, slightly curvilinear, decreasing slope angles.

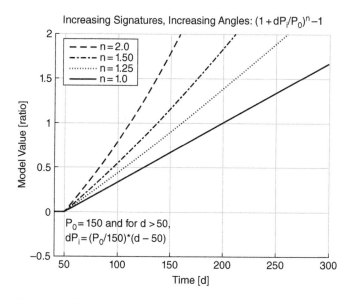

Figure 4.12 Power function #7: increasing curves, slightly curvilinear, increasing slope angles.

Power Function #9: Increasing Signatures, Vertically or Horizontally Asymptotic

The following model is for a power function that produces characteristic signatures that increase and are vertically or horizontally asymptotic, depending on the value of the power n (see Figures 4.15 and 4.16):

$$f(dP_i, P_0) = 1 - (1 - dP_i/P_0)^n \qquad (4.17)$$

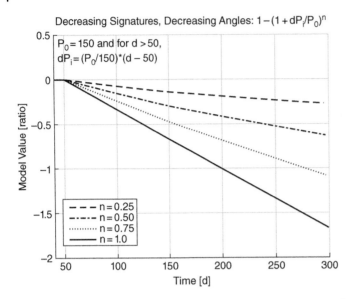

Figure 4.13 Power function #8: decreasing curves, slightly curvilinear, decreasing slope angles.

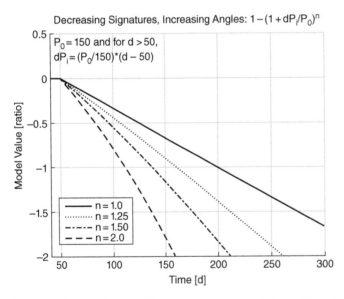

Figure 4.14 Power function #8: decreasing curves, slightly curvilinear, increasing slope angles.

The model implies the following:

- When $dP_i < P_0$, the signatures are increasing for all values of n.
- As $dP_i \to P_0$, the signatures become either vertically or horizontally asymptotic: the system approaches instability.
- When $dP_i > P_0$, the model is declared to be "not valid" and the system is defined as having ceased to function.

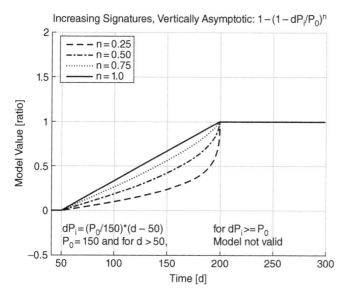

Figure 4.15 Power function #9: increasing curves, vertically asymptotic, $dP_i < P_0$.

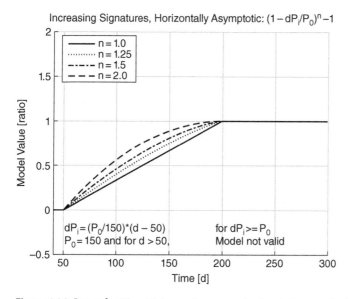

Figure 4.16 Power function #9: increasing curves, horizontally asymptotic, $dP_i < P_0$.

Power Function #10: Decreasing Signatures, Vertically or Horizontally Asymptotic

The following model is for a power function that produces characteristic signatures that decrease and are vertically or horizontally asymptotic, depending on the value of the power n (see Figures 4.17 and 4.18):

$$f(dP_i, P_0) = (1 - dP_i/P_0)^n - 1, \qquad (4.18)$$

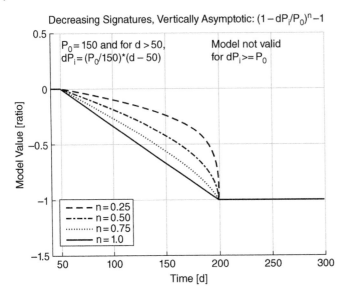

Figure 4.17 Power function #10: decreasing curves, vertically asymptotic, $dP_i < P_0$.

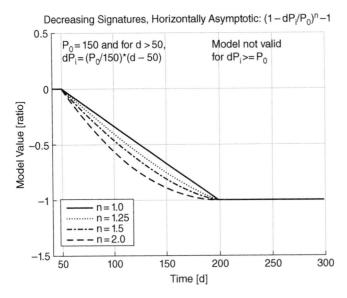

Figure 4.18 Power function #10: decreasing curves, horizontally asymptotic, $dP_i < P_0$.

This is the mirror image of Eq. (4.17) with respect to the horizontal axis. The model implies the following:

- When $dP_i < P_0$, the signatures are decreasing for all values of n.
- As $dP_i \rightarrow P_0$, the signatures become either vertically or horizontally asymptotic: the system approaches instability.

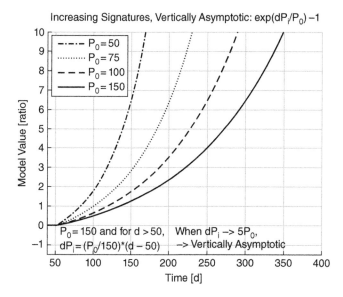

Figure 4.19 Exponential function #11: increasing curves, vertically asymptotic.

- When $dP_i > P_0$, the model is declared to be "not valid" and the system is defined as having ceased to function.

Exponential Function #11: Increasing Signatures, Approximately Vertically Asymptotic
The following is a model of an exponential function that produces characteristic signatures that increase and become increasingly vertically asymptotic (see Figure 4.19):

$$f(dP_i, P_0) = \exp(dP_i/P_0) - 1 \qquad (4.19)$$

As degradation increases, the signatures become approximately vertically asymptotic at $dP_i \geq 5P_0$. It is recommended that functional failure be defined to occur when or before degradation reaches a value of $dP_i = 2P_0$.

Exponential Function #12: Decreasing Signatures, Approximately Vertically Asymptotic
The following is a model of an exponential function that produces characteristic signatures that decrease and become increasingly vertically asymptotic slopes (see Figure 4.20):

$$f(dP_i, P_0) = 1 - \exp(dP_i/P_0), \qquad (4.20)$$

This is the mirror image of Eq. (4.19) with respect to the horizontal axis. As degradation increases, the signatures become approximately vertically asymptotic at $dP_i \geq 5P_0$. It is recommended that functional failure be defined to occur when or before degradation reaches a value of $dP_i = 2P_0$.

Exponential Function #13: Increasing Signatures, Horizontally Asymptotic
The following is a model of an exponential function that produces characteristic signatures that increase and become increasingly horizontally asymptotic (see Figure 4.21):

$$f(dP_i, P_0) = 1 - \exp(-dP_i/P_0) \qquad (4.21)$$

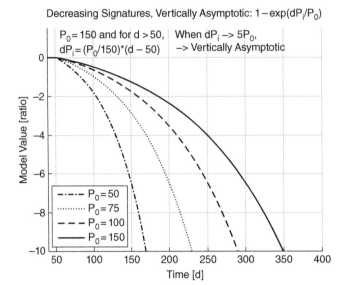

Figure 4.20 Exponential function #12: decreasing curves, vertically asymptotic.

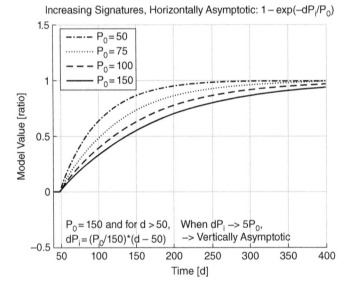

Figure 4.21 Exponential function #13: increasing curves, horizontally asymptotic.

As degradation increases, the signatures become horizontally asymptotic at $dP_i \geq 5P_0$. It is recommended that functional failure be defined to occur when or before degradation reaches a value of $dP_i = 2P_0$.

Exponential Function #14: Decreasing Signatures, Horizontally Asymptotic

The following is a model of an exponential function that produces characteristic signatures that decrease and become increasingly horizontally asymptotic (see Figure 4.22):

$$f(dP_i, P_0) = \exp(-dP_i/P_0) - 1, \tag{4.22}$$

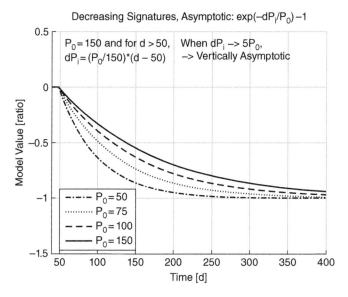

Figure 4.22 Exponential function #14: decreasing curves, horizontally asymptotic.

This is the mirror image of Eq. (4.21) with respect to the horizontal axis. As degradation increases, the signatures become horizontally asymptotic at $dP_i \geq 5P_0$. It is recommended that functional failure be defined to occur when or before degradation reaches a value of $dP_i = 2P_0$.

4.3.2 Example Plots of Representative FFP Degradation Signatures

Figures 4.3–4.22 present example plots of FFP signatures in response to increasing degradation.

4.3.3 Converting Decreasing Signatures to Increasing Signatures

The following degradation-signature models produce decreasing signatures:

Power function #2, Eq. (4.10):	$f(dP_i, P_0) = -(dP_i/P_0)^n$
Power function #4, Eq. (4.12):	$f(dP_i, P_0) = 1 - [1/(1 - dP_i/P_0)]^n$
Power function #6, Eq. (4.14):	$f(dP_i, P_0) = [1/(1 + dP_i/P_0)]^n - 1$
Power function #8, Eq. (4.16):	$f(dP_i, P_0) = 1 - (1 + dP_i/P_0)^n$
Power function #10, Eq. (4.18):	$f(dP_i, P_0) = (1 - dP_i/P_0)^n - 1$
Exponential-function #12, Eq. (4.20):	$f(dP_i, P_0) = 1 - \exp(dP_i/P_0)$
Exponential-function #14, Eq. (4.22):	$f(dP_i, P_0) = \exp(-dP_i/P_0) - 1$

To reduce the number of models, transform decreasing signatures into increasing signatures by doing the following:

$$f(\text{increasing}) = -f(\text{decreasing}) \tag{4.23}$$

Table 4.1 List of decreasing signatures and models and corresponding increasing models.

Signature name	Decreasing model	Increasing model
Power function #2	$-(dP_i/P_0)^n$	$(dP_i/P_0)^n$
Power function #4	$1 - [1/(1 - dP_i/P_0)]^n$	$[1/(1 - dP_i/P_0)]^n - 1$
Power function #6	$[1/(1 + dP_i/P_0)]^n - 1$	$1 - [1/(1 + dP_i/P_0)]^n$
Power function #8	$1 - (1 + dP_i/P_0)^n$	$(1 + dP_i/P_0)^n - 1$
Power function #10	$(1 - dP_i/P_0)^n - 1$	$1 - (1 - dP_i/P_0)^n$
Exponential function #12	$1 - \exp(dP_i/P_0)$	$\exp(dP_i/P_0) - 1$
Exponential function #14	$\exp(-dP_i/P_0) - 1$	$1 - \exp(-dP_i/P_0)$

Example 4.2 Suppose you have a decreasing FFP signature, and further suppose you determine the signature is modeled by Eq. (4.10). Use Eq. (4.23) to transform that signature into an increasing signature that is modeled by Eq. (4.9):

$$-f(dP_i, P_0) = -[-(dP_i/P_0)^n] = (dP_i/P_0)^n$$

Using the method of Example 4.2, you can convert all of the decreasing signatures to increasing signatures. Then you get the results shown in Table 4.1.

4.4 DPS Modeling: FFP to DPS Transform Models

In this section, we develop the models to transform FFP signature data into DPS data. To do that, we start with Eq. (4.4) and solve Eq. (4.5):

$$FFP_i = f(dP_i, P_0) \tag{4.24}$$

$$dP_i/P_0 = g(FFP_i) \tag{4.25}$$

A solution set $\{dP_i/P_0\}$ is a set of DPS data, $\{DPS_i\}$, that is a degradation-progression signature.

4.4.1 Developing Transform Models: FFP to DPS

We shall substitute FFP into Eq. (4.24), solve the result to get Eq. (4.25), and verify the solution. Verification shall consist of (i) producing FFP example data, (ii) performing a visual verification of the signature, (iii) transforming the example data to produce a DPS-based signature, and (iv) performing a visual verification that the DPS-based signature is an ideal transfer curve: a straight line.

Transform Model for Power Function #1
From Eq. (4.9), for the degradation-signal model for power function #1, and from Eq. (4.24):

$$f(dP_i, P_0) = (dP_i/P_0)^n = FFP_i$$

Since

$$DPS_i = dP_i/P_0$$

we have

$$DPS_i = (FFP_i)^{1/n} \tag{4.26}$$

Figure 4.23 shows plots of simulated FFP signatures, and Figure 4.24 shows plots of DPS data after transforming the FFP signatures: the modeling for transform Eq. (4.26) is verified.

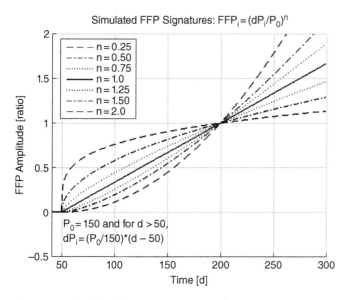

Figure 4.23 Simulated FFP signatures: power function #1.

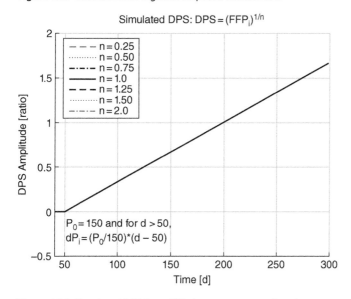

Figure 4.24 Simulated DPS from FFP signatures: power function #1.

Transform Model for Power Function #3

From Eq. (4.11), the degradation-signal model for power function #3, and from Eq. (4.24):

$$f(dP_i, P_0) = [1/(1 - dP_i/P_0)]^n - 1 = FFP_i$$

So

$$[1/(1 - dP_i/P_0)]^n = FFP_i + 1 \rightarrow [1/(1 - dP_i/P_0)] = (FFP_i + 1)^{1/n}$$

and therefore

$$(1 - dP_i/P_0) = 1/(FFP_i + 1)^{1/n}$$

Finally

$$DPS_i = dP_i/P_0 = 1 - 1/(FFP_i + 1)^{1/n} \qquad (4.27)$$

Figure 4.25 shows plots of simulated FFP signatures, and Figure 4.26 shows plots of DPS data after transforming the FFP signatures: the modeling for transform Eq. (4.27) is verified.

Transform Model for Power Function #5

From Eq. (4.13), the degradation-signal model for power function #5, and from Eq. (4.24):

$$f(dP_i, P_0) = 1 - [1/(1 + dP_i/P_0)]^n = FFP_i$$

So

$$[1/(1 + dP_i/P_0)]^n = 1 - FFP_i \rightarrow [1/(1 + dP_i/P_0)] = (1 - FFP_i)^{1/n}$$

and therefore

$$(1 + dP_i/P_0) = 1/(1 - FFP_i)^{1/n}$$

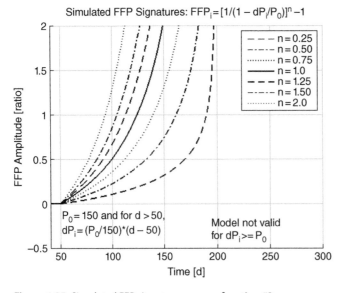

Figure 4.25 Simulated FFP signatures: power function #3.

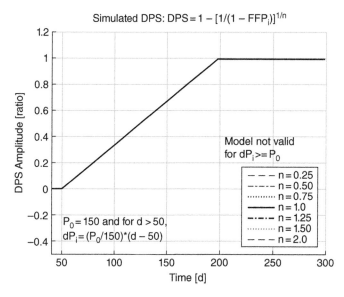

Figure 4.26 Simulated DPS from FFP signatures: power function #3.

Finally,

$$DPS_i = dP_i/P_0 = 1/(1 - FFP_i)^{1/n} - 1 \qquad (4.28)$$

Figure 4.27 shows plots of simulated FFP signatures, and Figure 4.28 shows plots of DPS data after transforming the FFP signatures: the modeling for transform Eq. (4.28) is verified.

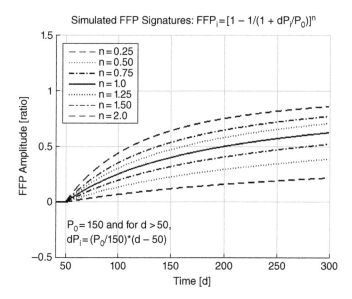

Figure 4.27 Simulated FFP signatures: power function #5.

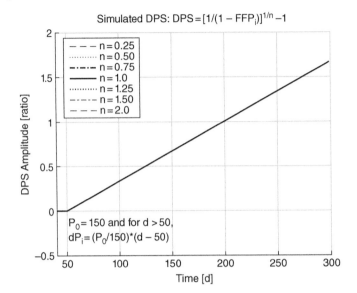

Figure 4.28 Simulated DPS from FFP signatures: power function #5.

Transform Model for Power Function #7

From Eq. (4.15), the degradation-signal model for power function #7, and from Eq. (4.24):

$$f(dP_i, P_0) = (1 + dP_i/P_0)^n - 1 = FFP_i$$

So

$$(1 + dP_i/P_0)^n = FFP_i + 1 \rightarrow (1 + dP_i/P_0) = (FFP_i + 1)^{1/n}$$

and therefore

$$DPS_i = dP_i/P_0 = (FFP_i + 1)^{1/n} - 1 \tag{4.29}$$

Figure 4.29 shows plots of simulated FFP signatures, and Figure 4.30 shows plots of DPS data after transforming the FFP signatures: the modeling for transform Eq. (4.29) is verified.

Transform Model for Power Function #9

From Eq. (4.17), the degradation-signal model for power function #9, and from Eq. (4.24):

$$f(dP_i, P_0) = 1 - (1 - dP_i/P_0)^n = FFP_i$$

So

$$(1 - dP_i/P_0)^n = 1 - FFP_i \rightarrow (1 - dP_i/P_0) = (1 - FFP_i)^{1/n}$$

and

$$DPS_i = dP_i/P_0 = 1 - (1 - FFP_i)^{1/n} \tag{4.30}$$

Figure 4.31 shows plots of simulated FFP signatures, and Figure 4.32 shows plots of DPS data after transforming the FFP signatures: the modeling for transform Eq. (4.30) is verified.

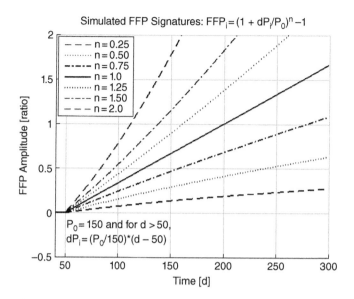

Figure 4.29 Simulated FFP signatures: power function #7.

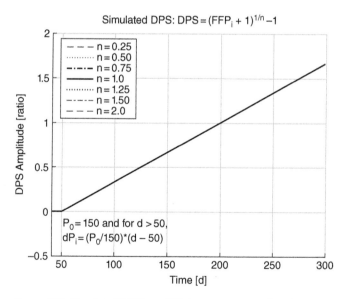

Figure 4.30 Simulated DPS from FFP signatures: power function #7.

Transform Model for Exponential Function #11

From Eq. (4.19), the degradation-signal model for exponential function #11, and from Eq. (4.24)

$$f(dP_i, P_0) = \exp(dP_i/P_0) - 1 = FFP_i$$

so

$$\exp(dP_i/P_0) = FFP_i + 1$$

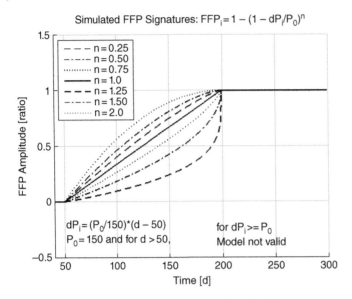

Figure 4.31 Simulated FFP signatures: power function #9.

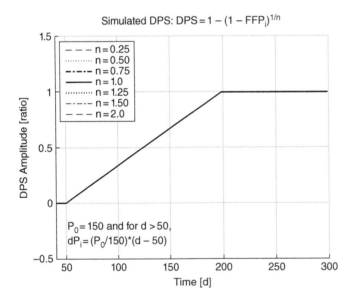

Figure 4.32 Simulated DPS from FFP signatures: power function #9.

and

$$DPS_i = dP_i/P_0 = \ln(FFP_i + 1) \tag{4.31}$$

Figure 4.33 shows plots of simulated FFP signatures, and Figure 4.34 shows plots of DPS data after transforming the FFP signatures: the modeling for transform Eq. (4.31) is verified.

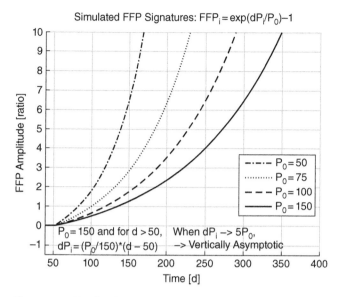

Figure 4.33 Simulated FFP signatures: exponential function #11.

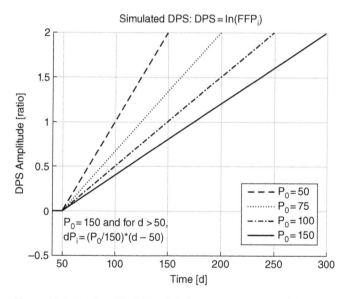

Figure 4.34 Simulated DPS from FFP signatures: exponential function #11.

Transform Model for Exponential Function #13

From Eq. (4.21), the degradation-signal model for exponential function #13, and from Eq. (4.24):

$$f(dP_i, P_0) = 1 - \exp(-dP_i/P_0) = FFP_i$$

So

$$\exp(-dP_i/P_0) = 1 - FFP_i \rightarrow -dP_i/P_0 = \ln(1 - FFP_i)$$

and therefore

$$dP_i/P_0 = -\ln(1 - FFP_i)$$

Finally,

$$DPS_i = dP_i/P_0 = \ln(1/(1 - FFP_i)) \qquad (4.32)$$

Figure 4.35 shows plots of simulated FFP signatures, and Figure 4.36 shows plots of DPS data after transforming the FFP signatures: the modeling for transform Eq. (4.32) is verified.

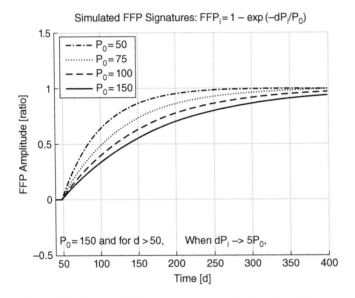

Figure 4.35 Simulated FFP signatures: exponential function #13.

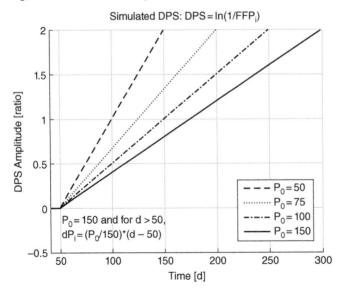

Figure 4.36 Simulated DPS from FFP signatures: exponential function #13.

4.4.2 Example Plots of FFP Signatures and DPS Signatures

Figures 4.23–4.36 present example plots of increasing FFP degradation signatures and their transforms to DPS.

4.5 FFS Modeling: Failure Level and Signature Modeling

Before you can transform a DPS into an FFS, you need to define and specify a threshold value of DPS data for functional failure. While it is possible to choose any such value, a better way is to define a threshold value of FFP signature data and then transform that value into a DPS_{BASED} value. An example of why using an FFP_{BASED} value is preferred is the following: it is always the case that if the value of a component is reduced to 0, the component has failed; similarly, if the value of a component has doubled in size, it would be reasonable to declare that component as failed. A qualitative assessment such as this is better made for FFP_{BASED} values. As a matter of practice and preference, this book recommends that functional failure be defined when or before the value of a degrading component is reduced by or increased by 70%, which is when or before an FFP_i value of 0.70.

4.5.1 Developing DPS-Based Failure Level (FL) Models Using FFP Defined Failure Levels

For each of the representative sets of example FFP signatures, we shall use Eqs. (4.4)–(4.6) to solve Eq. (4.7) to create FL_{DPS} models, which shall be verified by plotting and comparing FFP and DFP signatures using values for FL_{FFP} and FL_{DPS}:

$$FFP_i = f(dP_i, P_0), DPS_i = g(FFP_i), FL_{FFP} = FFP_i$$

and

$$FL_{DPS} = g(FL_{FFP})$$

DPS-Based FL Model for Power Function #1

From Eq. (4.26), to transform an FFP signature for power function #1, and from Eq. (4.7):

$$dP_i/P_0 = (FFP_i)^{1/n}$$

So

$$FL_{DPS} = (FL_{FFP})^{1/n} \qquad (4.33)$$

Example 4.3 Define FL_{FFP} as 0.65 for a power function #1 signature and $n = 2$, and calculate the value of FL_{DPS}. Verify the results by plotting the FFP signature, the DPS, and the failure levels:

$$FFP_i = (dP_i/P_0)^n$$

$$DPS_i = (FFP_i)^{1/n} \text{ and } FL_{DPS} = (FL_{FFP})^{1/n}$$

The calculated value of FL_{DPS} equals 0.81, and the plots in Figure 4.37 show that functional failure occurs at the same point in time for both FFP signature data and DPS data: Eq. (4.33) is verified.

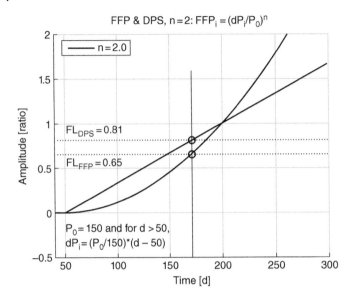

Figure 4.37 Simulated FFP and DPS signatures: power function #1 for $n = 2.0$.

DPS-Based FL Model for Power Function #3

From Eq. (4.27), to transform an FFP signature for power function #3, and from Eq. (4.7):

$$FL_{DPS} = 1 - 1/(FL_{FFP} + 1)^{1/n} \tag{4.34}$$

Example 4.4 Define FL_{FFP} as 0.75 for a power function #3 signature and $n = 1.5$, and calculate the value of FL_{DPS}. Verify your results by plotting the FFP signature, the DPS, and the failure levels:

$$FFP_i = [1/(1 - dP_i/P_0)]^n - 1 \text{ and } DPS_i = 1 - 1/(FFP_i + 1)^{1/n}$$

The calculated value of FL_{DPS} equals 0.31, and the plots in Figure 4.38 show that functional failure occurs at the same point in time for both FFP signature data and DPS data: Eq. (4.34) is verified.

DPS-Based FL Model for Power Function #5

From Eq. (4.28), to transform an FFP signature for power function #5, and from Eq. (4.7):

$$FL_{DPS} = 1/(1 - FL_{FFP})^{1/n} - 1 \tag{4.35}$$

Example 4.5 Define FL_{FFP} as 0.70 for a power function #5 signature and $n = 1.5$, and calculate the value of FL_{DPS}. Verify your results by plotting the FFP signature, the DPS, and the failure levels:

$$FFP_i = [1/(1 + dP_i/P_0)]^n - 1$$

$$DPS_i = 1/(1 + FFP_i)^{1/n} - 1$$

The calculated value of FL_{DPS} equals 1.23, and the plots in Figure 4.39 show that functional failure occurs at the same point in time for both FFP signature data and DPS data: Eq. (4.35) is verified.

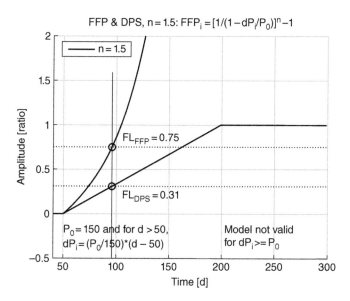

Figure 4.38 Simulated FFP and DPS signatures: power function #3 for $n = 2.0$.

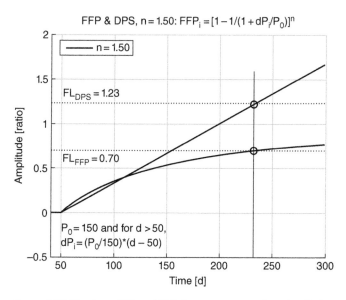

Figure 4.39 Simulated FFP and DPS signatures: power function #5 for $n = 1.5$.

DPS-Based FL Model for Power Function #7

From Eq. (4.29), to transform an FFP signature for power function #7, and from Eq. (4.7):

$$FL_{DPS} = (FL_{FFP} + 1)^{1/n} - 1 \tag{4.36}$$

Example 4.6 Define FL_{FFP} as 0.85 for a power function #7 signature and $n = 0.75$, and calculate the value of FL_{DPS}. Verify your results by plotting the FFP signature, the DPS, and the failure levels:

$$FFP_i = \left(1 + \frac{dP_i}{dP_0}\right)^n - 1$$

$$DPS_i = (FFP_i + 1)^{1/n} - 1$$

The calculated value of FL_{DPS} equals 1.27, and the plots in Figure 4.40 show that functional failure occurs at the same point in time for both FFP signature data and DPS data: Eq. (4.36) is verified.

DPS-Based FL Model for Power Function #9

From Eq. (4.30), to transform an FFP signature for power function #9, and from Eq. (4.7):

$$FL_{DPS} = 1 - (1 - FL_{FFP})^{1/n} \tag{4.37}$$

Example 4.7 Define FL_{FFP} as 0.40 for a power function #9 signature and $n = 0.25$, and calculate the value of FL_{DPS}. Verify your results by plotting the FFP signature, the DPS, and the failure levels:

$$FFP_i = 1 - (1 - dP_i/P_0)^n$$

$$DPS_i = 1 - (1 - FFP_i)^{1/n}$$

The calculated value of FL_{DPS} equals 0.87, and the plots in Figure 4.41 show that functional failure occurs at the same point in time for both FFP signature data and DPS data: Eq. (4.37) is verified.

DPS-Based FL Model for Exponential Function #11

From Eq. (4.31), to transform an FFP signature for exponential function #11, and from Eq. (4.7):

$$FL_{DPS} = \ln(FL_{FFP} + 1) \tag{4.38}$$

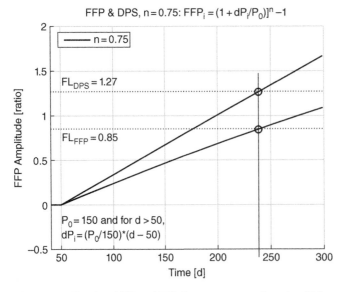

Figure 4.40 Simulated FFP and DPS signatures: power function #7 for $n = 0.75$.

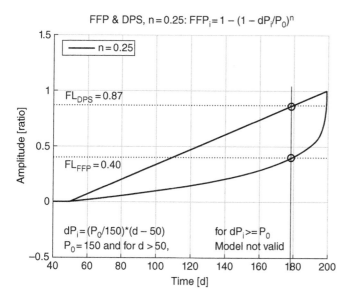

Figure 4.41 Simulated FFP and DPS signatures: power function #9 for $n = 0.25$.

Example 4.8 Define FL_{FFP} as 3.0 for an exponential function #11 signature and $P_0 = 100$, and calculate the value of FL_{DPS}. Verify your results by plotting the FFP signature, the DPS, and the failure levels:

$$FFP_i = \exp(dP_i/P_0) - 1$$

$$DPS_i = \ln(FFP_i + 1)$$

The calculated value of FL_{DPS} equals 1.39, and the plots in Figure 4.42 show that functional failure occurs at the same point in time for both FFP signature data and DPS data: Eq. (4.38) is verified.

DPS-Based FL Model for Exponential Function #13
From Eq. (4.32), to transform an FFP signature for exponential function #11, and from Eq. (4.7):

$$FL_{DPS} = \ln(1/(1 - FL_{FFP})) \tag{4.39}$$

Example 4.9 Define FL_{FFP} as 3.0 for an exponential function #13 signature and $P_0 = 100$, and calculate the value of FL_{DPS}. Verify the results by plotting the FFP signature, the DPS, and the failure levels:

$$FFP_i = 1 - \exp(-dP_i/P_0)$$

$$DPS_i = \ln(1/(1 - FFP_i))$$

$$FL_{DPS} = \ln(1/(1 - FL_{FFP}))$$

The calculated value of FL_{DPS} equals 1.39, and the plots in Figure 4.43 show that functional failure occurs at the same point in time for both FFP signature data and DPS data: Eq. (4.39) is verified.

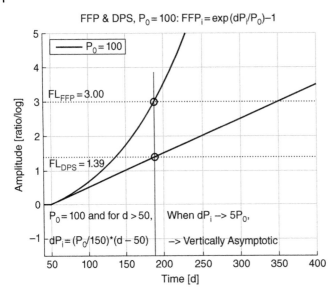

Figure 4.42 Simulated FFP and DPS signatures: exponential function #11 for $P_0 = 100$..

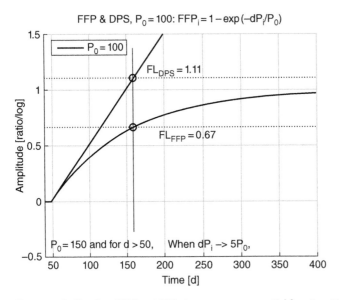

Figure 4.43 Simulated FFP and DPS signatures: exponential function #13 for $P_0 = 150$.

4.5.2 Modeling Results for Failure Levels: FFP-Based and DPS-Based

The results of specifying an FFP-based failure level (FL_{FFP}) and converting that value to a DPS-based failure level (FL_{DPS}) should result in the two signatures reaching the threshold level for functional failure at the same point in time. The plots in Figures 4.37–4.43 confirm the expected results for various degradation signatures for various levels of functional failure.

4.5.3   Transforming DPS Data into FFS Data

There is a final signature transform we need to perform: from DPS data to FFS data. Using DPS data as input to a prediction system would cause that system to acquire knowledge about the value of DPS data at which functional failure occurs. We transform DPS data into DPS-based FFS data using Eq. (3.18):

$$FFS_{DPSi} = 100\, DPS_i/FL_{DPS}$$

The advantage of this transform is the presence of a single model for all failure levels, which greatly simplifies prediction-algorithm processing to produce prognostic information based on which a prognosis of system health can be made. We provide two examples as illustrations.

Example 4.10   Transform the DPS data used in Example 4.3, plot the FFS data, and comment on the results.

The FFS plot on the top plot of Figure 4.44 shows that functional failure at 100% amplitude occurs at the same point in time as shown in the FFP and DPS plots on the bottom plot of Figure 4.44. This is the expected result.

Example 4.11   Transform the DPS data used in Example 4.5, plot the FFS data, and comment on the results.

The FFS plot on the top plot of Figure 4.45 shows functional failure at 100% amplitude occurs at the same point in time as shown in the FFP and DPS plots on the bottom plot of Figure 4.45. This is the expected result.

4.6   Heuristic-Based Approach to Modeling of Signatures: Summary

This chapter presented a rationale for transforming CBD signatures into other signatures: FFP, DPS, and FFS. You learned that both an FFP and a DPS can be transformed into an FFS. You saw that an FFP is a normalized version of a CBD and that an FFP greatly reduces the number of signature models; and we showed that transforming an FFP into an FFS produces a very amenable signature to use as input for prediction algorithms.

The heuristic-based approach to modeling signatures presented in this chapter is summarized as follows:

- Transform CBD signatures into FFP signatures.
- Transform FFP signatures into DPS data.
- Transform DPS data into FFS data.

The last transform requires the specification of an FFP-based threshold level at which functional failure occurs, followed by the calculation of a DPS-based threshold level of failure: this threshold value is used to transform DPS data into FFS data.

It is possible to use knowledge and analysis based on, for example, PoF and FMEA to derive an accurate DPS-based model as a function of a change in system, an assembly, or a component parameter that is degrading: $f(dP_i, P_0)$. In practice, however,

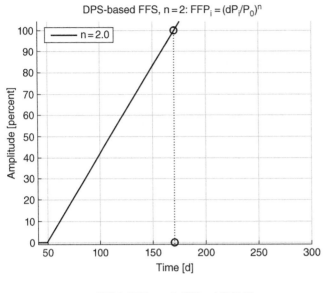

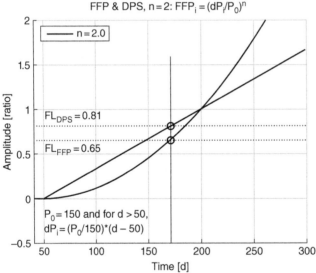

Figure 4.44 Example plots: FFS (top) and FFP and DPS (bottom): power function #1.

highly accurate models are not required unless the system, the data sensor, and the data-conditioning and transform-processing routines produce noiseless data: data that is absolutely flat in the absence of degradation, data that varies with 100% correlation to degradation, and degradation that causes data to increase or decrease monotonically as degradation progresses.

This chapter described how, given a set of DPS models, you can transform CBD-based data into an FFS that is amenable to processing by prediction algorithms to produce prognostic information. The objective is to produce prognostic information having

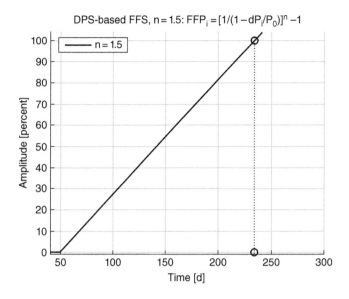

DPS-based FFS, n = 1.5: $FFP_i = [1/(1 - dP_i/P_0)]^n - 1$

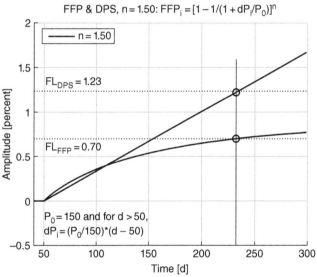

FFP & DPS, n = 1.50: $FFP_i = [1 - 1/(1 + dP_i/P_0)]^n$

$FL_{DPS} = 1.23$

$FL_{FFP} = 0.70$

$P_0 = 150$ and for d > 50,
$dP_i = (P_0/150)*(d - 50)$

Figure 4.45 Example plots: FFS (top) and FFP and DPS (bottom): power function #5.

an accuracy consistent with the signal-to-noise ratio (SNR) of that CBD-based data; the sampling rate used to acquire that CBD-based data; and the measurement accuracy of the system, especially the sensor and data converters (digital-to-analog and analog-to-digital) used to process and transform data.

Table 4.2 lists increasing FFP functions and the model, FFP-to-DPS transform, and FL(FFP) to FL(DPS) model. Not shown in the table is the DPS-to-FFS transform:

$$FFS_{DPSi} = 100\, DPS_i / FL_{DPS}$$

Table 4.2 Models: decreasing signature, increasing signature, FFP-to-DPS, and FL(FFP) to FL(DPS).

Increasing signature function	FFP signature model	FFP-to-DPS transform model	FFP-based FL to DPS-based FL model
Power function #1	$(dP_i/P_0)^n$	$(FFP_i)^{1/n}$	$(FL_{FFP})^{1/n}$
Power function #3	$[1/(1 - dP_i/P_0)]^n - 1$	$1 - 1/(FFP_i + 1)^{1/n}$	$1 - 1/(FL_{FFP} + 1)^{1/n}$
Power function #5	$1 - [1/(1 + dP_i/P_0)]^n$	$1/(1 - FFP_i)^{1/n} - 1$	$1/(1 - FL_{FFP})^{1/n} - 1$
Power function #7	$(1 + dP_i/P_0)^n - 1$	$(FFP_i + 1)^{1/n} - 1$	$(FL_{FFP} + 1)^{1/n} - 1$
Power function #9	$1 - (1 - dP_i/P_0)^n$	$1 - (1 - FFP_i)^{1/n}$	$1 - (1 - FL_{FFP})^{1/n}$
Exp. function #11	$\exp(dP_i/P_0) - 1$	$\ln(FFP_i + 1)$	$\ln(FL_{FFP} + 1)$
Exp. function #13	$1 - \exp(-dP_i/P_0)$	$\ln(1/(1 - FFP_i))$	$\ln(1/(1 - FL_{FFP}))$

In the next chapter, you will learn that signatures are not ideal: they contain, for example, offset errors, distortion, and noise – including signal variations due to feedback effects, multiple failure-mode effects, and so on. Included in the next chapter are methods for mitigating and/or ameliorating such non-ideality.

References

Carr, J.J. and Brown, J.M. (2000). *Introduction to Biomedical Equipment Technology*, 4e. Upper Saddle River, NJ: Prentice Hall.

Erickson, R. (1999). *Fundamentals of Power Electronics*. Norwell, MA: Kluwer Academic Publishers.

Hofmeister, J., Goodman, D., and Wagoner, R. (2016). Advanced anomaly detection method for condition monitoring of complex equipment and systems. 2016 Machine Failure Prevention Technology, Dayton, Ohio, US, 24–26 May.

Hofmeister, J., Wagoner, R., and Goodman, D. (2013). Prognostic health management (PHM) of electrical systems using conditioned-based data for anomaly and prognostic reasoning. *Chemical Engineering Transactions* 33: 992–996.

Hofmeister, J., Szidarovszky, F., and Goodman, D. (2017). An approach to processing condition-based data for use in prognostic algorithms. 2017 Machine Failure Prevention Technology, Virginia Beach, Virginia, US, 15–18 May.

Judkins, J.B. and Hofmeister, J.P. (2007). Non-invasive prognostication of switch mode power supplies with feedback loop having gain, IEEE Aerospace Conference 2007, Big Sky, Montana, US, 4–9 Mar.

Judkins, J.B., Hofmeister, J., and Vohnout, S. (2007). A prognostic sensor for voltage regulated switch-mode power supplies. IEEE Aerospace Conference 2007, Big Sky, Montana, US, 4–9 Mar, Track 11–0804, 1–8.

Kwon, D., Azarian, M.H., and Pecht, M. (2010). *Degradation of Digital Signal Characteristics Due to Intermediate Stages of Interconnect Failure*, 281–296. College Park, MD: Center for Adv. Life Cycle Eng. (CALCE), Univ. of Maryland.

Texas Instruments. (1995). Understanding data converters. Application Report SLAA013.

Viswanadham, P. and Singh, P. (1998). *Failure Modes and Mechanisms in Electronic Packages*, Chapter 7, 283–307. New York, NY: Chapman and Hall.

Further Reading

Filliben, J. and Heckert, A. (2003). Probability distributions. In: *Engineering Statistics Handbook*. National Institute of Standards and Technology. http://www.itl.nist.gov/div898/handbook/eda/section3/eda36.htm.

Tobias, P. (2003a). Extreme value distributions. In: *Engineering Statistics Handbook*. National Institute of Standards and Technology. https://www.itl.nist.gov/div898/handbook/apr/section1/apr163.htm.

Tobias, P. (2003b). How do you project reliability at use conditions? In: *Engineering Statistics Handbook*. National Institute of Standards and Technology. https://www.itl.nist.gov/div898/handbook/apr/section4/apr43.htm.

5

Non-Ideal Data: Effects and Conditioning

5.1 Introduction to Non-Ideal Data: Effects and Conditioning

Condition-based data (CBD) contains feature data (FD) that forms signatures that are highly correlated to failure, damage, and degradation – especially fatigue damage due to the cumulative effects of stresses and strains induced by temperature, voltage, current, shock, vibration, and so on. Those signatures form curves that are not ideal, such as that shown in Figure 5.1; instead, CBD-based signatures contain noise, they are distorted, and they change in response to that noise. That non-ideality, when fault-to-failure progression (FFP) signatures are transformed into degradation progression signatures (DPS) and then into functional failure signatures (FFS), results in a non-ideal transfer curve and errors in prognostic information. Those errors include the following: (i) an offset error between the time when degradation begins and the time of detection of the onset of degradation; and (ii) nonlinearity errors that reduce the accuracy of estimates of remaining useful life (RUL), state of health (SoH), and prognostic horizon (PH) – or end of life (EOL).

5.1.1 Review of Chapter 4

Chapter 4 presented a set of seven signature models that resulted in FFP signatures having ideal, characteristic curves, such as those plotted on the top of Figure 5.2. Those ideal signatures were transformed into ideal DPS transfer curves of straight lines starting at an origin of 0 amplitude and passing through another data point having an amplitude of 1, such as those plotted on the bottom of Figure 5.2. Two signatures are shown on the top of Figure 5.3: a curvilinear FFP signature and its transform into a linear DPS. That DPS was transformed into the FFS shown on the bottom of Figure 5.3; that FFS is a transfer curve that is very amenable to processing by a prediction system to produce prognostic information.

5.1.2 Data Acquisition, Manipulation, and Transformation

In this chapter, we continue an offline phase of developing models and methods: those oriented toward addressing nonlinearity effects due to noise. We use examples and descriptions to illustrate commonly encountered causes and effects of noise on the quality of data and on the nonlinearity of the final transfer curve – an FFS – of a sensing system. We present methods you might use to mitigate and/or ameliorate nonlinearity

Prognostics and Health Management: A Practical Approach to Improving System Reliability Using Condition-Based Data, First Edition. Douglas Goodman, James P. Hofmeister and Ferenc Szidarovszky.
© 2019 John Wiley & Sons Ltd. Published 2019 by John Wiley & Sons Ltd.

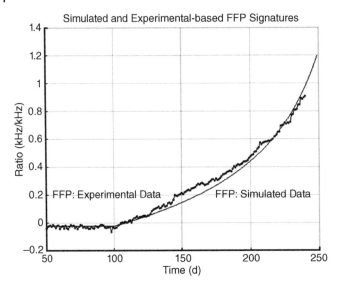

Figure 5.1 Example of a non-ideal FFP signature and an ideal representation of that signature.

that prevents your sensing solution from meeting accuracy, resolution, and precision requirements.

In this book, we are not interested in how to prevent or avoid noise when designing and developing systems; rather, we are concerned with mitigating the effects of noise after data is collected by sensor frameworks, especially the effects of such noise on the linearity of the transfer curve of FFS data. Non-ideality of that transfer curve, not the prediction algorithms, is the largest factor in the relative accuracy of prognostic information. We shall focus on the following three major processes:

- Acquiring and manipulating data to improve the quality of data: removing and mitigating noise to a level sufficient to meet requirements related to prognosis accuracy.
- Transforming data to form signatures that are amenable to prediction processing to produce prognostic information within accuracy requirements: attaching a sensing system to a node with the expectation that in the absence of degradation, data acquired and manipulated at that node will show neither a decreasing nor an increasing signature.
- Modeling degradation signatures and signal-conditioning methods in an offline phase, and then using those models and methods in an online phase to support the transformation of data into an FFS: transforming curvilinear, noisy CBD-based signatures into linear, almost noiseless straight-line transfer curves.

In Chapters 3 and 4, we used the resonant frequency of a damped-ringing response of a switched-mode power supply (SMPS) as the feature of interest. In this chapter, we shall discuss the amplitude of the ripple voltage at the output node of a SMPS as the feature of interest. The base set of models and methods to process signatures was completed in Chapter 4.

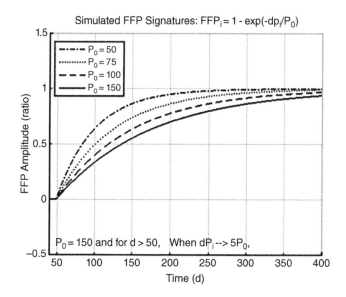

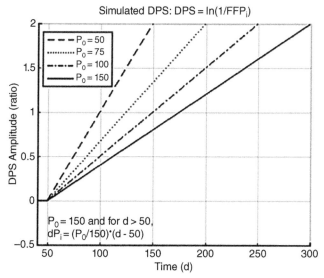

Figure 5.2 Plots of a family of FFP signatures and DPS transfer curves.

5.1.3 Chapter Objectives

CBD-based signatures are not ideal: instead, they contain noise, they are distorted, they contain errors, and so on. Although we may have, for example, an intelligent sensor that is robust enough to attach to the output node of an SMPS – to sample data; perform Bessel, Butterworth, or Chebyshev filtering; digitize analog data; perform digital signal processing; and extract a condition indicator such as the amplitude of the ripple voltage (V_R, a type of FD) at the output node of that SMPS – we might discover that our signature is so noisy, it cannot be used without further signal conditioning.

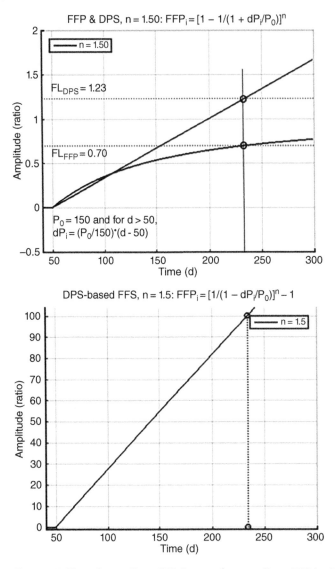

Figure 5.3 Plots of a curvilinear FFP, the transform to a linear DPS (top), and the transform to an FFS (bottom).

A heuristic-based approach to non-ideal CBD signatures for prognostics and health management/monitoring (PHM) can be described as comprising four major processes in an offline and an online phase, as illustrated in Figures 5.4 and 5.5 (IEEE 2017; Medjaher and Zerhouni 2013):

1. Acquire and manipulate data to improve the quality of the data.
2. Transform the data to form signatures that are amenable to prediction processing to produce prognostic information within accuracy requirements.

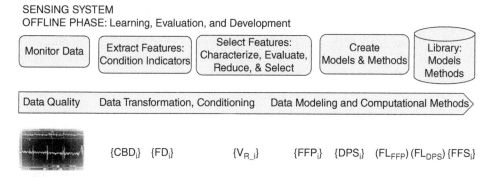

Figure 5.4 Offline phase to develop a prognostic-enabling solution of a PHM system.

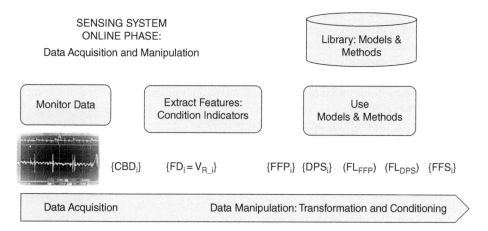

Figure 5.5 Diagram of an online phase to exploit a prognostic-enabling solution.

3. Model degradation signatures and signal-conditioning methods in an offline phase, and then use those models and methods to support the transformation of data into an FFS.
4. Process transformed data to produce prognostic information.

In this chapter, we shall use the models (process #3) developed in Chapter 4 to transform exemplary data (process #2) to demonstrate causes and effects that reduce the quality of signatures (process #1), which reduces the accuracy of prognostic information (process #4). The objectives are to present and discuss sources of errors, significant effects of those errors on signatures, and methodologies and techniques that, when employed, ameliorate and/or mitigate errors by removing, reducing, and/or by avoiding such causes and effects. The goal is to improve the linearity and accuracy of FFS data used as input to the prediction system of a PHM system and, in doing so, improve the accuracy of prognostic information used to provide a prognosis of the health of the system being monitored and managed.

We shall apply a heuristic-based approach to an example of a prognostic-enabled component of an assembly in a system that, when subjected to degradation leading to failure, produces noisy CBD. That example will be an SMPS in which the capacitance of the output filter is reduced as degradation proceeds. The purposes include the following: (i) illustrate nonlinearity errors related to noise; (ii) identify, list, explain, and demonstrate some of the more common causes and effects leading to nonlinearity errors; and (iii) present and show methodologies that ameliorate and/or mitigate non-ideality and/or the effects of non-ideality.

5.1.4 Chapter Organization

The remainder of this chapter is organized to present and discuss topics related to the cause and effect of non-ideality in signatures that result in nonlinearity errors. We present and discuss various methods and techniques to ameliorate and mitigate those effects:

5.2 Heuristic-Based Approach Applied to Non-Ideal CBD Signatures
    This section summarizes a heuristic-based approach for application to CBD signatures through the use of examples. Noise is identified as an issue in achieving high accuracy in prognostic information.
5.3 Errors and Non-Ideality in FFS Data
    This section presents and discusses topics related to nonlinearity, such as noise margin and offset error; measurement error, uncertainty, and sampling; other sources of noise; and data smoothing and non-ideality in FFS data.
5.4 Heuristic Method for Adjusting FFS Data
    This section describes a method for adjusting FFS data, adjusted FFS data, and data-conditioning another example data set.
5.5 Summary: Non-Ideal Data, Effects, and Conditioning
    This section summarizes the material presented in this chapter.

5.2 Heuristic-Based Approach Applied to Non-Ideal CBD Signatures

CBD signatures are not ideal: they contain offset errors, distortion, and noise – including signal variations due to, for example, feedback effects and multiple failure-mode effects. Included in this chapter are methods for mitigating and/or ameliorating such non-ideality. Other errors are introduced in the processing of CBD signatures and the transformation of those signatures into FFS used as input to a prediction system.

Even if CBD was totally absent of noise and variability not related to degradation, attaching a sensor to a node to collect data and then processing that data introduces error into the data. Sensors may perform noise filtering or data sampling, or act as analog-to-digital data converters (ADCs), and digitized data is processed using digital signal processing (DSP) methods and techniques to include additional filtering, data fusion, data and domain transforms, data storage, and data transmission, all of which introduce errors into the data. Operational and environmental variability, such as voltage and temperature variability, are also sources of signal variability that result in non-ideal data (Texas Instruments 1995; Jenq and Qiong 2002; Hofmeister et al. 2013, 2016, 2017).

5.2.1 Summary of a Heuristic-Based Approach Applied to Non-Ideal CBD Signatures

Chapter 4 presented a representative set of increasing FFP signatures: the negative of a decreasing signature produces an increasing signature. This chapter uses an approach that is summarized as follows and that we shall apply to non-ideal CBD signatures (an example is shown in Figure 5.6):

1. Select a target within a system to prognostic enable, such as a component or assembly. The component should have a sufficiently high rate of failure and sufficient negative consequences related to failure to justify the costs of prognostic-enabling that target.
2. Design and perform experiments that replicate degradation leading to the failure of interest and that produce CBD from which one or more candidate features of interest can be isolated and extracted. Evaluate those candidates, and reduce the set, preferably to three or less.
3. Characterize those CBD features as condition indicators, leading to one or more CBD signatures correlated to increasing degradation. Evaluate the signatures, and reduce the number of features to one or two candidates.
4. Convert an exemplary decreasing CBD signature to an increasing CBD signature.
5. Transform an exemplary increasing CBD signature into an FFP signature.
6. Analyze and characterize the exemplary FFP signature as one or more degradation signatures of a set of known signatures in a library of models:
   - As necessary, develop a model for a new degradation-signature model using a method similar to that used in Chapter 3.
   - In this book, we will not develop any degradation-signature models beyond the seven developed in Chapter 4.
7. Transform the exemplary FFP signature data into a DPS.
8. Transform the exemplary DPS into an FFS.

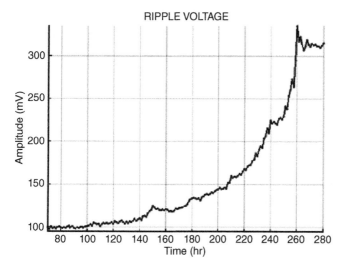

Figure 5.6 Plot of a non-ideal CBD signature data: noisy ripple voltage, output of a switched-mode regulator.

9. Perform experiments, collect new data, apply a set of sensing solutions, and evaluate the nonlinearity of the exemplary FFS. Choose the most promising set of solutions.

10. Select appropriate data-conditioning methods to sufficiently reduce nonlinearity to meet requirements regarding accuracy, resolution, and precision of any prognostic estimates produced using the FFS. Assume the prediction system of the PHM system will not introduce further nonlinearity errors.

11. Adapt and implement data-conditioning steps, such as those just described, to provide a solution in the frameworks of the sensing system of the PHM system. The solution includes the sensor and all firmware and/or software computational routines necessary to acquire and manipulate CBD signatures used to detect, isolate, and identify states related to health of the system; see Figures 1.1 and 1.2 in Chapter 1 (Carr and Brown 2000; Hofmeister et al. 2013; CAVE3 2015; IEEE 2017).

5.2.2 Example Target for Prognostic Enabling

For illustrative purposes, the capacitance of the output filter of an SMPS is selected as the prognostic target, and the output ripple voltage is selected as the CBD feature to be extracted as a condition indicator and processed as a signature. That power supply uses a design incorporating a sufficiently large value of output filtering capacitance that the amplitude of the ripple voltage is dependent on the resistivity of the load and not on the amount of filtering capacitance. Degradation effects that result in the loss of filtering capacitance will cause the amplitude of the output ripple voltage to increase (Erickson 1999; Judkins et al. 2007; Singh 2014).

Acquire Data, Transform It, and Evaluate It Qualitatively

We begin the process of prognostic-enabling the SMPS by running experiments that inject faults and capture data. We verify that degrading the filter capacitance results in ripple voltage like that shown in Figure 5.6: the signature is noisy and clearly not an ideal curve. The exemplary set data is a derived from actual experimental data; additional data points have been added, times were changed from minutes (an accelerated test) to hours, and noise amplitudes were changed for illustrative purposes. The signature results from a square root ($n = 0.5$) power function #3 type of degradation.

Example 5.1 Apply the modeling we have developed to the exemplary data shown in Figure 5.6, and comment on the results:

- From the shape of the curve, and using Table 4.2 in Chapter 4:

FFP-signature model:	$[1/(1 - dP_i/P_0)]^n - 1$
FFP-to-DPS transform model:	$1 - 1/(FFP_i + 1)^{1/n}$
FL transform from FFP-based to DPS-based:	$1 - 1/(FL_{FFP} + 1)^{1/n}$

- The nominal value of the ripple voltage is 99.3 mV ($FD_0 = 99.3$ mV): calculated average of 10 samples prior to onset of degradation.
- Use 0 for the noise margin ($NM = 0$ mV).
- Define functional failure as $FL_{CBD} = 250$ mV, when CBD equals 250 mV.
- Injected degradation for the experiment began when time equaled 98 hours.

- Experimental data showed that CBD amplitude of 250 mV occurred when time equaled 153 hours.

First, transform the CBD-based signature into an FFP signature, combining Eqs. (3.8) and (4.3) with the result shown on the top of Figure 5.7:

$$FFP_i = (CBD_i - FD_0 - NM)/FD_0 \tag{5.1}$$

Then transform the FFP signature data into DPS data using the FFP-to-DPS transform model, and plot it as shown on the bottom of Figure 5.7. From the specification of

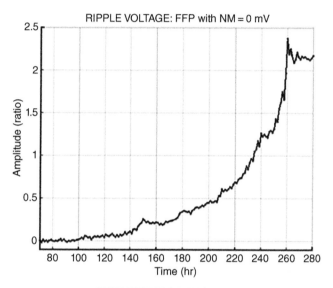

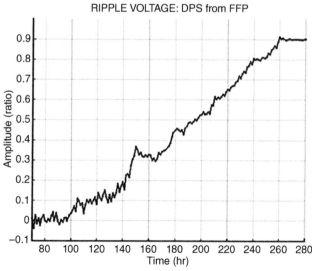

Figure 5.7 Example plots: non-ideal FFP signature data (top) and transformed DPS data (bottom).

$FL_{CBD} = 250$ mV (functional failure), calculate FL_{FFP}

$$FL_{FFP} = (FL_{CBD} - FD_0)/FD_0 = 1.55$$

which you use to calculate FL_{DPS}:

$$FL_{DPS} = 1 - 1/(FL_{FFP} + 1)^{1/n} = 1 - 1/2.55^2 = 0.85$$

to create FFS data that is plotted with the result shown on the top of Figure 5.8. An ideal FFS plot is also included on the bottom of Figure 5.8. A qualitative comparison of the FFS data to the plotted ideal FFS appears favorable – although the data for times <100 hours is questionable, since that data indicates the presence of degradation.

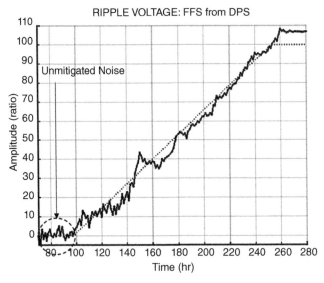

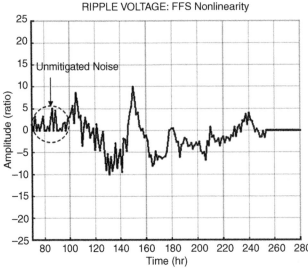

Figure 5.8 Example plots: non-ideal and ideal FFS data (top) and FNL (bottom).

Table 5.1 FFS nonlinearity procedure.

Procedure name	Procedure expression
Calculate time to failure (TTF)	$TTF = t_{FAILURE} - t_{ONSET}$
Create set of ideal data {$IDEAL\_FFS_i$}	$IDEAL\_FFS_i = 100\ (t_i - t_{ONSET})/TFF$
Calculate point-by-point nonlinearity {FNL_i}	$FNL_i = FFS_i - IDEAL\_FFS_i$
Calculate positive nonlinearity (FNL_P)	$FNL_P = \max(\{FNL_i\})$
Calculate negative nonlinearity (FNL_N)	$FNL_N = \min(\{FNL_i\})$
Calculate total nonlinearity error (FNL_E)	$FNL_E = FNL_P - FNL_N$

Evaluate Data Quantitatively

Because an ideal FFS is a straight-line transfer curve, a quantitative evaluation of results is possible by employing the FFS nonlinearity (FNL) method from Chapter 3. You decide it would be informative to calculate the positive nonlinearity, negative nonlinearity, and total nonlinearity:

$$FNL_P = \max(\{FNL_i\}) \tag{5.2}$$

$$FNL_N = \min(\{FNL_i\}) \tag{5.3}$$

$$FNL_E = FNL_P - FNL_N \tag{5.4}$$

Table 5.1 lists the steps in a procedure to quantitatively evaluate the nonlinearity of an FFS.

Example 5.2   Perform an FNL procedure as outlined in Table 5.1, and then plot and evaluate the results. Assume the following:

- The desired FFS nonlinearity is to be within $\pm 10\%$ ($PD_\alpha = PD_{10}$), on or before 75% of the maximum prognostic distance (PD), which is known from the experiment to be 155 hours starting at the onset of degradation (98 hours) and ending at functional failure (253 hours).
- The desired FFS nonlinearity is to be within 5% ($PD_\alpha = PD_{10}$), on or before 50% of the maximum PD.
- No positive FFS nonlinearity prior to onset of degradation.

The nonlinearity of the FFS data on the left in Figure 5.8 is calculated and plotted on the right in Figure 5.8. Table 5.2 lists the quantitative calculations, specifications, and results and evaluation. The evaluation is, "requirements are not met" – as indicated in the far-right column of the table.

You conclude that the results did not meet requirements because of the effects of unmitigated noise in the data (see Figure 5.8) prior to the onset of degradation at 100 hours.

Example 5.3   Since the CBD-based signature indicates the presence of noise of about 2 mV, it would be reasonable to use an NM of 3.0 mV, which equals the noise plus a 50% margin (contingency). Perform the transforms, the FNL procedure, and evaluations described in Examples 5.1 and 5.2.

Table 5.2 List of calculations, specifications, and results for Example 5.2.

Name	Calculation/Specification	Required	Result → Criteria
Maximum PD	PD = TTF = 155 hours	N/A	
PD within 10%	$PD1_{10\%} = 100\,(PD - PD_{10})/PD$	75%	PD_{10} at 34.5 h 77.7% → met
PD within 5%	$PD_{5\%} = 100\,(PD - PD_5)/PD$	50%	PD_5 at 109.5 h 29.4% → did not meet
FNL_P	$FNL_P = \max(\{FNL_i\})$	N/S	10.0%
FNL_N	$FNL_N = \min(\{FNL_i\})$	N/S	−10.0%
FNL_E	$FNL_E = FNL_P - FNL_N$	N/S	20.0%
FNL_P	Before onset of degradation	0%	~5% → did not meet

N/A, not applicable; N/S, not specified.

Table 5.3 List of calculations, specifications, and results for Example 5.3.

Name	Calculation/Specification	Required	Result	Chg.
Maximum PD	PD = TTF = 151.25 h	N/A		
PD within 10%	$PD1_{10\%} = 100(PD - PD_{10})/PD$	75%	PD_{10} at 65 h 57.7% → did not meet	N
PD within 5%	$PD_{5\%} = 100\,(PD - PD_5)/PD$	50%	PD_5 at 105.0 h 30.6% → did not meet	P
FNL_P	$FNL_P = \max(\{FNL_i\})$	N/S	8.4%	P
FNL_N	$FNL_N = \min(\{FNL_i\})$	N/S	−14.0%	N
FNL_E	$FNL_E = FNL_P - FNL_N$	N/S	22.4%	N
FNL_P	Before onset of degradation	0%	0% → met	P

N/A, not applicable; N/S, not specified.

Using $NM = 3.0$ mV leads to one significant improvement: degradation is no longer detected ($FFS_i > 0$) prior to the actual time of the onset of degradation (98 hours). Instead of an early-detection error, degradation is now detected at a later time (103.75 hours), and functional failure is detected at 253.75 hours. The offset error of 5.75 hours in the detection of the onset of degradation causes other changes in calculations: compare the plots in Figure 5.8 ($NM = 0$ mV) to the plots in Figure 5.9 ($NM = 3.0$ mV). Table 5.3 lists the results and indicates whether the result was a positive improvement (P) or a negative change (N).

5.2.3 Noise is an Issue in Achieving High Accuracy in Prognostic Information

As we have just discussed, noise is an issue in achieving high accuracy in prognostic information: errors in the FFS data that is input to a prediction system are very likely to cause corresponding errors in the prognostic information. Example 5.2 illustrated two

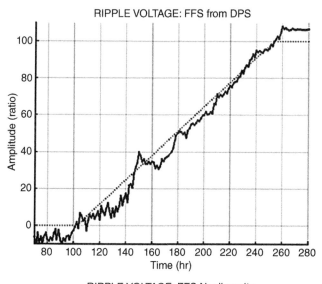

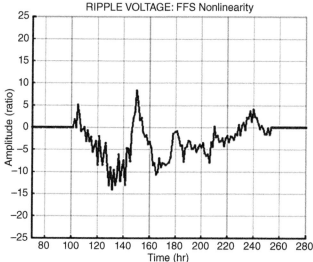

Figure 5.9 Example plots: non-ideal and ideal FFS data (top) and FNL (bottom) after using a NM of 3.0 mV.

types of error: (i) false detection of degradation in the absence of detection and (ii) failure to meet PD requirements at a 5% level of accuracy. Example 5.3 illustrated that using a mitigation method, such as NM, introduces other forms of error such as a reduction in PD and an increase in nonlinearity.

Before making further attempts to ameliorate and/or mitigate the effects of noise, you need to understand the sources and effects of noise. Further, note that by "ameliorate and/or mitigate the effects of noise," we are referring to solutions to be exploited after data is collected and processed by a sensor framework within the sensing system – we are not referring to solutions to reduce noise within the monitored system.

5.3 Errors and Non-Ideality in FFS Data

In this section, we present topics related to noise: its sources and effects, and methods to ameliorate and/or mitigate errors and noise in FFS data. *Noise*, in the context of this book, is any unwanted signal or measurement that is not related to degradation: a specific failure mode or increasing level of damage. The topics presented include but are not limited to the following:

- Noise margin and offset error
- Measurement error, uncertainty, and sampling
- Operating and environmental variability
- Nonlinear degradation
- Multiple modes of degradation

Most errors and noise in FFS data are the result of noise in the original signal at monitored nodes. Some errors and noise are introduced by hardware, firmware, and software used to acquire, manipulate, and store and retrieve data; other errors and noise are introduced by the solutions employed to ameliorate and/or mitigate noise and the effects of noise (Stiernberg 2008; Vijayaraghavan et al. (2008).

It should be understood that due to expense, in terms of both time and money, it is not possible to ameliorate all forms of noise. In addition to expense, there are operational limitations: often, if not invariably, ameliorating noise results in an increase in the weight of the solution(s) and an increase in the power to drive the solution(s). Therefore, mitigation methodologies are often used when, for whatever reason, amelioration methodologies are deemed too expensive and noise is evaluated as being unacceptably high.

A most important consideration is the following: it is not necessary to eliminate and/or mitigate all sources and effects of noise. Instead, eliminate, reduce, and/or mitigate appropriate and sufficient sources and effects of noise to meet the accuracy requirements related to the prognostic target.

The topics presented in this section include commonly encountered sources of variability that contribute to noise, such as the following:

- Variability due to quantization error and other errors related to digitization.
- Variability due to measurement errors related to sampling.
- Variability due to the operating and ambient environments, including switching effects (spikes, glitches, and notches) and environmental effects (temperature, pressure, airflow, and so on).

We could, but will not, devote this entire book to noise and errors that ultimately result in noise, the sources of noise, and actions to address issues due to noise. Instead, we shall focus on noise that results in accuracy issues you are likely to encounter.

5.3.1 Noise Margin and Offset Errors

A commonly used mitigation method is an NM, such as that in Eq. (5.1), but using an NM or other mitigation method often introduces other errors. A primary error related to NM is an offset error in detecting the time of the onset of degradation:

$$OFFSET_E = t_{DETECT} - t_{ONSET} \qquad (5.5)$$

This can be expressed as a percentage error related to estimated prognostic distance,

$$OFFSET_{PDE} = 100(OFFSET_E/PD_{EST}) \tag{5.6}$$

$$PD_{EST} = TTF_{EST} = t_{FF} - t_{DETECT} \tag{5.7}$$

where t_{FF} is the time when functional failure is detected, and t_{DETECT} is when degradation is detected

$$PD_{MAXIMUM} = TTF_{TRUE} = t_{EOL} - t_{ONSET} \tag{5.8}$$

where t_{EOL} is the true end of life or time of functional failure, and t_{ONSET} is the true time of the onset of detection. The error in prognostic distance is given by

$$PD_E = PD_{MAXIMUM} - PD_{EST} \tag{5.9}$$

Example 5.4   Using the data and results of Examples 5.2 and 5.3, calculate the offset and PD errors introduced by the NM value used in Example 5.3. From the data and from Eq. (5.8),

$$PD_{MAXIMUM} = t_{EOL} - t_{ONSET} = 253 - 98 = 155 \text{ hours}$$

After the NM is used, the estimated PD and error are calculated using Eqs. (5.7) and (5.9)

$$PD_{EST} = t_{FF} - t_{DETECT} = 253.75 - 102.5 = 151.25 \text{ hours}$$

$$PD_E = PD_{MAXIMUM} - PD_{EST} = 155 - 151.25 = 3.75 \text{ hours}$$

resulting in a 2.5% error. The offset error is calculated using Eqs. (5.5) and (5.6):

$$OFFSET_E = t_{DETECT} - t_{ONSET} = 103.75 - 98.0 = 5.75 \text{ hours}$$

$$OFFSET_{PDE} = 100(OFFSET_E/TTF_{TRUE}) = 100(5.75/155) = 3.7\%$$

An important objective in achieving high accuracy and reliability in prognostic information is to reduce NM to the smallest value necessary to reliably prevent false detection of degradation and, at the same time, reduce the level of introduced errors. Invariably, to do so, you need to consider other methods to ameliorate and/or mitigate noise and its effects, and that means you must be aware of commonly encountered sources of noise and effects introduced by the sensing system and the pros and cons of solutions you might consider employing in the sensing system. The system being monitored is not subject to being improved by the sensing or prediction systems of a PHM system.

5.3.2   Measurement Error, Uncertainty, and Sampling

Measurement error, uncertainty, and sampling involve the monitoring, acquisition, manipulation, and storage and retrieval of data at points where noise and the effects of noise can be amplified and/or injected into signals. You need to take this into consideration when designing, developing, and evaluating a sensing system.

We shall use two cases as examples: one is a sensor that samples a signal at a node, digitizes the voltage, performs DSP to extract FD, transforms the FD value from a digital value to a scalar value, and then transmits that value to a hub for collection and processing as a data point in a signature; the second is a resistive-temperature detector

(RTD) type of sensor across which a known voltage is applied and from which the current flowing through the temperature-sensitive element can be measured, digitized, and converted to a scalar value that is used as a table-lookup input to extract a temperature value (ITS 1990). Both cases are replete with data-acquisition and -manipulation points of processing for noise to be injected, amplified, and/or transformed:

- Variability due to power sources, the operating environment, and degradation
- Quantization and other errors introduced by data converters
- Transformation errors due to simplifying of expressions
- Delays in time between measurements
- Inappropriate sampling rates
- ADC: input range and sampling rate
- Other sources of noise

Variability Due to Power Sources, the Operating Environment, and Degradation

A major source of noise is the power used in a system: energy sources such as voltage and current supplies, generators, and so on. The output of such sources is not constant: their outputs vary due to variability in their inputs and because of loading effects at their output nodes; they vary in response to their environment, such as temperature, humidity, altitude, and pressure; and they vary in response to degradation within the sources themselves. That variability often results in noise: signal variations that are not related to a failure/degradation mode of interest.

Example 5.5   Suppose, for example, you wish to prognostic enable the coil of a starter relay that is known to exhibit a change in resistance as the coil degrades due to stresses induced by the high currents required to activate the relay. Due to costs and other considerations associated with removal and replacement of the relay to perform a direct measurement of resistance, you are limited to measuring the voltage source for the relay, the current driving the relay, and the ambient temperature of the location of the relay.

You decide to use data fusion, $R_{RELAY} = V_{RELAY}/I_{RELAY}$, to cancel out the effects of variability in the voltage source. But R_{RELAY} is temperature dependent – as temperature changes, R_{RELAY} changes. You decide to fuse the calculated value of R_{RELAY} with measured temperature data using the following

$$R_{RELAY} \approx R_0(1 + TC_R T_M) \rightarrow R_0 \approx (V_{RELAY}/I_{RELAY})/(1 + TC_R T_M)$$

where R_0 is the nondegraded value of the resistance of the relay coil, TC_R is a temperature coefficient of resistance, and T_M is a measured temperature. But the calculated value of R_0 is subject to the following variability: V_{RELAY} is a measurement, I_{RELAY} is a measurement, and T_M is a measurement – and measurement uncertainty applies to all three. Further, the resistance-to-temperature expression provides only an approximation, and the value of TC_R is not constant across the entire range of temperature (ITS 1990). In summary, although the two data fusions are significant solutions to reducing noise amplitude in this example, it is not possible to eliminate the effects of voltage and temperature variations; nor is it possible to totally eliminate measurement uncertainty.

An effective way to ameliorate the effects of noise due to variations in power sources is to take at least one other measurement that exhibits similar noise effects due to common sources of noise, and then use differential techniques to cancel out that noise. An obvious drawback is that an additional sensor, additional weight, additional power input to the sensing system, and additional points of potential failure are added to the sensing system.

Quantization and Other Errors Introduced by Data Converters

Quantization error refers to a case where measurement values change in discrete increments rather than continuously. Both of the exemplary cases in this section use measurement-related methods that introduce quantization errors into data. The first uses an ADC that digitizes input, and the second uses digitized input data, processes that data to create a scalar resistance value, and then uses the scalar value as table-lookup input to extract a temperature value from a table – another form of digitization that results in quantization errors.

Other errors introduced by data converters include step errors, offset errors, gain errors, nonlinearity errors, and so on, all of which contribute to measurement uncertainty and measurement error. Quantization errors also occur when lookup tables are used to convert data (Baker 2010).

Example 5.6 Suppose you are using a voltage sensor that employs an 8-bit digital output having a 6-bit effective number of bits (ENOB). Further suppose that the voltage reference for the ADC is 3.0 VDC; and the ADC exhibits an offset error, a step error, and a gain error so that the ADC transfer curve is as shown in Figure 5.10. If you measure an ideal current, such as that shown on the top of Figure 5.11, the sensor output will exhibit quantization, nonlinearity, and gain errors as shown on the bottom of Figure 5.11.

The quantization error leads to the step-like output of the sensor rather than straight-line output; the nonlinearity error is the unequal step widths, and the gain error is the difference between the input current amplitude (2.5 mA maximum) and the value of the digitized sensor output (< 2.5 mA).

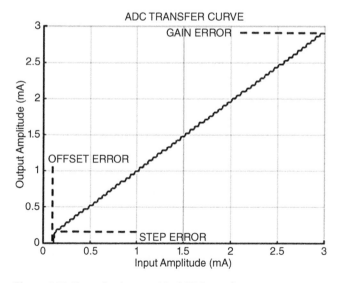

Figure 5.10 Example plot: non-ideal ADC transfer curve.

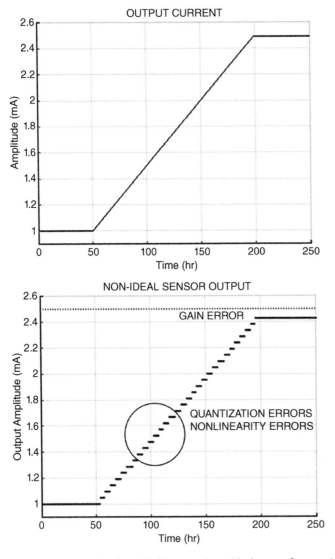

Figure 5.11 Example plots: ideal input and non-ideal output from an ADC.

In addition to quantization errors, offset errors, step errors, gain errors, and nonlinearity errors, other errors can be injected into digital data. Causes of those errors are many and include the following: clock jitter, aperture errors, voltage droop during the hold period of sample-and-hold input stages, thermal noise because of large capacitance values used in sample-and-hold input stages, and nonlinear gain in the various stages used in data converters.

Example 5.7 Suppose the ADC from the previous example is used to measure the current flowing through an RTD sensor. Also suppose a second instantiation of that ADC is used to measure the voltage applied across the same RTD, and both measurements

Table 5.4 Example of lookup values for resistance-temperature, platinum RTD (ITS 1990).

T											
−10	96.09	95.69	95.30	94.91	94.52	94.12	93.73	94.34	92.95	92.55	92.16
0	100.0	99.61	99.22	98.83	98.44	98.04	97.65	97.26	96.87	96.48	96.09
0	100.00	100.39	100.78	101.17	101.56	101.95	102.34	102.73	103.12	103.51	103.90
10	103.90	104.29	104.68	105.07	105.46	105.85	106.24	106.63	107.02	107.40	107.79

are used to calculate the resistance of the RTD sensing element, $R_{RTD} = V_{RTD}/I_{RTD}$. Then the calculated resistance is the result of fusing noisy measurements consisting of quantization, nonlinearity, gain, and other errors.

That noisy value of resistance is then used as an input to a table lookup (see Table 5.4) to select a temperature value: each column represents a one-degree change in temperature. The temperature is an indirect measurement involving two analog-to-digital conversions, a data fusion, and then a table lookup that is a form of digitization coupled with a transform from one data type (resistance) to another data type (temperature). The effects of noise can be compounded and become unacceptably high.

In cases where you determine that an active data converter, such as an ADC, is a major contributor to noise issues that require remediation, you could replace that data converter with one that has a higher ENOB and/or less nonlinearity, and/or a faster sampling rate, and/or a higher input-frequency range, and so on. Again, though, you are invariably faced with higher exploitation costs, more weight, and more power input to drive your sensing system.

When noise issues are traced back to quantization errors because of the use of lookup tables, you might consider solving a complex expression – tables often are used to replace such processing. Consider, however, the complexity of such solutions: for example, lookup tables for platinum-wire RTD applications replace solving a complex Callendar-Van Dusen expression (ITS 1990),

$$R/R_0 = 1 + A\,T + B\,T^2 - C\,(T - 100)T^3$$

for

$$-200°C \leq T \leq 0°C$$

$$R/R_0 = 1 + A\,T + B\,T^2$$

for

$$-0°C \leq T \leq 661°C$$

where

T = temperature in °C, R = resistance at temperature T, R_0 = resistance at 0 °C
$A = \alpha(1 + \delta/100)$, $B = -(\alpha\,\delta)\,/\,(100^2)$, $C = -(\alpha\,\delta)\,/\,(100^4)$
$\alpha = (R_{100} - R_0)\,/\,(100 + R_0)$
$\delta = (R_0\,[1 + \alpha\,260] - R_{200})\,/\,(4.16\,R_0\,\alpha)$

Solving more complex expressions to improve accuracy and reduce noise usually requires more memory to store variables and parameter values, more processing speed, and higher exploitation costs. An alternative is to use a simpler but less accurate expression, such as the linear expression used in Example 5.5,

$$R/R_0 = 1 + A\,T$$

where A is a temperature coefficient of resistance. There are many excellent sources of information on fitting using a linear, quadratic, cubic, or rational polynomial function: the Mosaic Industries website, for example, produces documentation on calibrating resistance.

You can also employ interpolation, extraction, and data-smoothing methods and techniques to reduce the magnitude of the effects of noise.

Measurement Delays

Measurement of multiple variables, such as voltage and current, are not exactly simultaneous: there is a time difference between the acquisition, manipulation, and storing of measurement data. In Example 5.5, voltage and current measurements were fused to create temperature-dependent values of resistance

$$R = V/I$$

that were then transformed into temperature-independent values using a linear resistance-temperature expression and a measured value of temperature:

$$R_0 \approx (V/I)/(1 + TC_R T_M)$$

An experiment was designed and performed using a test bed to simulate degradation using large variations in supplied voltage and temperature to evaluate the effectiveness of the method described in Example 5.5: example plots of the measurements of temperate, voltage, and current are shown Figure 5.12, and the calculated resistances are plotted in Figure 5.13. The plot of the temperature-independent resistance values becomes more and more noisy as degradation progresses. That noise was subsequently attributed to three effects: (i) differences in time between the measurements, (ii) conversion errors introduced using the linear expression for relating resistance and temperature, and (iii) an amplification effect as the resistance increased.

Evaluation: (i) attempting to reduce noise by changing the data-acquisition and -manipulation methods would be very costly; (ii) such noise does increase measurement uncertainty, but the error is usually insignificant; and so (iii) the level of inaccuracy is acceptable. Although no changes were required or made, the following changes were available: (i) use a higher sampling rate, thereby decreasing the time differential between measurements; (ii) use faster computations to also decrease the time differential between measurements; and (iii) use parallel rather than serial measurements.

Sampling Rates

Sampling is a form of low-pass filtering that has an important relationship with respect to resolution and accuracy of a sensing system. In this book, we refer to three different sampling rates: the rate at which the system performs signal sampling at nodes, the rate at which features are extracted from a node (extraction sampling), and the rate at which FD is sampled (data sampling).

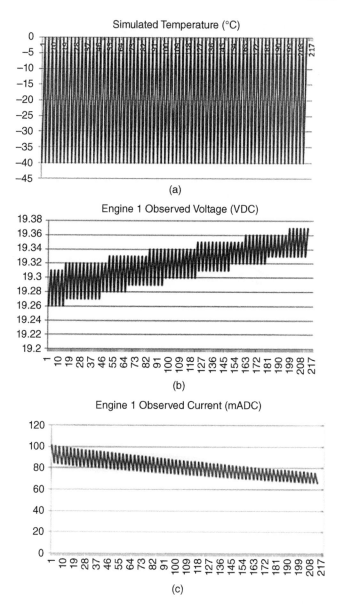

Figure 5.12 Temperature (a), voltage (b), and current (c) plots.

The Nyquist-Shannon sampling theorem states that the sampling rate must be at least twice that of the frequency of interest (Smith 2002). In addition to determining how often we need to sample a node to meet resolution requirements, we need to determine the following: (i) how many feature samples to extract when we sample a node, (ii) how many samples of FD to extract from each feature sample, and (iii) at what rate we will sample FD.

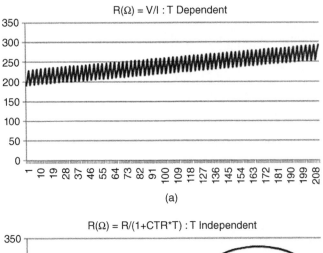

(a)

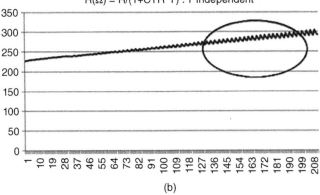

(b)

Figure 5.13 Temperature-dependent (a) and temperature-independent (b) plots of calculated resistance.

Example 5.8 You are given the task of specifying a set of engineering specifications for prognostic-enabling a SMPS using a method based on frequency response instead of the ripple-voltage method you have been working on. You are informed that the customer specifications are as follows:

- Advance notice of functional failure must be at least 72 days.
- Within 20 days after initial detection of a pending functional failure, RUL estimates must converge to within 10% accuracy with a precision of ±1 day; and within 48 days, RUL estimates must be accurate to within 5%.
- At this point in the design of the sensing system, you decide you have sufficient information to perform initial experiments to formulate a set of draft engineering specifications.

You design and perform an initial set of experiments and confirm the SMPS produces a response you are familiar with, an example of which is shown in Figure 5.14. You analyze the response and conclude the following:

- The response is noisy, and the sinusoidal periods are approximately equal but do vary: $T1 \neq T2 \neq T3 \neq T4 \neq T5$

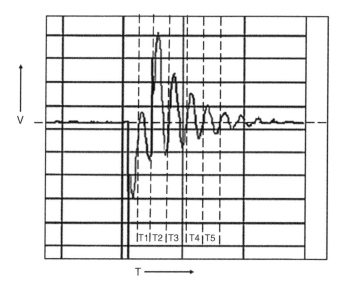

Figure 5.14 Example of a damped-ringing response.

- The nominal frequency is approximately
  $1 \, \text{MHz} \, (\overline{T} \cong 1 \, \mu s)$
- The frequency at functional failure is approximately
  1.4 MHz.

The ± 1 day requirement leads you to the following design specification:

- The node-sampling rate shall be once per 12 hours (see Figure 5.15).
  The noisy signal, the 10% and 5% accuracy requirements, and the experimental data lead you to the following design specifications:
- Each time a node is sampled, 10 samples of the response shall be extracted.
- Five periods shall be used to calculate an average frequency per response;
- The average frequency of each of the 10 responses per node sample shall be averaged, providing the sample average for node.
- Each response shall be windowed (data window : 1 μs delay + 6 μs sampling period).
- The ADC used in the sensor shall have a 12-bit output.
  You review and evaluate your design, taking into consideration such factors as settling between frequency responses, which leads you to another design specification:
- Feature sampling shall occur at intervals of 1 ms.

These initial specifications become input to the next phase of a design to prognostic-enable the SMPS: the hardware design and/or selection of the sensor; the selection of an ADC; the design and/or selection of the controller for the sensor; the software design of the control flow and data management for the sensing system; the design and/or selection of the data and communications protocols, formats, and transceivers; the firmware design and programming of microcontrollers; and the design and programming of the software elements of the sensing system.

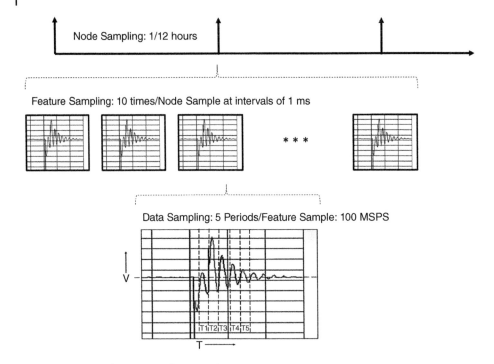

Figure 5.15 Sampling diagram for Example 5.8.

In this section, we have briefly touched on many aspects of dealing with measurement: errors, uncertainty, and sampling rates:

- Direct measurements leading to changes in a parameter of interest are not always feasible. An effective method is to fuse two or more measurements to transform data into a data type that exhibits characteristic signatures of the parameter of interest – for example, fusing voltage and current to transform data to resistance.
- Measurements often are not independent; instead, they are dependent on multiple factors and effects. An effective method for dealing with dependent measurements is to fuse multiple measurements in a manner that cancels or significantly reduces noise. An example is fusing resistance values with an expression that relates resistance at different temperature states other than 0 °C and at 0 °C.
- Noise can be introduced into a sensing system due to the manner in which sensed data is acquired, manipulated, stored, and retrieved. Such noise can be eliminated or otherwise minimized by careful design regarding concurrency of data acquisition, the rate at which data is acquired, the filtering methods used in the sensing system, and increased computational speeds to reduce processing time and latency.
- Sampling is a form of low-pass filtering and data smoothing through averaging. There are three basic modes of sampling: the rate at which nodes are sampling, the number of times features are extracted during a given sampling of a node, and the number of data examples per extracted feature from a node.

ADC: Input Range and Sampling Rate

A source of error related to digitization is the relationship of a reference voltage used by data converters, the ENOB of data converters, and the maximum amplitude of the data being converted: the input range. Another source of error is the relationship of the ADC sampling rate and the rate of change of the input signal.

Example 5.9   Suppose your input system uses an appropriate transducer: for example, one that converts amperes to voltage, torr to voltage, force to voltage, and so on. You decide to use sensor specifications from Example 5.9. Further suppose you are using a voltage sensor that uses a one-volt reference and has a 12-bit output (as specified) with an ENOB of 10 bits (not specified). The voltage value of the least-significant bit (LSB) of that ADC becomes

$$LSB = V_{REF}/2^{(\#bits)} = 1/2^{(12)} = 0.24\,mV$$

which might lead you to believe you can measure a 10 mV signal with 2.4% accuracy. However, because the ENOB is 10, the effective LSB is only

$$LSB_{EFF} = 1/2^{(10)} = 1.0\,mV$$

which means you can measure a 10 mV signal with only 10% accuracy. You run some experiments and discover the measurement inaccuracy is worse than 10% (see Figure 5.16): the sampling rate of the ADC (about 1.6 MSPS) is too low for the rate of change of the input, as indicated by the staircase appearance of the value of the digital output in Figure 5.1. The value is obtained by a digital-to-analog converter (DAC) operation. If you want to measure with an accuracy of 5% or better, you need to:

- Use a sensor with an ADC sampling rate of at least 20 SPS – 40 MSPS would be better.
- Use a sensor with an ADC ENOB of at least 11 bits and/or an ADC with a lower reference voltage.

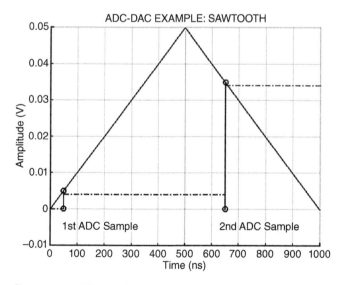

Figure 5.16 ADC example: saw-tooth input, sampling, digital output value.

5.3.3 Other Sources of Noise

There are many other sources of noise in signatures. Among them, the following are significant enough to describe in further detail:

- White and thermal noise
- Nonlinear degradation: feedback and multiple modes of degradation
- Large amplitude perturbations

White and Thermal Noise

White noise is one of the many noises commonly referred to as *background noise*. It is random noise that has an even distribution of power (Vijayaraghavan et al. 2008) and is caused by, for example, random motion of carriers due to temperature and light (see the top of Figure 5.17).

Background noise can be mitigated by using low-pass filtering techniques, including sample averaging and data smoothing. Instead of continuously sampling data, periodically take a number of consecutive samples: for example, a burst of 10 samples in one second – and calculate the average. Repeat at a suitable sampling rate such as, for example, once an hour (see the bottom of Figure 5.17). Sample averaging, when performed in the sensor framework, also reduces the amount of data that needs to be transmitted and collected for processing by a vector framework.

You could use such filtering in the sensor hardware and firmware and/or in post-processing software to condition the collected data. In applications involving rotating equipment such as engines, shafts, drive trains, and so on, this form of filtering can be accomplished through the use of any number of time-synchronous averaging (TSA) algorithms (Bechhoefer & Kingsley 2009).

Nonlinear Degradation: Feedback and Multiple Modes of Degradation

Chapter 3 used an example of a square root degradation in which the frequency component of a damped-ringing response changed as the filtering capacitance degraded; a plot of the data in that example is shown in Figure 5.18. The signature exhibits both a linear and a nonlinear curve: the linear curve is illustrated by the dashed line between the data points at time 100 days and at time 200 days.

Nonlinearity in signatures does not refer to whether the shape is linear or curvilinear: instead, it refers to a significant change in the shape of the curve. The source of the nonlinearity in Chapter 3 was determined to be feedback between the output of the power supply and the input of the pulse-width modulator in the supply. Such changes in shape can also be due to multiple modes of degradation. For example, the output of a sub-assembly might exhibit the effects of multiple assemblies, such as a power supply that uses power-switching devices, an H bridge controller, and a capacitive output filter. The capacitive output filter degrades, which reduces the deliverable power, which causes the H bridge controller to change the pulse width and/or the pulse rate of the modulator to compensate for the loss of capacitance, which results in the linear portion of the signature. Eventually, the degradation becomes severe enough that the feedback control in the power supply is no longer able to compensate. It is not unusual that the combined loss of capacitance coupled with a high power demand causes the switching devices and/or their gate drivers to be overdriven and become permanently damaged – and a second mode of degradation ensues.

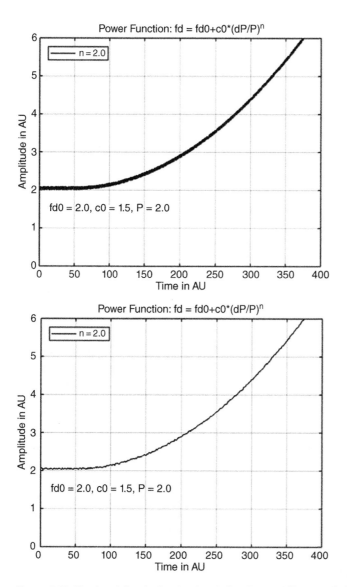

Figure 5.17 Simulated data before (top) and after (bottom) filtering of white (random) noise.

We believe that rather than attempt to mitigate, filter, and/or condition such changes in the shape of the signature, it is better to view a change in shape as evidence that the prognostic target – a device, component, assembly, or subsystem – has functionally failed and is no longer capable of operating within specification. Such changes in shape are indicative of either another failure mode and/or a phase change in the properties of a device or component: perhaps the crystalline structure of a semiconductor material has changed, or a rotating shaft has changed from an inelastic phase to a plastic phase, or vibration caused by a spalled tooth in a gear has caused assembly mounts to loosen.

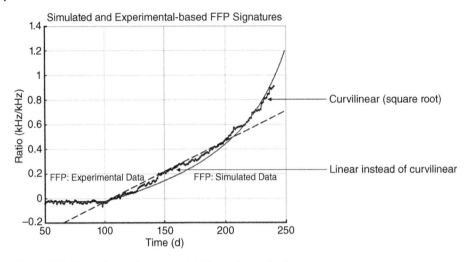

Figure 5.18 Degradation signature exhibiting a change in shape.

Large-Amplitude Perturbations

Large perturbations in amplitude are particularly vexing in prognostic solutions: special conditioning is often required to meet accuracy requirements. Consider, for example, the set of temperature data plotted on the top of Figure 5.19. Although it is obvious that the signature is increasing in value, it is difficult to determine the time of the onset of degradation and the time when functional failure occurs – defined as, for example, when there is a 10-degree error.

Suppose the temperature measurements are those taken by a sensor at the compressor inlet of a jet engine, and there is another temperature sensor, such as one to measure ambient temperature or the compressor inlet temperature of other engines. Further suppose there is an inconsequential time difference between those other sets of data. Given those conditions, several assumptions can be made: (i) abrupt changes in data are not associated with degradation; (ii) noise in each data set might be common-mode noise; (iii) when only one set of data exhibits an increasing signature, then the failure mode is isolated to that engine. An example of this case is shown in the plots on the bottom of Figure 5.19.

Chapter 2 introduced the concept of calculating the distance between sample vectors (see section "Mahalanobis Distance Modeling of Failure of Capacitors"). By subtracting two vectors, we obtain the differential distance and thereby mitigate noise because we subtract common-mode components of noise. The result, plotted in Figure 5.20, is a much less noisy signature.

The general procedure for an ith set out of M sets of data taken essentially at the same time is expressed as follows:

$$X_{Ri} = \left(\frac{1}{M-1}\right)\left[\sum_{1}^{n=M} X_n\right] - X_i \text{ mean reference for an } i\text{th data point} \qquad (5.10)$$

$$X_{Di} = X_i - X_{Ri} \text{ } i\text{th differential distance} \qquad (5.11)$$

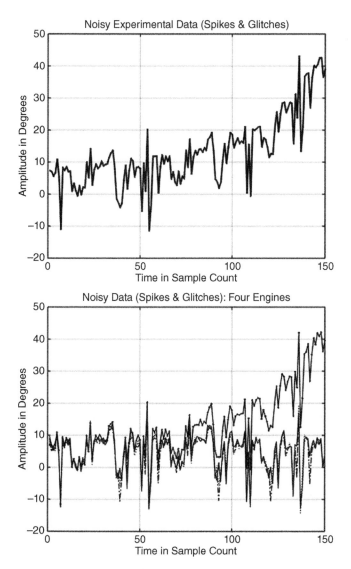

Figure 5.19 Experimental data: temperature measurements for a jet engine.

$\{X_{Di}\}$ is a CBD signature. When we apply this method to, for example, temperature data from each of four engines on the same aircraft (top plots in Figure 5.21), we obtain the differential signatures shown in the bottom plots in Figure 5.21.

Although one differential-distance signature in Figure 5.21 is visually increasing, a computational routine would use the largest amplitude value of all four signatures to create a single differential-degradation signature, as shown in Figure 5.22: compare that to the original temperature data shown in Figure 5.19. In Chapter 7, we will provide an example of how to further condition this signature using an algorithm to mitigate large-amplitude perturbations and data smoothing.

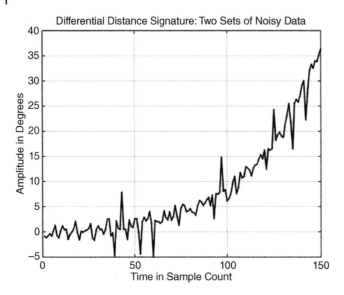

Figure 5.20 Differential signature from temperature measurements for each of two engines.

5.3.4 Data Smoothing and Non-Ideality in FFS Data

Topics of discussion related to non-ideal data in this chapter have included noise margin and offset error and measurement error, uncertainty, and sampling; and noise issues related to variability of power sources, converters, and temperatures. Even after due diligence in the design and exploitation of the sensing hardware and firmware, you may discover noise issues you need to address, as exemplified by the data from Chapter 3 plotted in Figure 5.23: an FFP signature without using any NM.

Example 5.10   Evaluation indicates that a manufacturing tolerance of 10% translates into an NM of 5% to account for variability in the nominal value, 24.0 kHz, of the FD (frequency). Signal variability due to measurement uncertainty, noise, and operating and environment conditions translates into an additional NM of 2.5%: a sum of 7.5%, to which you add a 33% safety margin. This translates into a further NM of 2.5%: a total NM of 10.0% = 2.4 kHz. The resulting FFP signature is shown in Figure 5.23.

Example 5.11   You decide to use a dynamically calculated value for FD_0 and incorporate a calibration step in the design of your sensing system: (i) acquire the frequency value for 10 node samples; (ii) calculate the average frequency; and (iii) define and use a node definition that includes the node name, sensor type and serial number, nominal frequency, and noise margin.

The average of the measured nominal frequencies is 23.9 kHz. Now, since you know what the nominal value is, you can reduce the noise margin to 4%. You obtain an updated FFP signature, which is plotted in Figure 5.24. Detection of degradation (amplitude crossover at 0%) is improved from an offset error of about 30 days to an offset error of about 15 days: a 50% error reduction.

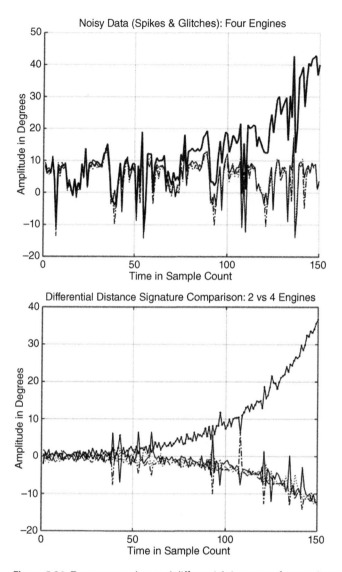

Figure 5.21 Temperature data and differential signatures: four engines on an aircraft.

Example 5.12 Careful examination of the signature at the crossover amplitude of 0% reveals that there are two crossover points, due to, you conclude, the noise in the data. You decide to employ a moving-average method of data smoothing: (i) use a rolling average of three data points; (ii) load a three-slot first-in-first-out (FIFO) data queue with the very first data point acquired during the initialization stage; (iii) store the FIFO queue and a pointer to a next slot in the queue; (iv) when the next data point is acquired, store that value in the data queue; (v) update and store the data pointer; (vi) calculate a rolling average of the data queue; (vii) use the rolling average as the current amplitude to calculate the FFP signature; (viii) use the node definition area to store the data queue

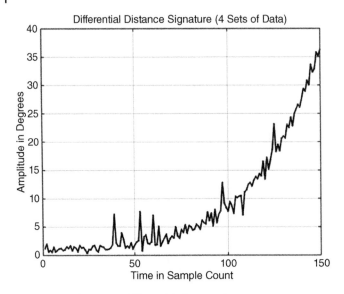

Figure 5.22 Composite differential-distance signature.

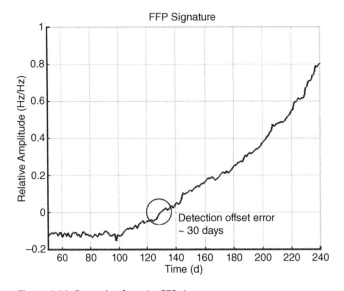

Figure 5.23 Example of a noisy FFP signature.

and pointers; and (ix) reduce the NM from 4% to 2%. Figure 5.25 shows the plot of the smoothed FFP signature: the error in detecting degradation is improved from 15 to 5 days, which is a 67% error reduction.

Example 5.13 You evaluate the characteristic curve of the data plotted in Figure 5.25 and decide the signature is probably the result of a power function #3 with a power of 1/2. You then transform the FFP signature into a DPS using the following expression

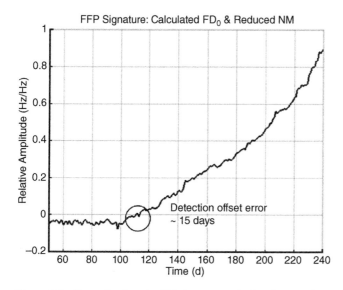

Figure 5.24 Example of a noisy FFP signature: calculated nominal FD value, reduced NM value.

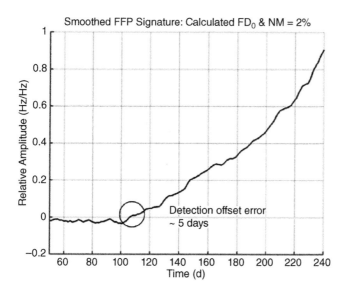

Figure 5.25 Example of a smoothed (3-point moving average) FFP signature.

(from Chapter 4)

$$DPS_i = 1 - 1/(FFP_i + 1)^{1/n}$$

which results in the plot shown in Figure 5.26.

Example 5.14 You define functional failure to occur when the FFP signature reaches a value 0.65, which is transformed into a DPS-based FL:

$$FL_{FFP} = 0.65 \rightarrow FL_{DPS} = 1 - 1/(FL_{FFP} + 1)^2 = 0.63$$

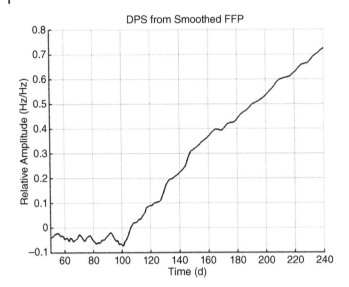

Figure 5.26 Example of a DPS from a smoothed FFP signature.

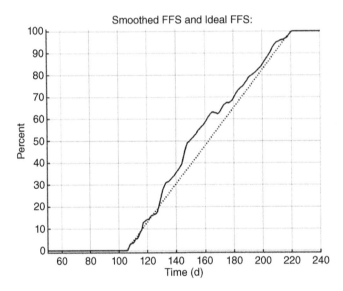

Figure 5.27 Example of an FFS from a smoothed FFP signature.

The FL_{DPS} value is used to transform DPS data into FFS data, as plotted in Figure 5.27.

Example 5.15 You calculate and plot the point-by-point FNL using the procedure listed in Table 5.1:

$$TTF = t_{FAILURE} - t_{ONSET}$$

$$IDEAL\_FFS_i = 100(t_i - t_{ONSET})/TFF$$

$$FNL_i = FFS_i - IDEAL\_FFS_i$$

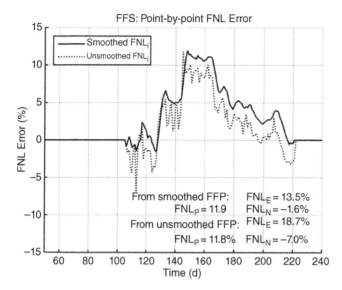

Figure 5.28 Example of a {FNLi} plot from a smoothed FFS.

$$FNL_E = FNL_P - FNL_N = \max(\{FNL_i\}) - \min(\{FNL_i\})$$

From the data,

$$t_{ONSET} = 105 \text{ days and } t_{FAILURE} = 220 \text{ days} \rightarrow TTF = 115 \text{ days}$$

The plot of the nonlinearity, $\{FNL_i\}$, is shown in Figure 5.28 (bold). Also shown is a plot of the nonlinearity for an FFS from using an unsmoothed FFP signature. The total nonlinearity error is reduced from 18.7% to 13.5% through the use of a four-point moving-average method of data smoothing.

5.4 Heuristic Method for Adjusting FFS Data

In the examples in the previous section, data smoothing using a four-point moving average results in less-abrupt changes in linearity and in the following error reductions:

- Error in detecting the onset of degradation is reduced from 30 days to 5 days: an 83% error reduction.
- Total nonlinearity error is reduced from 18.5% to 13.5%: a 27% reduction.

In this section, we present a heuristic method to improve FFS linearity by adjusting a model to received FFS data, one data point at a time, and then using the adjusted model to change the amplitude of the received data point. The model is a computed path along which FFS data is presumed to prefer to travel : see Figure 5.29.

5.4.1 Description of a Method for Adjusting FFS Data

The heuristic method for adjusting FFS data consists of a concept of an FFS model that defines an area having a diagonal that represents a preferred path along which FFS data travels from the onset of degradation to functional failure. Because of noise, FFS data is

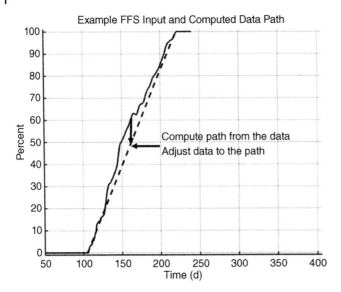

Figure 5.29 Example random-walk paths and FFS input.

not an ideal straight line from the lower-left corner of the area to the upper-right corner of the area.

The objective of the heuristic method is to (i) compute a preferred path, taking into account changes in amplitude (vertical deviations) and different rates (horizontal deviations); (ii) adjust the length of the model (time to failure, TTF); and then (iii) adjust the amplitude of the input data toward the preferred path: the diagonal of the model. This method includes the following significant design points (refer back to Figure 5.29):

- The model and data-point adjustments are performed one data point at time, without any knowledge of when degradation begins or when functional failure occurs.
- The input FFS data points are used to detect the onset of degradation and the time of functional failure.
- Model adjustments are made using geometry-based computations to solve a random-walk problem in which an FFS travels on a path that starts in the lower-left corner and continues to the upper-right corner of a rectangular area: from the onset of detectable degradation to functional failure.
- The height of the area is understood to be 100% – from the definition of an FFS transfer curve.
- The length of the area (horizontal axis) is calculated using the input data.
- The amplitude of each data point is adjusted toward the adjusted path and then averaged with up to three previous data points to create a smoothed FFS curve.

5.4.2 Adjusted FFS Data

When the FFS data shown in Figure 5.27 is input, one data point at a time, into the described heuristic method for adjusting FFS data, the result is the adjusted FFS data plotted in Figure 5.30; the input FFS is also plotted in Figure 5.30. The FNL plots for the adjusted FFS and for the smoothed FFS are shown in Figure 5.31. The total FNL error

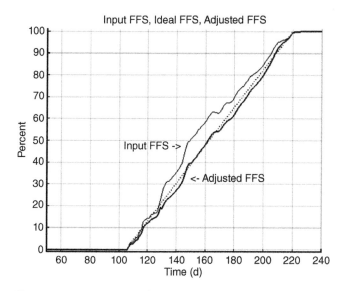

Figure 5.30 Example plots of input FFS data and adjusted FFS data.

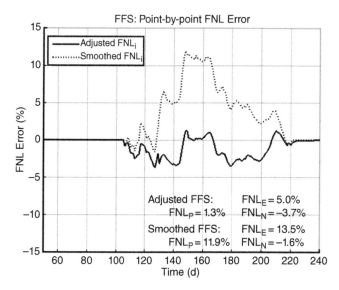

Figure 5.31 Example of a {FNLi} plot from an adjusted FFS.

is reduced: prior to any conditioning, total nonlinearity was 18.6%; after conditioning, total nonlinearity is 8.0%.

5.4.3 Data Conditioning: Another Example Data Set

We used an example data set to create the FFP signature shown in Figure 5.23, to demonstrate the effectiveness of using a four-point moving average to smooth FFP signatures and reduce offset errors in detecting the onset of degradation and reduce nonlinearity in FFS data (Figure 5.28). We then demonstrated a heuristic method for improving the linearity of FFS data to further reduce nonlinearity (Figure 5.31). We reduced total nonlinearity from 13.5% to 8.0%.

In this section, we apply the same data-conditioning methods to the input data shown in Figure 5.6 to demonstrate that the methodology is extendable to other sets of data. Data similar to that shown in Figure 5.6 is first transformed into an FFP signature as shown on the top of Figure 5.32.

Conditioning FFP Signature Data

We perform the first step in a conditioning procedure, dynamically measuring an average value of the feature rather than using a nominal manufactured value: this lets us

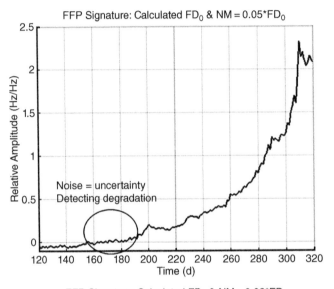

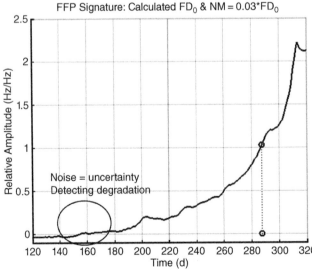

Figure 5.32 Ripple voltage: plots of an unsmoothed (top) and smoothed (bottom) FFP signature.

use a smaller NM value. But noise is a problem, even though we used a relatively large margin (5% of the nominal feature value); there is an area of uncertainty in detecting the onset of degradation, as indicated in the plot. Rather than increase the NM, it is better to smooth the FFP data and then evaluate how much NM is required.

Example 5.16   You apply a four-point rolling average method, evaluate the result, and decide that instead of increasing NM, you can reduce it. The result is the FFP signature plotted on the bottom of Figure 5.32. The FFP signature is much smoother, which allows you to reduce the NM value, resulting in a reduced offset error in detecting degradation.

You transform the FFP signature into DPS and then define functional failure to occur when the FFP signature exceeds 1.0: at that threshold, the ripple voltage has doubled in amplitude compared to the nominal value. You convert the FL_{FFP} value to an FL_{DPS} value of 0.75,

$$FL_{FFP} = 1.0 \rightarrow FL_{DPS} = 1 - 1/(FL_{FFP} + 1)^2 = 0.75$$

and use the 0.75 threshold to transform the DPS data into FFS data. A plot of the DPS from an unsmoothed FFP signature is shown on the top of Figure 5.33, and the DPS from a smoothed FFP signature is shown on the bottom of Figure 5.33. The corresponding plots of the FFS data are shown in Figure 5.34; there is a clear difference in the appearance of the two sets of plots.

Analysis of the data indicates that functional failure is detected when a data point reaches or exceeds the specified failure threshold, as indicated in the plots.

Reevaluating NM

It is a good idea, after employing data smoothing, to reevaluate your choice of the value for NM. It will often be the case that you can lower the NM, which will typically reduce any offset error, such as the 5.7 hours encountered in Example 5.3.

Example 5.17   After evaluating the plot shown in Figure 5.33, you decide to change the NM from 3.0 mV to 2.0 mV and rerun the FFP, the DPS, the FL, and the FFS models. The FFS plots for an NM of 3.0 mV and an NM of 2.0 mV are shown in Figure 5.34. The detection of onset of degradation (when FFS > 0) for a NM of 3.0 mV occurs at time = 153.7 days; and for a NM of 2.0 mV, detection occurs at time = −151.3 day: the offset error is reduced from 5.75 days (3.7%) to 3.25 days (2.1%).

Conditioning FFS Data

The FFP signature is transformed into a DPS, a functional-failure level is defined and converted to a DPS-based FL, and then the DPS data is transformed into FFS data. As each FFS data point is created, it is subjected to the heuristic method from Section 5.4 to further linearize the FFS data.

5.5   Summary: Non-Ideal Data, Effects, and Conditioning

This chapter presented topics related to non-ideal data. You learned that noise is an issue in achieving high accuracy in prognostic data: those issues become evident when

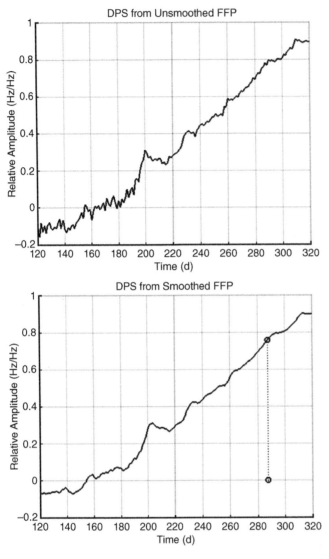

Figure 5.33 Ripple voltage: plots of an unsmoothed (top) and smoothed (bottom) FFS data.

you acquire, transform, and evaluate data. Noise is defined as any variability in data not related to a particular mode of failure; it must be sufficiently reduced and/or mitigated to meet the accuracy requirements of a PHM system for each prognostic-enabled device, component, or assembly.

Methods to reduce the effects of noise and nonlinearity in data were presented related to noise margin, measurement errors and uncertainty, operating and ambient environment, nonlinear degradation, and multiple modes of degradation. You also learned a heuristic method for adjusting FFS data and how to evaluate the end result of data conditioning and model adjustment.

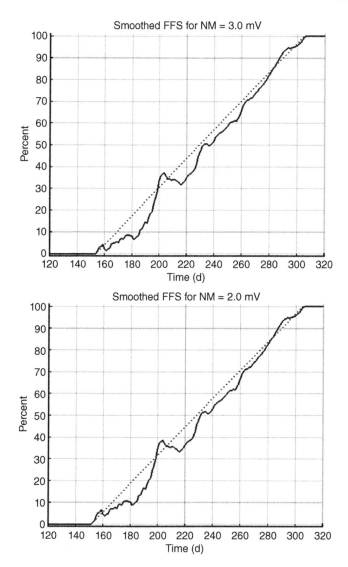

Figure 5.34 Smoothed FFS: for NM = 3.0 mV (top) and for NM = 2.0 mV (bottom).

In the next chapter, you will learn about important characteristics and metrics related to prognostic information output by a prediction framework in response to FFS data produced by a vector framework.

References

Baker, R.J. (2010). *CMOS Circuit Design, Layout, and Simulation*, 3e. Wiley-IEEE Press.

Bechhoefer, E. and Kingsley, M. (2009). A review of time synchronous average algorithms. 2009 PHM Conference, San Diego, California, US, 27 Sep. – 1 Oct.

Carr, J.J. and Brown, J.M. (2000). *Introduction to Biomedical Equipment Technology*, 4e. Upper Saddle River, New Jersey: Prentice Hall.

Erickson, R. (1999). *Fundamentals of Power Electronics*. Norwell, MA: Kluwer Academic Publishers.

Hofmeister, J., Goodman, D., and Wagoner, R. (2016). Advanced anomaly detection method for condition monitoring of complex equipment and systems. 2016 Machine Failure Prevention Technology, Dayton, Ohio, US, 24–26 May.

Hofmeister, J., Szidarovszky, F., and Goodman, D. (2017). An approach to processing condition-based data for use in prognostic algorithms. 2017 Machine Failure Prevention Technology, Virginia Beach, Virginia, US, 15–18 May.

Hofmeister, J., Wagoner, R., and Goodman, D. (2013). Prognostic health management (PHM) of electrical systems using conditioned-based data for anomaly and prognostic reasoning. *Chemical Engineering Transactions* 33: 992–996.

IEEE. (2017). Draft standard framework for prognosis and health management (PHM) of electronic systems. IEEE 1856/D33.

ITS. (1990). International temperature scale. National Institute of Science and Technology.

Jenq, Y.C. and Li, Q. (2002). Differential non-linearity, integral non-linearity, and signal to noise ratio of an analog to digital converter. Portland, Oregon: Department of Electrical and Computer Engineering, Portland State University.

Judkins, J.B., Hofmeister, J., and Vohnout, S. (2007). A prognostic sensor for voltage regulated switch-mode power supplies. IEEE Aerospace Conference 2007, Big Sky, Montana, US, 4–9 Mar, Track 11–0804, 1–8.

Medjaher, K. and Zerhouni, N. (2013). Framework for a hybrid prognostics. *Chemical Engineering Transactions* 33: 91–96. https://doi.org/10.3303/CET1333016.

National Science Foundation Center for Advanced Vehicle and Extreme Environment Electronics at Auburn University (CAVE3). (2015). Prognostics health management for electronics. http://cave.auburn.edu/rsrch-thrusts/prognostic-health-management-for-electronics.html (accessed November 2015).

Singh, S.P. (2014). Output ripple voltage for buck switching regulator. Application Report SLVA630A, Texas Instruments, Inc.

Smith, S.W. (2002). *Digital Signal Processing: A Practical Guide for Engineers and Scientists*, 1e. Newnes Publishing.

Stiernberg, C. (2008). Five tips to reduce measurement noise. National Instruments.

Texas Instruments. (1995). Understanding data converters. Application Report SLAA013.

Vijayaraghavan, G., Brown, M., and Barnes, M. (2008). Electrical noise and mitigation. In: *Practical Grounding, Bonding, Shielding and Surge Protection*. Elsevier.

Further Reading

Filliben, J. and Heckert, A. (2003). Probability distributions. In: *Engineering Statistics Handbook*. National Institute of Standards and Technology. http://www.itl.nist.gov/div898/handbook/eda/section3/eda36.htm.

O'Connor, P. and Kleyner, A. (2012). *Practical Reliability Engineering*. Chichester, UK: Wiley.

Pecht, M. (2008). *Prognostics and Health Management of Electronics*. Hoboken, New Jersey: Wiley.

Tobias, P. (2003a). Extreme value distributions. In: *Engineering Statistics Handbook*. National Institute of Standards and Technology. https://www.itl.nist.gov/div898/handbook/apr/section1/apr163.htm.

Tobias, P. (2003b). How do you project reliability at use conditions? In: *Engineering Statistics Handbook*. National Institute of Standards and Technology. https://www.itl.nist.gov/div898/handbook/apr/section4/apr43.htm.

6

Design: Robust Prototype of an Exemplary PHM System

6.1 PHM System: Review

The design and development of a system to support prognostics and health management/monitoring (PHM) is complex, and there are many approaches to do so. Even though each such system (with the possible exception of demonstrations, test beds, and experiments) is unique, there are important design considerations. For example, (i) whether to use node-based, assembly-based, or some other architecture to sample and acquire data from nodes; (ii) data-handling requirements, such as conditioning, transforming, and fusing; (iii) resolution, precision, and accuracy requirements for predicting information; and (iv) checkpoint/restart (stop and resume operations). A robust PHM system needs to address these issues.

For illustrative purposes, this chapter is written directly to you, the reader, in the role of a lead design engineer of prognostic solutions who has been chosen to lead the design and development of a robust prototype PHM system. That system is to serve as a proof of feasibility for a customer and as a demonstration of the capabilities and expertise you and your team possess. This chapter is based on historical and case-study data.

Before introducing new issues and solutions, we will review all of the previous chapters, because you will need to appropriately select, adapt, and apply other approaches; a large number (but not all) of these have already been presented in this book.

6.1.1 Chapter 1: Introduction to Prognostics

In Chapter 1, you learned that prognostics involves accurately detecting and reporting future failures in systems by detecting degradation (diagnostics) and creating information such as state of health (SoH) and remaining useful life (RUL). We defined core frameworks for a PHM system: a sensor framework, a feature-vector framework; and a prediction framework as shown in Figure 1.1 and repeated here in Figure 6.1. The primary focuses of that chapter were the foundation of reliability theory, failure distributions under extreme stress, uncertainty measures in parameter estimation, system reliability and prognostic health management, and prognostic information: we presented over 130 equations dealing with those topics for modeling.

Section 1.6 included a framework diagram for PHM, repeated in Figure 6.2. Section 1.7 included two figures that are repeated in Figures 6.3 and 6.4.

Prognostics and Health Management: A Practical Approach to Improving System Reliability Using Condition-Based Data, First Edition. Douglas Goodman, James P. Hofmeister and Ferenc Szidarovszky.
© 2019 John Wiley & Sons Ltd. Published 2019 by John Wiley & Sons Ltd.

CORE PROGNOSTIC FRAMEWORKS for PROGNOSIS in a PHM SYSTEM

CONTROL & DATA FLOW FRAMEWORK

SENSOR FRAMEWORK	FEATURE-VECTOR FRAMEWORK		PREDICTION FRAMEWORK
Sense	Acquire	Analyze	Analyze

Sense
Physical Sensors
Soft Variables
Microprocessors

S

Acquire
Data
• Capture
• Processing
• Storage
• Management
• Communication

DA   DM

Analyze
Diagnostics
• Detection
• Isolation
• Identification

SD

Analyze
Assessment
• Health
• State

Prognostics
• Health
• Remaining Life
• Horizon (End of Life)

HA   PA

Figure 6.1 Core prognostic frameworks in a PHM system. Source: after IEEE (2017).

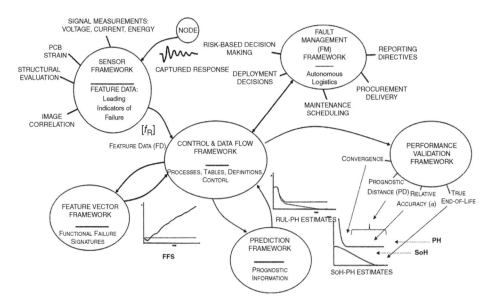

Figure 6.2 A framework for CBM-based PHM. Source: after CAVE3 (2015).

6.1.2 Chapter 2: Prognostic Approaches for Prognosis and Health Management

Chapter 2 described classical approaches to PHM: model-based, data-driven, and hybrid (see Figure 6.5). You learned that a model-based approach includes, for example, deductive and inductive analytical modeling, distribution modeling, physics of failure and reliability modeling, and other aspects – all of which lead to complexity. Two major data-driven approaches were presented – statistical and machine learning – along with a hybrid approach that used both and was the most complex. You learned the primary

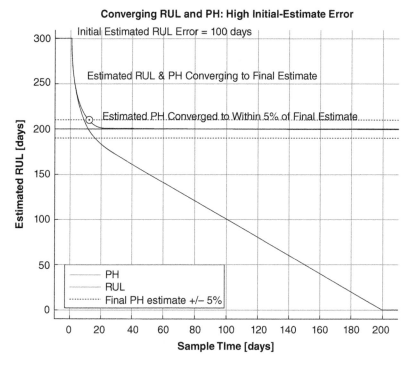

Figure 6.3 Random walk with Kalman-like filtering solution for a high-value initial-estimate error.

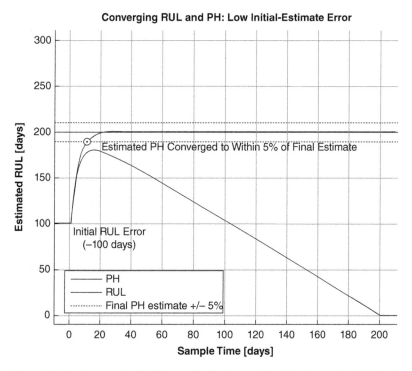

Figure 6.4 Random walk with Kalman-like filtering solution for a low-value initial-estimate error.

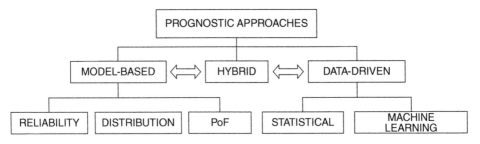

Figure 6.5 Block diagram showing three approaches to PHM.

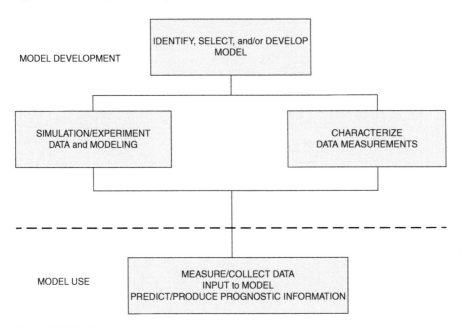

Figure 6.6 Model development and use.

disadvantages of those classical approaches: they are not applicable to a specific prognostic target in a system; and/or they are nondeterministic and not suitable for applying to prognostic targets; and/or it is complex to adapt them to sensor data.

You also learned that modeling is a two-phase process, model development and model use, and that leading indicators of failure can be extracted from sensor data and collected to form condition-based data (CBD) signatures. Figure 6.6 illustrates an approach for modeling development and use for CBD, and Figure 6.7 compares the approaches.

A heuristic-based approach for CBD was introduced: identify characteristic curves of failure modes, and then develop and use signature-based models. That approach is diagramed in Figure 6.8, and differences in focus of model-based and heuristic-based approaches to PHM are listed in Table 6.1. Keep in mind that although there are differences in focus, the approach often employs analysis and modeling techniques such as reliability modeling, physics of failure (PoF) analysis, and failure mode and effect analysis (FMEA) (Hofmeister et al. 2013, 2016, 2017, 2018a,b; Medjaher and Zerhouni 2013; Pecht 2008).

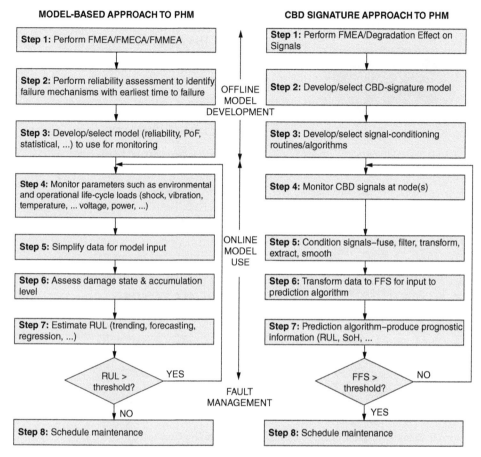

MODEL-BASED APPROACH TO PHM

Step 1: Perform FMEA/FMECA/FMMEA

Step 2: Perform reliability assessment to identify failure mechanisms with earliest time to failure

Step 3: Develop/select model (reliability, PoF, statistical, ...) to use for monitoring

Step 4: Monitor parameters such as environmental and operational life-cycle loads (shock, vibration, temperature, ... voltage, power, ...)

Step 5: Simplify data for model input

Step 6: Assess damage state & accumulation level

Step 7: Estimate RUL (trending, forecasting, regression, ...)

RUL > threshold? YES

NO

Step 8: Schedule maintenance

CBD SIGNATURE APPROACH TO PHM

Step 1: Perform FMEA/Degradation Effect on Signals

Step 2: Develop/select CBD-signature model

Step 3: Develop/select signal-conditioning routines/algorithms

Step 4: Monitor CBD signals at node(s)

Step 5: Condition signals–fuse, filter, transform, extract, smooth

Step 6: Transform data to FFS for input to prediction algorithm

Step 7: Prediction algorithm–produce prognostic information (RUL, SoH, ...

FFS > threshold? NO

YES

Step 8: Schedule maintenance

OFFLINE MODEL DEVELOPMENT

ONLINE MODEL USE

FAULT MANAGEMENT

Figure 6.7 Diagram: model-based and CBD-signature approaches to PHM.

6.1.3 Chapter 3: Failure Progression Signatures

Chapter 3 provided a more in-depth look at CBD signatures and the desirability of transforming those signatures (CBD to fault-to-failure progression [FFP] signature to degradation progression signature [DPS]) into functional failure signature (FFS) input to prediction algorithms that produce prognostic information. A procedural diagram is shown in Figure 6.9.

The chapter described and showed how those signatures are used for reliable condition-based monitoring (RCM): detection of the onset of degradation; the increasing progression of damage due to degradation; prognostic estimates for when damage is likely to reach a level defined as functional failure; and detection of functional failure. Especially significant is the linearization of curvilinear data that is transformed to DPS data and DPS-based FFS; see Figure 6.10.

Chapter 3 described how to assess the linearity of FFS data that is input to a set of prediction algorithms. A set of four equations are used to calculate FFS nonlinearity (FNL). For example, given an FFS produced using FFP and one produced using DPS data (Figure 6.11) the FNL (Figure 6.12) shows the DPS-based FFS is less nonlinear. Extensive

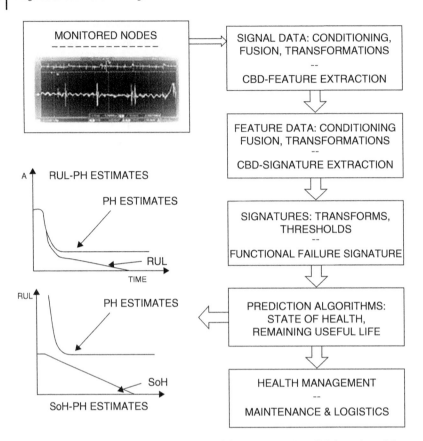

Figure 6.8 Example diagram: heuristic-based CBM system using CBD-based modeling.

Table 6.1 Differences in focus of model-based and heuristic-based approaches to PHM.

Step	Model-based focus	Heuristic-based focus
1	Identify failure modes, effects analysis	Identify failure modes, effects analysis; identify nodes and signatures, precursors to failure
2	Identify failure modes, earliest TTF	Characterize basic curves (signatures) related to failure
3	Develop models to use for failure prediction	Select and/or develop models to transform signatures: CBD to FFP to DPS to FFS
4	Monitor environmental, usage, and operational loads Model the inputs for prediction	Monitor signals, environmental, and operational loads as required to condition the data
5	Simplify and condition data for input to prediction models	Condition and transform signal data, using environmental and operational loads to condition the data (instead of modeling inputs to prediction models)
6	Assess state and level of accumulated damage	Use FFS data to detect damage and as input data to prediction algorithms rather than prediction models
7	Produce prognostic information	Same
8	Perform fault management	Same

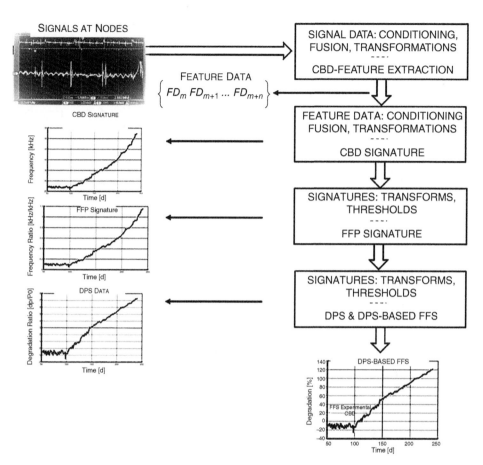

Figure 6.9 Procedural diagram for producing a DPS-based FFS.

experience with many different types of curves leads to the conclusion that the smaller the value of FNL is, the more accurate the prediction information produced is when the FFS is processed.

6.1.4 Chapter 4: Heuristic-Based Approach to Modeling CBD Signatures

Chapter 4 presented a set of seven signature models that resulted in FFP signatures having ideal, characteristic curves, such as those plotted on the left in Figure 6.13. Those ideal signatures were transformed into ideal DPS transfer curves of straight lines starting at an origin of 0 amplitude and passing through another data point having an amplitude of 1, such as those plotted on the right in Figure 6.13.

6.1.5 Chapter 5: Non-Ideal Data: Effects and Conditioning

Chapter 5 introduced two diagrams showing an offline and an online phase for developing a prognostic solution in a PHM system (Figures 6.14 and 6.15, Medjaher and

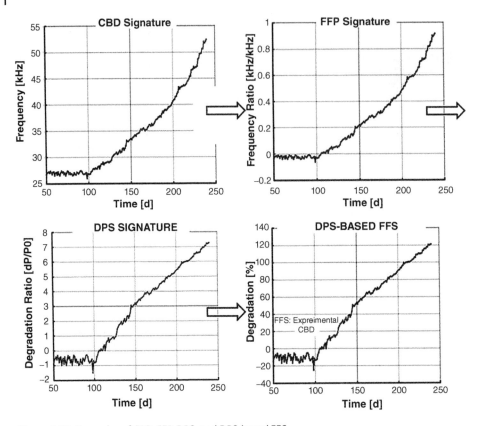

Figure 6.10 Examples of CBD, FFP, DPS, and DPS-based FFS.

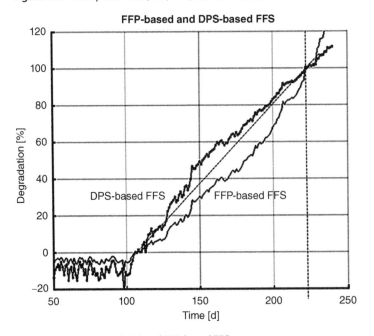

Figure 6.11 DPS-based FFS and FFP-based FFS.

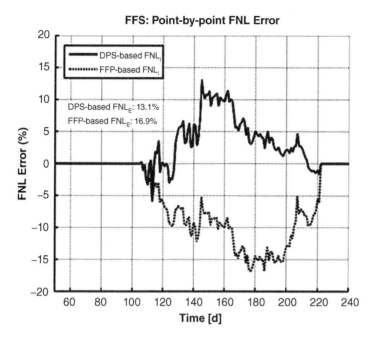

Figure 6.12 FNL plots for the FFS shown in Figure 6.11.

Zerhouni 2013; IEEE 2017). Those diagrams illustrate significant steps in the design and development of a prognostic solution for a node in a system.

The focus of that chapter was causes and effects that reduce the quality of signatures: sources of errors, significant effects of those errors on signatures, and methodologies and techniques that, when employed, ameliorate and/or mitigate errors by removing, reducing, and/or by avoiding such causes and effects. A heuristic-based approach accomplishes the following: (i) illustrates nonlinearity errors related to noise; (ii) identifies, lists, explains, and demonstrates some of the more common causes and effects leading to nonlinearity errors; and (iii) presents methodologies that ameliorate and/or mitigate non-ideality and/or the effects of non-ideality.

An example of noisy data is the multiple-input set of four CBD signatures on the bottom plots of Figure 6.16; one method of noise conditioning and mitigation is to calculate a mean-based differential distance for each noisy signal, as shown by the plots on the right side of that figure. From the set of four conditioned signatures, a single signature (Figure 6.17) can be isolated; further conditioned; transformed into an FFP signature, a DPS, and an FFS; and used as input to a prediction framework.

The mean-differential distance method

1. Calculates the difference between each signal and the mean of the other signals, using data fusion to mitigate common-mode noise.
2. Compares each result and chooses the highest value. Using a second data fusion creates a mean-based differential distance of multiple input signals to extract feature data (FD: change caused by degradation) to create a single signature related to degradation.

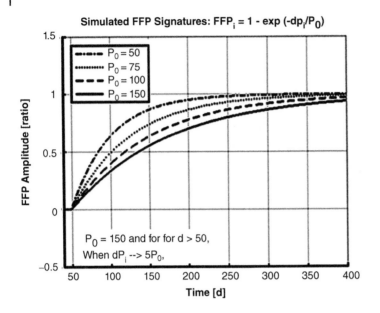

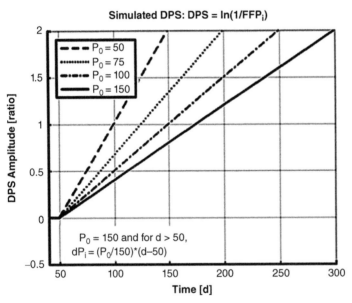

Figure 6.13 Plots of a family of FFP signatures and DPS transfer curves.

This approach is an example of an important goal of conditioning and transforming CBD: improving the linearity and accuracy of signature data used as input to the prediction system of a PHM system and, in doing so, improving the accuracy of prognostic information used to provide a prognosis of the health of the system being monitored and managed.

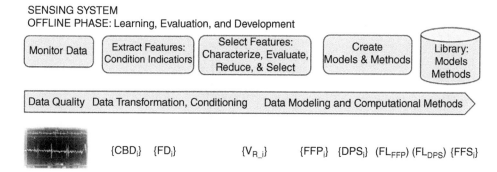

Figure 6.14 Offline phase to develop a prognostic-enabling solution for a PHM system.

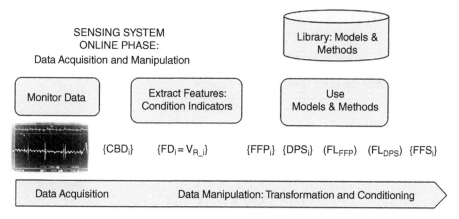

Figure 6.15 Online phase to exploit a prognostic-enabling solution.

6.1.6 Chapter Objectives

The methods presented in the previous chapters provide a solid foundation for designing and developing individual prognostic solutions. This chapter illustrates important aspects and considerations in the design and development of a robust prototype of an exemplary PHM system; this information is presented as though you are the lead designer of such a PHM system.

Although single-node PHM solutions are useful in investigating, characterizing, developing, and demonstrating single- or limited-sensing system solutions of simple, limited capability, they are impractical in a real, complex system with multiple nodes, different types of sensors, single and multivariate signal processing, and varying sample rates at different sampling frequencies. And for prognostics, system solutions need a robust architecture of frameworks, control and data flow, and memory, to do the following: (i) sufficiently remember the past – what has previously transpired at each node; (ii) sense the present – collect measurement data from each node; (iii) process the present and combine it with the past to project the future – produce and act on prediction information for each node. A very important consideration is that a PHM system is not going to run forever: it will be intentionally and unintentionally paused

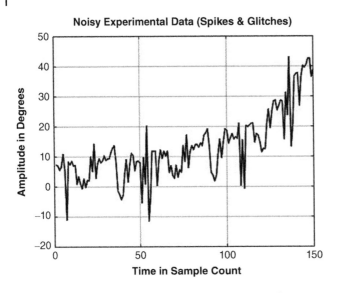

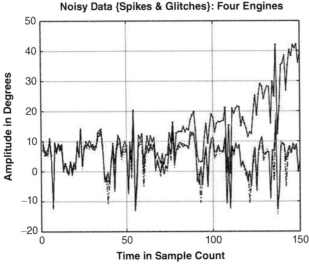

Figure 6.16 Multiple temperature signals: before and after differential-distance conditioning.

and resumed, gracefully ended (shut down) and restarted; and not-so-gracefully ended (crashed) and started over. A PHM system is required to support a sufficiently robust checkpoint (save the present for use as the past) and restart (restore the past and resume operation) that meets customer requirements regarding recovery and continuation of prognostic monitoring and prediction accuracy.

Referring back to Figures 6.1 and 6.2, this chapter will be based on an exemplary PHM system[1] that is directed to you in a role of a designer of a PHM system that

---

1 The design of a prototype, exemplary PHM system is patterned after one developed by Ridgetop Group, Inc., Tucson, AZ, as a developer kit. That PHM system was used to process data and produce example illustrations and results in this book – especially in this chapter.

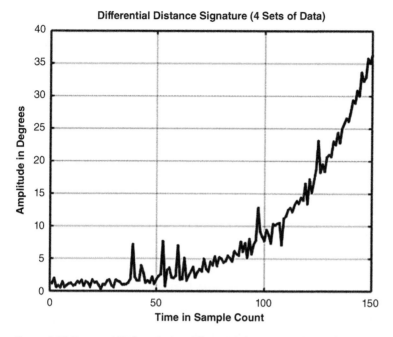

Figure 6.17 Extracted FD from fusing differential-distance conditioned CBD.

incorporates the following:

- A sensing framework comprising multiple nodes.
- A feature-vector framework comprising multiple program modules.
- A prediction framework comprising two program modules to provide prediction information.
- A performance-validation framework, offline mode.
- A control- and data-flow framework that supports the following: (i) ring sensing of nodes, (ii) different sampling rates and frequencies, (iii) alerts, and (iv) a checkpoint/restart.

6.1.7 Chapter Organization

The remainder of this chapter is organized to present and discuss topics related to an exemplary PHM system using frameworks, control and data flow, and a checkpoint/restart:

6.2 Design Approaches for a PHM System

This section describes major approaches to designing a PHM system: select and evaluate candidate targets and their failure modes for prognostic enabling; select models and computational routines from an offline phase; and select an architecture for online use.

6.3 Sampling and Polling

This section describes practical considerations and aspects related to continual sampling, periodic sampling, periodic-burst sampling, and polling. Those considerations are important in the design and development of a PHM system.

6.4 Initial Design Specifications

This section describes important considerations in the initial design for a PHM system: operation in test/demonstration mode versus actual; the design and use of a test bed; and the results of using a test bed in design and development.

6.5 Special RMS Method for AC Phase Currents

This section describes the need for a special method of calculating root mean square (RMS) values of phase currents to support prognostic enabling of the power-switching transistors in the controller of an electro-mechanical actuator (EMA).

6.6 Diagnostic and Prognostic Procedure

This section describes the diagnostic and prognostic procedure for the power supply and for the EMA.

6.7 Specifications: Robustness and Capability

This section describes the node-based architecture and the node-based definition files used to define and control the functional operation of the PHM system.

6.8 Node Specifications

This section provides details on specifying system node definitions and node definitions to manage processing and flow of control and data in a PHM system.

6.9 System Verification and Performance Metrics

This section provides details on an important aspect of verifying a system – the accuracy of prognostic information – by calculating and evaluating performance metrics. Those metrics include degradation detection, prognostic distance (PD), convergence of prognostic horizon (PH) estimates, RUL, PH, and SoH estimates.

6.10 System Verification: Advanced Prognostics

This section describes how using an advanced version of a prediction program, ARULEAV, results in improved performance metrics compared to a basic prediction program, DXARULE.

6.11 PHM System Verification: EMA Faults

This section describes the results of using the advanced program, ARULEAV, to process signatures related to three different types of EMA faults.

6.12 PHM System Verification: Functional Integration

This section describes the results of concurrent processing of multiple nodes.

6.13 Summary: A Robust Prototype PHM System

This section summarizes the material presented in this chapter.

6.2 Design Approaches for a PHM System

Examples of major considerations you face when designing a PHM system include the following:

- What failure modes are candidate targets for prognostic enabling?
- What are the constituent subsystems of those failure modes that are to be prognostic enabled?
- What are the assemblies and subassemblies, often referred to as line-replaceable units (LRUs), within each subsystem to be prognostic enabled? Such LRUs are often considered a logical node.

- Are the target failure modes due to degradation in a shop-replaceable unit (SRU), such as a circuit-card assembly (CCA) within an air data unit or a gear within a transmission?
- Can an existing sensor or sensors be used to detect degradation related to the failure mode? If so, good; otherwise, will one or more sensors be added to support prognostic enablement? If so, good; otherwise, that particular failure mode needs to be dropped from the candidate list.
- What is the cost to design, develop, test, verify, field, and maintain each candidate target for prognostic enablement? Costs include, but are not limited to, monetary, calendar time to field a solution, weight and power required for a solution, required personnel skills, and costs to sustain and service a fielded solution.

6.2.1 Selecting and Evaluating Targets and Their Failure Modes

This chapter assumes that your team and the customer have created and winnowed a list of candidate targets to the actuator subsystem shown in Figure 6.18, which will be used as a prognostic target for an exemplary PHM system. The actuator subsystem comprises the following major blocks: a motor drive (1) with H bridge output (2) and a brushless DC motor (3) (BLDC). The EMA drives a positional wing surface (4), is powered by a switch mode power supply (SMPS) (5), and is controlled by positioning commands (6).

Your prognostic-enabling team is to prognostic enable four major failure modes: (i) tantalum oxide capacitors in the output filter of the SMPS are prone to failing short and burning open; (ii) the power-switching metallic oxide semiconductor field-effect transistors (MOSFET) in the H bridge are prone to becoming damaged and exhibiting

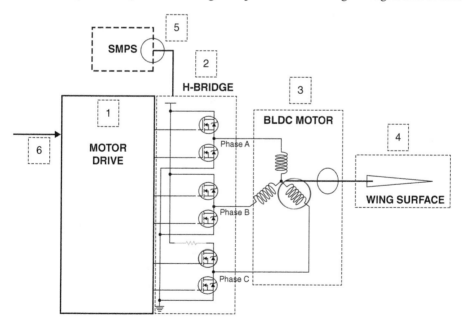

ELECTROMECHANICAL ACTUATOR

Figure 6.18 Block diagram of an example EMA subsystem.

high-value resistance when switched on; (iii) the coil windings of the EMA motor are prone to damage with resulting loss of motor power; and (iv) the load on the shaft of the EMA motor increases. The goal is to design and develop a PHM system that collects data from all the sensors; processes that data, one data point at time; and makes a prognosis of the prognostic-enabled system for each data point. The resultant PHM system is not simply a demonstration in which each node progresses from a state of no degradation to a state of failure before any prognosis is performed; rather, each LRU is to be processed as a complete system in which there is no a priori knowledge of which LRU will degrade or when, which failure mode is the source of degradation, and when a failure mode is likely to result in functional failure.

6.2.2   Offline Prognostic Approaches: Selecting Results

As explained in previous chapters, one or more offline approaches need to be employed to complete the design and development of the models and computational routines required to support prognostic enabling. A first step is to evaluate the sensors and sensor data that will be available. Referring back to Figure 6.18, there are three sets of sensor data you can use:

- Positioning commands (6).
- Output voltage of the SMPS (5).
- Phase-current measurements from the H bridge (2): phase A, phase B, and phase C.

Your team decides to use a CBD approach for the offline phase (Figure 6.19) in the design and development of the prototype PHM system: to use historical, experimental, test, and simulation data and FMEA to create signature models; and to design and develop a set of computational routines to process collected sensor data (Figure 6.20). Those computational routines are to condition and transform sensor data into a set of FFS data as input to prediction algorithms in a program module.

6.2.3   Selecting a Base Architecture for the Online Phase

The operation of the prototype PHM system is the online phase for prognostics is the operation where your models are used to process sensor data to produce prognostic information. Evaluation of the offline phase results for the LRUs, sensors, sensor data,

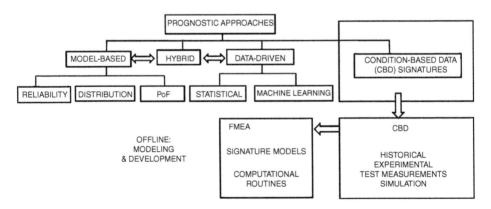

Figure 6.19 Offline modeling and development diagram.

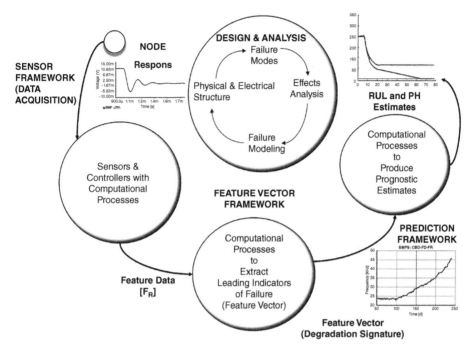

Figure 6.20 Design and analysis diagram.

and models leads you and your team to propose using a framework architecture based on that shown in Figure 6.19; and the customer approves. Now that you have an approved architecture, you need to prepare, review, and obtain approval for a set of initial design specifications.

6.3 Sampling and Polling

Sampling is important consideration in the design of a PHM system: you need to (i) evaluate, select, and specify continual versus periodic sampling (Figure 6.21); and (ii) determine sampling rates, sampling frequencies, and sampling duration (Figure 6.22). You need specifications for each node from which you intend to collect data.

6.3.1 Continual – Periodic Sampling

As a general rule, you should use continual sampling (see Figure 6.21, top) to detect unexpected faults that transition rapidly from a state of zero damage to a state of

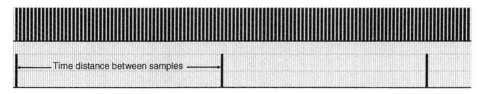

Figure 6.21 Continual (top) and periodic (bottom) sampling.

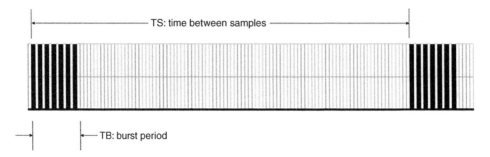

Figure 6.22 Period-burst sampling: sampling period (TS) and burst period (TB).

failure – fast enough that detecting damage and estimating a future time of failure is, practically speaking, not viable. Sampling is performed to support diagnostics by detecting damage as soon as possible, with the intent to mitigate further damage to the system or a catastrophe due to failure to detect the fault. Use periodic sampling (see Figure 6.21, bottom) when the expected distance in time between the onset of damage and functional failure is sufficiently long: (i) there is no urgency to detect a fault within the distance in time between samples, and (ii) your customer wishes to defer maintenance to a future point in time – provided the customer is sufficiently confident in the accuracy of the prediction information from the PHM system.

6.3.2 Periodic-Burst Sampling

You should be aware that only using periodic sampling usually results in insufficient accuracy. It is highly likely that to support prognostics, you should use periodic-burst sampling (Figure 6.22). To do so, you need to (i) select the sampling period; (ii) use burst-mode sampling; (iii) select the sampling frequency; and (vi) select a burst period. Those design decisions must be made to meet the accuracy requirements of the prognostic information.

Example 6.1 The SMPS you are to prognostic enable is a DC-DC converter that supplies power to EMAs (Figure 6.23). The power supply has an output of 24 VDC with an output filtering circuit of 4800 μF capacitance comprising 22 parallel-connected tantalum oxide capacitors that are known to fail short and burn open so that each failure of a capacitor reduces the filtering capacitance by 220 μF. The SMPS has a ripple-voltage

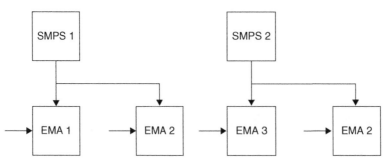

Figure 6.23 Example of power supplies and EMA subsystems.

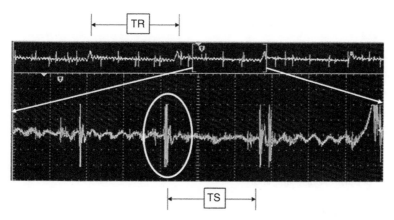

Figure 6.24 SMPS output showing ripple period (TR) and sampling period (TS).

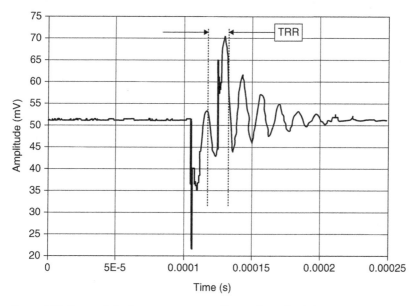

Figure 6.25 Damped-ringing response caused by an abrupt load change.

frequency of 400 Hz (nominal). You conduct an experiment and obtain a captured image of the output as shown in Figure 6.24: the box labeled TR illustrates the ripple-voltage period (2.5 ms for 400 Hz), and the circled outputs are damped-ringing responses caused by abrupt perturbations injected by a sensor into the output load. In this experiment, you are using a sampling rate of 2000 per second (0.5 ms sampling period).

You zoom in, capture, and evaluate one of the damped-ringing responses (Figure 6.25): the resonant frequency is about 70 kHz (the box labeled TRR shows a period of 14.29 μs). Your analysis leads you to decide to program the sensor to sample the ringing response for a 100 μs period – about seven sinusoids (rings). You further decide to program the sensor to calculate the period of the first seven rings, discard the first and last, and calculate and output the average of the middle five rings as the sampled frequency. That

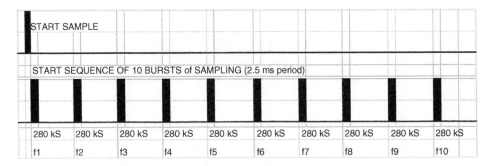

Figure 6.26 Example of burst of burst sampling.

averaging technique mitigates noise caused by, for example, clock jitter, ADC digitization errors, ripple-voltage amplitude variations, and switching and thermal noise.

You realize you are not done yet: you need to determine how many samples the sensor will take during that 100 μs burst period. Your customer stated that they would like overall accuracy of 5% with occasional accuracy of 10% within 72 hours before functional failure. This leads you to settle on a target accuracy of 2.5% when measuring frequency, which means the sensor needs to sample each ring at least 40 times. For seven rings, the sensor needs to sample the output 280 times during the burst-sampling period, which is a sampling frequency of 2.8 MHz – this is much higher than anticipated and, compared to a lower sampling frequency, means higher costs for the sensor, greater power consumption, and greater use of storage and transmission bandwidth.

A solution to the problems associated with a high sampling frequency is to use a lower sampling frequency and increase the number of sampling periods. So, you decide to use a burst-sampling design. Referring to Figure 6.26, you (i) start a sample period; (ii) inject a burst of 10 perturbations and data sampling spaced 2.5 ms apart for a total burst period of 25 ms (a sampling rate of 40 S/S); (iii) during each of the 10 perturbation and data-sampling periods, use a 280 kHz sampling frequency; (iv) calculate and save the average frequency, f1, f2, …f10; and (v) calculate the average of the 10 averages as a single scalar value as the sampled data.

6.3.3 Polling

An important architectural decision is the basic method of polling you will design and develop for the PHM system: (i) interrupt-driven, where sensors issue a signal when they have data to be collected; (ii) call polling, where a controller queries a sensor (ready-to-send data) and waits for a response (data ready) to collect data; or (iii) hub polling, where a controller asks for data (*send data*) from a sensor on the hub and either collects the data (*data sent*) or moves on to another sensor (*no data*). There are also other architectures that might be suitable, including mixed-polling mode.

A significant problem with an interrupt-driven architecture is that the data-sampling cycle is determined by the sensor and is not suitable when different duty periods for data sampling are used, depending on, for example, the criticality of the node to which the sensor is attached and the estimated SoH of that node. For a low-criticality, healthy node, sampling should occur less often than for, say, another node having an identical sensor, higher criticality, and a detected state of degradation.

Your team decides to use call polling (with wait for response) for a number of reasons, including the fact that the sensors in the system are *smart sensors*. Upon being polled, they wake up and do the following: (i) sense the node; (ii) perform a series of samples (burst-mode); (iii) convert analog data to digital data; (iv) compute a scalar value of the average of the burst-mode samples; (v) register that scalar value; (vi) inform the calling PHM system (*data ready*); and then (vii) go into quiescence (*sleep*). The advantages of this type of polling include the following: (i) the PHM system determines when data is to be collected and thereby supports the concept of determining the rate of monitoring as specified by node criticality and SoH; (ii) less sensor power is required; (iii) data traffic is reduced; and (iv) a multiplicity of nodes operating at different rates of sampling is supportable.

6.4   Initial Design Specifications

Before you can design and develop a prototype PHM system, a set of initial design specifications needs to be prepared, reviewed, and approved. Those specifications include your plans to do the following: (i) use a model-use phase, as illustrated in Figure 6.6; (ii) use results of an initial design and analysis phase (Figures 6.19 and 6.20) to prognostic enable a number of EMA subsystems (Figure 6.18); and (iii) prognostic enable a system configured as previously shown in Figure 6.23. The system includes the following: (i) a voltage sensor at the output of each SMPS; (ii) three AC current sensors in each EMA; and (iii) positioning commands at the input of each sensor.

The requirements for the prototype include the following:

1. Prognostic-enable each SMPS for a single mode of failure: output filter, degraded capacitance.
2. Prognostic-enable each EMA for a multiplicity of failure modes:
   a. Power-switch switching transistors, degraded on-resistance.
   b. Motor transformer, degraded winding.
   c. Motor shaft, excessive load.
3. A single system comprising support for all prognostic-enabled units (as opposed to an individual system for each prognostic-enable unit).
4. Checkpoint/restart support for both planned and unplanned unit and system shutdowns.
5. Fault management and services are not required in the prototype.

6.4.1   Operation: Test/Demonstration vs. Real

To test and/or demonstrate your prototype PHM system, it is impractical to use real devices in real time – too much time is required, even if you use injected faults. You should not, for example, change a value every second to simulate degradation; instead, you should to use a larger interval, perhaps five minutes, to allow the test device, component, assembly, and/or subsystem to equilibrate. And you should inject faults that will generate a large number of data points, such as 200 or more. For a five-minute interval, for a single prognostic-enabled mode, your test/demonstration would take more than 15 hours to complete – your customer is not going to wait that long. You need to specify how your team will test and verify the design and operation of a prototype PHM system.

TEST Mode

One approach to solve a time issue related to demonstrating realistic-degradation profiles in a short period of time is to do the following:

1. Acquire, save, and use historical data and/or experimental/test data collected over a long period of time.
2. Read the acquired data from a saved file, one data point at time.
3. For each such data point, translate the acquisition time to a system clock time.
4. Run your prototype PHM system in a TEST mode wherein the PHM system clock is accelerated. For example, each second of real clock time is translated into a number of minutes of sampling time.

Your customer accepts your recommendation that the prototype PHM system shall support two modes of operation: (i) sampling performed using real time and (ii) sampling performed using collected and saved data acquired during an accelerated test.

TEST Data: Saved

When you operate your prototype PHM system in TEST mode, you need to decide how to handle the difference in time between when the data was acquired and when the data is processed during a test and demonstration – especially in a multiple-node, multiple-degradation test setup. For example, the test data for a first node might have been acquired in a relative real clock time of 1.0, 2.0, ... hours, whereas the test data for a second node might have been acquired in a relative real clock time of 20, 40, ... minutes. If you employ a simple ring type of polling without employing a realistic test clock, then after four sets of samples, your prototype PHM system will have processed 8 hours of data in, for example, 4 seconds for the first node, and 80 minutes (1.33 hours) of data for the second node. Information and alerts issued by your prototype PHM system would be nonsensical: for example consider how your customer might react when seeing a third alert following the first two alerts in Figure 6.27. Although you might be able to convince the customer to accept the seemingly nonsensical times as being an artifact of running the prototype PHM system in a TEST mode, we recommend you employ a more robust TEST clock – especially were the customer to point out that the test data was derived and obtained based on original historical data in which degradation was known to be reached in about 50 days, not 2.0 hours.

TEST Data: Real Time

When you operate your prototype PHM system in TEST mode using an actual, fault-injected prognostic target, two considerations are important: (i) the resolution of changes in value of an injected fault compared to an equilibration or settling time; and (ii) mixed-mode testing. You need to always use a real system clock when monitoring actual prognostic targets, and you need to use a simulated clock when processing data with saved collection times.

One method of simulating a clock is to use the data-collection time as the node time, regardless of the actual real clock or a simulated system clock. This means each node

NODE 1 is 25% degraded at time	2.00 hours, esimated RUL is 36 hours
NODE 1 is 50% degraded at time	3.00 hours, esimated RUL is 35 hours
NODE 2 is 25% degraded at time	1.00 hours, esimated RUL is 200 hours

Figure 6.27 Example of alerts issued using an unrealistic PHM system clock and/or polling method.

may have times that are different from any other node, as shown in Figure 6.27, in which the times for NODE 1 are different from the times for NODE 2.

Another method is to create a system TEST clock and translate the data-collection time to that TEST clock. You need to ensure that you and the customer are in full agreement as to the meaning of and specification of the clock used for full-function operation, the clock used to process historical/test data, and the clock used for demonstrations.

Data: Real-Time Acquisition with Batch-Mode Processing

Acquisition of data from sensors in real time, whether during test or during actual operation, requires a design that addresses design points such as the following: (i) Will each data point be processed from data acquisition through prediction processing in a single series of procedural steps? (ii) Will a number of data points be batched and saved for processing at a future time? (iii) In addition to, for example, active filtering and analog-to-digital data conversion, will other data conditioning be performed in real time (as opposed to, for example, near-real-time or batch-mode processing? (iv) Is fusion of multiple-measurement data necessary to create a useful feature that can be extracted? If so, what algorithms and/or computational routines are required? (v) What transforms, such as time-frequency domain or signature-signature, are needed?

It is not uncommon to acquire and collection data in real time – for example, when an aircraft is in flight – and then download the collected data for subsequent processing. When the time between when the data was acquired and when the data is processed is sufficiently long, such processing is said to be *near real time.* How long is sufficiently long? That is a matter for you and your customer to agree on, to avoid issues: for example, even if the time delay was less than one second, your customer might assert that was too long and that you need to fix the system so the delay is less than 1 millisecond. Another consideration is seemingly incongruous differences between node-data times, such as the 3.00 hours at NODE 1 followed by the 1.00 hour at NODE 2 shown in Figure 6.27; the decrease in time is because the data for NODE 1 was collected after the data for NODE 2.

Data: Fault Injected

Experimental data obtained using fault-injection can exhibits steps, such as those shown in Figure 6.28, because either there was insufficient time between changes in injected faults and/or the incremental change in values used for injecting faults was too large. Steps not due to real degradation might result in either or both of the following: (i) an incorrect analysis/result from FMEA; and (ii) an overly aggressive method to mitigate noise. Processing of experimental injection of faults can result in skewed data and times.

6.4.2   Test Bed

Even though you have access to historical data, you need to design and develop a test bed to meet the following objectives: (i) design and run an experiment to verify and demonstrate that you can replicate the failure mode, detect degradation, and prognose a future time of failure. In this section, we describe a suitable, two-part test bed (Ridgetop 2013): one part for the power supply and one part for the EMA. You need to review and approve the design and development of the test bed and how you intend to use test data. For example, you need to specify that test-bed data will be used in model development, model verification, and testing and demonstrating PHM capability. Remember: your

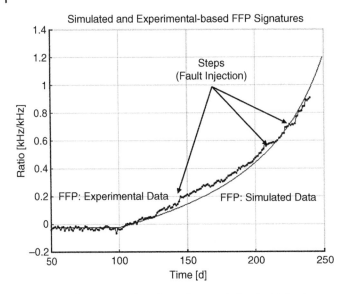

Figure 6.28 Steps in test data due to low-resolution fault injection.

customer needs to accept both the test bed and your intended use of test-bed data as being necessary and sufficient.

Test Bed: Power Supply Section
Figure 6.29 is an example of a test bed your team might design and develop for the power supply |1|. The power supply and the EMA |2| are examples of those shown in Figure 6.18. The diagram illustrates the design methods employed to address acquisition

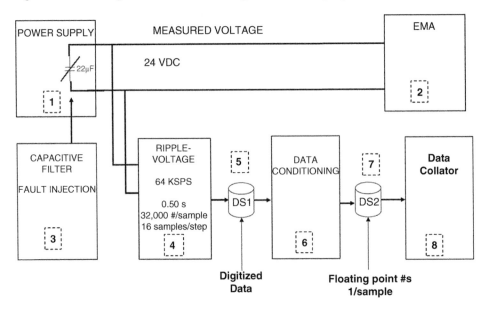

Figure 6.29 Diagram of a test bed to inject faults into a power supply.

of data in real time with final processing of data at a later time. The design solutions are summarized here:

|3| is a programmable method of fault injection. At selectable time intervals, the value of the filtering capacitance of the power supply is reduced.

|4| is a sensor that samples the output of the power supply 16 times during the interval between each fault injection. The sensor samples the output in a burst of 32 000 samples in a half-second interval. The output voltage is filtered to isolate the AC component ripple voltage and noise, which are digitized, transformed to scalar values, and written to an output file |5|.

|6| is a computational routine that reads the 32 000 scalar values, calculates a mean value, and writes a single floating-point number. That number and the time of the sample are written as a data point to an output file |7|.

|8| is a data collator for data acquired from both the power supply and the EMA.

Test Bed: EMA Section

Figure 6.30 is an example of a test bed your team might design and develop for the EMA |2| that is powered by power supply |1|. The diagram illustrates the design methods employed to address acquisition of data in real time with final processing of data at a later time. The design solutions are summarized here:

|9| is a sensor to detect the position of a movable surface driven by the EMA |2|.

|10| is three current sensors to measure the three phase currents (A, B, and C) in the windings of the BLDC motor in the EMA |2|.

Programmable method of fault injection: at selectable time intervals, the value of the filtering capacitance |11| is the controller in the EMA |2|.

|12| is three sets of two power-switching transistors that are used to provide AC power to the BLDC motor.

|13| is a winding (one of three) in the BLDC motor.

|14| is the shaft of the BLDC motor to position a movable surface.

|15| is positioning commands that are input to a test-bed controller |16| connected to the controller |11| for the BLDC motor.

|17| is the currents in the windings of the BLDC motor. Large-amplitude currents occur when the movable surface is being actively positioned (moved).

|18| is a programmable switch used to inject resistance |19| into the drain of any one of the power-switching transistors. This simulates a known PoF mechanism (Celaya et al. 2010). The failure mode causes measurable changes in the current flow in the BLDC windings.

|20| is a window control that uses positioning information to create a sampling window |21| to acquire data (see Figure 6.31) that is digitized and written to an output file |22|. The windowing technique avoids analysis issues caused by the extreme variations in amplitude at the start and end of moving a surface.

|23| is a program module comprising algorithms and computational routines to condition, transform, and extract feature data (FD) to an output file |24|.

Not shown in Figure 6.30 are two other fault-injection methods: one to simulate a damaged motor winding and the other to increase the load placed on the motor.

You design and conduct experiments that use the test bed to inject faults and acquire CBD. You apply methods and techniques from Chapters 1–5 to extract, condition, fuse,

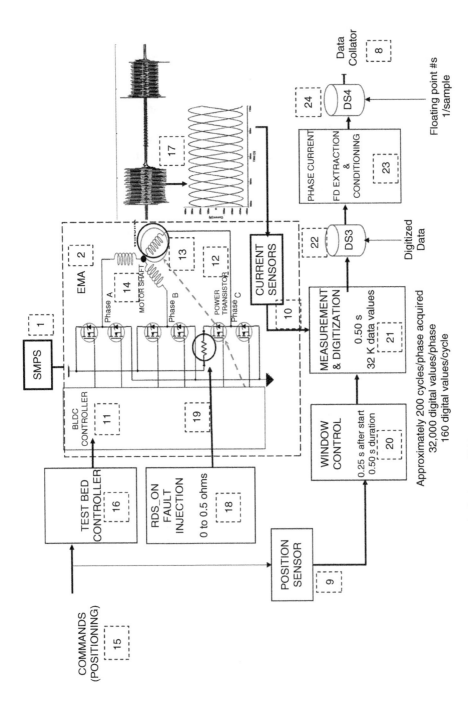

Figure 6.30 Diagram of a test bed to inject faults into an EMA.

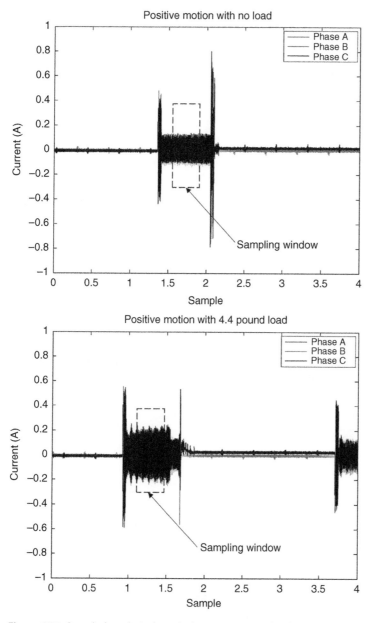

Figure 6.31 Sampled- and windowed-phase currents: no load (top) and extra load (bottom).

transform, and characterize FD as signatures. Your EMA experiments yield very noisy data that hides current changes related to degradation; so, you employ a windowing technique such as that shown in Figure 6.31. The right side of the right-hand plot exhibits little, if any, increase – this and other experiments confirm that for prognostic purposes, sampling should be done when the EMA is moving a load upward rather than downward.

6.4.3 Test Bed: Results

You use the test bed during the initial design and test phase for the following failure modes by Pareto ranking: (i) power supply, loss of filtering capacitance; (ii) EMA, switching power transistor; (iii) EMA, excessive loading on the motor shaft; and (iv) EMA, motor winding.

Power Supply

The fault-injection experiments using your test bed are successful, as evidenced by Figure 6.32.

EMA

The EMA experiments yield mixed results: in the absence of any degradation in an EMA, the measured phase currents are, with the exception of noise, unchanging and equal in magnitude in the middle half of actuator movement. Examination of Figure 6.33, and comparing raising a normally loaded surface (top plot) and an excessively loaded surface (bottom plot), exhibit an increase in phase current.

However, despite inserting hundreds of milliohms of resistance into drains of the power-switching transistors, you are unable to detect any change in the measured phase currents. You simulate the power supply with degraded transistors and confirm that despite a half-ohm degradation, there is very little change in current amplitude (see Figure 6.34): less than 1.5% (9 mA) from a nominal root mean square (rms) value of 650 mA. The latter observation leads you to more closely examine the experimental data – especially data for a high level of degradation.

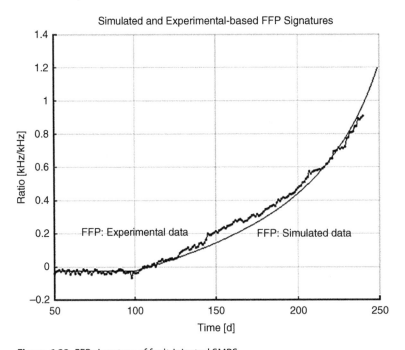

Figure 6.32 FFP signature of fault-injected SMPS.

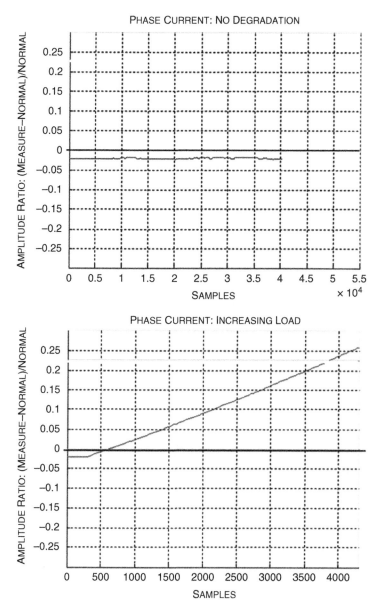

Figure 6.33 Sampled-phase currents: no load (top) and extra load (bottom).

6.5 Special RMS Method for AC Phase Currents

You run an experiment in which you introduce a severe degradation into one of the six power-switching transistors, which results in a level shifting of the reference for the three phase currents (Figure 6.35). You observe the following: (i) a very small change in the measured magnitudes of the phase currents and (ii) significant shifting in the reference levels of the phase currents. PoF analysis and FMEA confirm that because

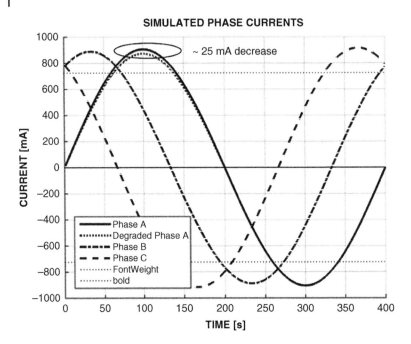

Figure 6.34 Current magnitude: no degradation (high), degraded transistor (reduced amplitude).

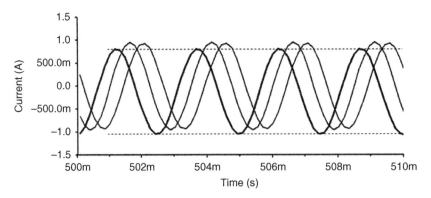

Figure 6.35 Shifted levels due to a degraded power-switching transistor.

the three windings of the motor are tied to a floating reference point, and due to the design of the H bridge (refer back to Figure 6.30), the observed shifting is an expected result.

Examination of the EMA circuitry (Figure 6.30) and PoF analysis confirms that level shifting of the reference (zero crossover) is real rather than an error in experiment or measurement: the level shift is caused by the necessity to maintain a zero-sum current, even when current imbalances are caused by degradation of the power-switching transistors. You solve the issue of low-amplitude measurement values by splitting the calculation of rms into two: a positive rms and a negative rms. You only use current values above or below a selected threshold: the dark samples seen in Figure 6.36.

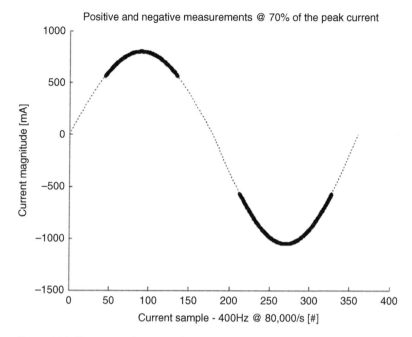

Figure 6.36 Illustration of using peak positive and negative threshold values.

6.5.1 Peak-RMS Method

You investigate and develop a special peak-RMS method: set a threshold level, such as 0.80, of the peak amplitude; and calculate only the rms value of the current samples at or above a high threshold (positive peaks) and at or below a low threshold. Furthermore, you truncate the values of the sampled data as illustrated in Figure 6.37. Only the peak phase current values are used in the calculation, as shown by the curve segments in Figure 6.37.

The special peak-RMS method is particularly advantageous in two respects: (i) a tremendous reduction in data transmitted (wirelessly) and (ii) emphasis of current differences. Sampling occurs 0.25 seconds after an EMA is activated to move upward, and sampling is ended 0.50 seconds later. Four sets of 32 000 digital samples are acquired: one for the power supply and one for each of three phase currents. Each analog sample is converted by its sensor into 12 bits of digital data: over 1.5 million bits of data. Sensor firmware reduces that to 128 000 scalar values and then into 10 scalar values: one for the power supply measurement; three values of rms (one for phase current); and six values of special peak-RMS calculations (two values – positive and negative – for each phase current).

6.5.2 Special Peak-RMS Method: Base Computational Routine

You design and develop a computational routine that is simple and straightforward:

Let $m(s)$ = measured amplitude of the phase current.
Let P_S = maximum positive amplitude of the phase current in a burst of 32 000 samples.

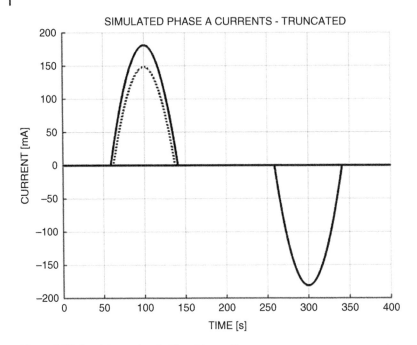

Figure 6.37 Special rms: threshold and truncation.

Let N_S = minimum negative amplitude of the phase current in a burst of 32 000 samples. Then

$$p(s) = m(s) \text{ when } m(s) \geq 0.80 \, P_S, \text{ else } p(s) = 0$$

$$n(s) = m(s) \text{ when } m(s) \leq 0.80 \, N_S, \text{ else } n(s) = 0$$

$$P_{RMS} = (1/P) \sum_{s=1}^{s=n} p(s) \tag{6.1}$$

$$N_{RMS} = (1/N) \sum_{s=1}^{s=n} n(s) \tag{6.2}$$

6.5.3 Special Peak-RMS Method: FFP Computational Routine

Let 120 = *nominal value* for phase current and AT = 0.80. Then for each of the three phase currents, the sample value for an FFP is calculated using Eq. (6.3) for each data point. The results are plotted in Figure 6.38:

$$P_{PK} = (P_{RMS} - (AT \, P_{RMS})) \text{ positive feature data} \tag{6.3}$$

$$N_{PK} = (N_{RMS} + (AT \, N_{RMS})) \text{ negative feature data} \tag{6.4}$$

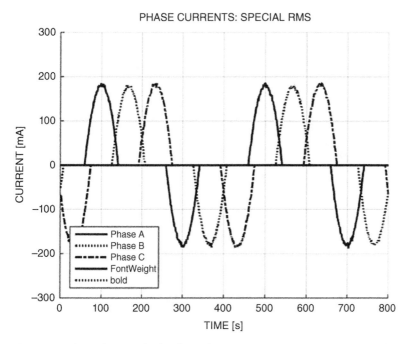

Figure 6.38 Special rms applied to three phase currents.

6.5.4 Peak-RMS Method: EMA

You design and run experiments with and without fault injection for the following failure modes:

- Excessive loading on the motor resulting from binding, friction, or excess weight on whatever surface the EMA is moving.
- Degrading motor winding resulting from, for example, excessive heat, leading to damaged winding insulation with resulting shorting of motor windings.
- Degrading power-switching MOSFETs in the H bridge controller for the EMA, with resulting decreased voltage and power delivered to the motor.

Excessive Loading: FFP Signature Computational Routine
You obtain data with and without excessive loading, apply Eq. (6.1)–(6.4), and then fuse and transform the feature data using Eq. (6.5) for each of the three phase currents to obtain and plot FFP data points as shown in Figure 6.39:

$$FFP = (P_{PK} - N_{PK} + 240)/120 \text{ positive feature data} \tag{6.5}$$

Excessive Loading: FFS Computational Routine
You analyze the results of your FFP computational routine and conclude the following:

1. Although there is a slight curvature, the signature is sufficiently linear to meet requirements. FFP-to-DPS transform will not be performed.

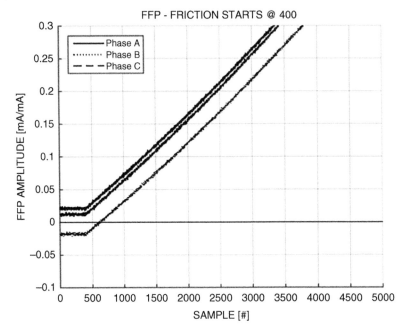

Figure 6.39 FFP signatures due to loading.

2. You will need to apply a noise margin (NM) of about 0.025, because in the absence of degradation, an FFP signature is above the zero threshold.
3. You should apply noise mitigation in the form of data smoothing. You decide to use point averaging.
4. You decide to use a failure threshold of 0.25 – approximately a 25% increase in current due to extra loading due to weight and/or friction.

You apply the NM (Eq. (6.6)); you use data smoothing to create the three FFP signatures shown in Figure 6.40. Next, you select the highest amplitude FFP; you select a failure threshold; and you use Eq. (6.7) to transform that FFP to an FFS (Figure 6.41).

$$FFP = (FFP - 0.025)\ FFP\ \text{with applied NM} \tag{6.6}$$

$$FFS = 100\ (FFP/0.25)\ \text{Obtain FFS for 0.25 failure threshold} \tag{6.7}$$

Motor Winding: FFP and FFS Signatures
You obtain data with and without injected shorted-winding faults, and you apply the same methods you used for the loading experiments. The FFP signatures are shown in Figure 6.42. You observe the following:

1. Only one FFP signature – that for Phase C – is increasing, instead of all three.
2. You can use the same FFS computational routine as for the EMA loading case (Figure 6.43).
3. The different behavior of the signatures: one signature changes versus three signatures increasing, for diagnostic purposes (fault isolation).
4. There is an unavoidable offset error (see Figure 6.43) due to amplitude differences in the phase currents in the absence of degradation.

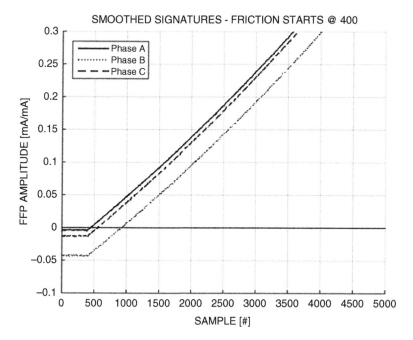

Figure 6.40 Smoothed FFP signatures due to loading.

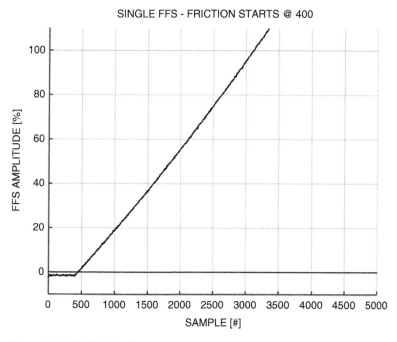

Figure 6.41 FFS: EMA loading.

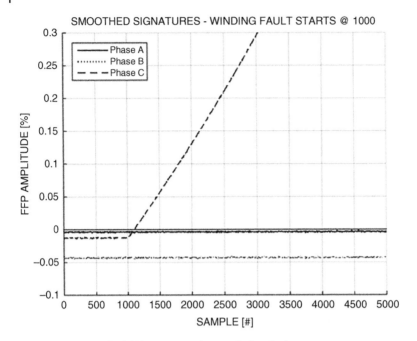

Figure 6.42 Smoothed FFP signatures due to winding faults.

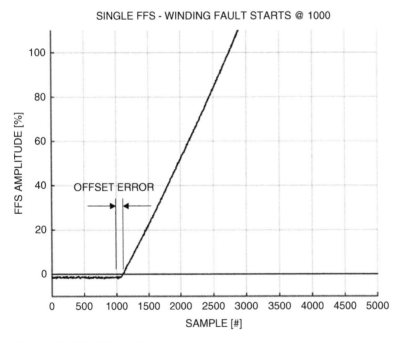

Figure 6.43 FFS: EMA winding.

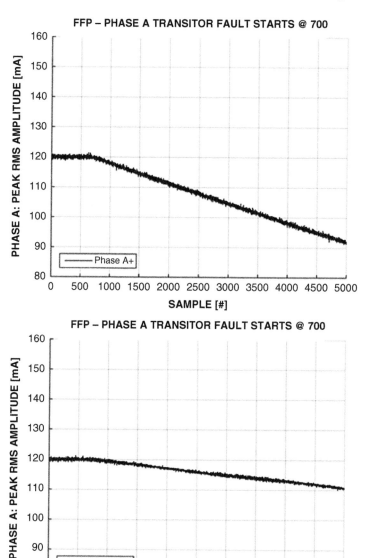

Figure 6.44 Phase A currents: transistor fault in the positive Phase A portion.

H-Bridge Controller: FFP and FFS Signatures

You obtain data with and without injected on-resistance faults in an H bridge controller. To do that, you cause the on-resistance of the power-switching transistor to increase over time in the positive half of the Phase A section of the H bridge. You plot the special peak RMS values for Phase A, Phase B, and Phase C as shown in Figures 6.44–6.46.

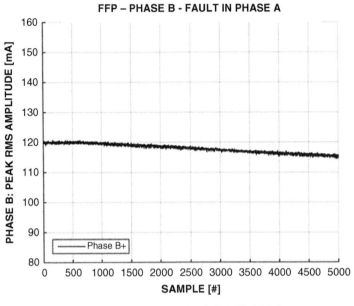

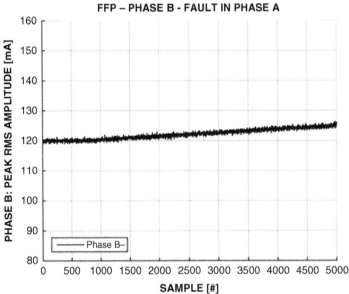

Figure 6.45 Phase B currents: transistor fault in the positive Phase A portion.

The results are not what you expected: rather than a decrease only in the positive half of the Phase A current, there is a change in both the positive and negative halves of all three phase currents; also see Figure 6.47. Before proceeding, you need to understand what is happening. Examination of the motor circuit (Figure 6.48) leads you to conclude that the common node for the three motor windings is floating, which means a change in amplitude of one phase current will change the reference level for all three phase currents.

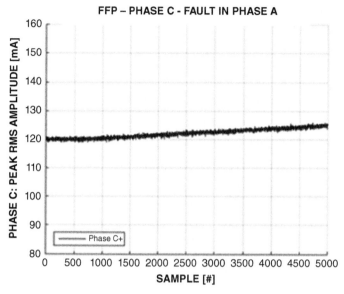

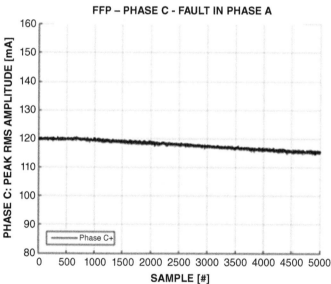

Figure 6.46 Phase C currents: transistor fault in the positive Phase A portion.

You then conclude that you can use the results as follows:

1. Sum the absolute value of each half of all three phase currents.
2. Use the highest-valued sum as an FFP data point.
3. The location of the fault is in the section of the H bridge corresponding to the largest decrease in magnitude. In this case, that is the positive half of the H bridge for the Phase A current (refer back to the top plot of Figure 6.44).
4. When you do, the resulting FFP is shown in Figure 6.49.

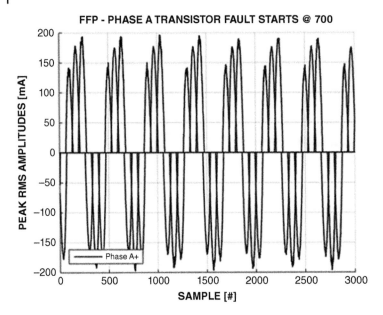

Figure 6.47 Peak-RMS currents: transistor fault in the positive Phase A portion.

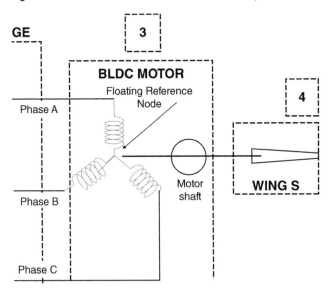

Figure 6.48 Close-up of the EMA motor winding.

After examining Figure 6.49 and reviewing your experiment and experimental results, you conclude the following:

1. You have a prognostic solution to diagnose and prognose a failing power-switching transistor in an H bridge driver in an EMA.
2. You use the following to convert feature data to FFP signature data:

$$FFP = (ABS(P_{PK}) + ABS(N_{PK}) + 240)/120 \text{ sum absolute peak values} \quad (6.8)$$

where 120 = nominal current value in the absence of degradation.

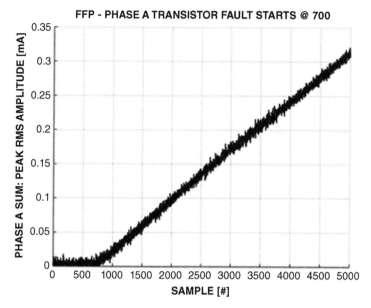

Figure 6.49 FFP: H bridge fault, sum of both halves of the Phase A current.

3. You use a four-point averaging method to smooth the FFP signature data (see Figure 6.50).
4. You can use Eqs. (6.6) and (6.7) to transform FFP signature data to FFS data (see Figure 6.51).

• In Eq. (6.6), change 0.025 to 0.10 for an NM as a further step toward noise mitigation.

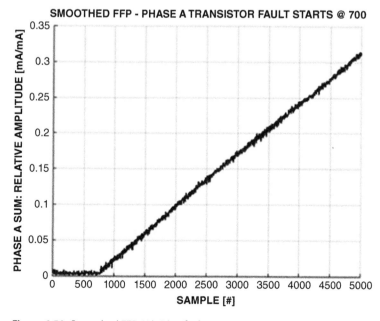

Figure 6.50 Smoothed FFP: H bridge fault.

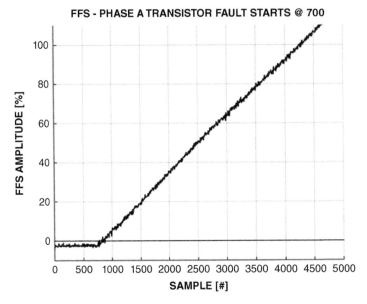

Figure 6.51 Smoothed FFS: H bridge fault.

- In Eq. (6.7), continue to use 0.25 for functional failure (about a 25% reduction in current).
- Apply five-point averaging to the FFS data as a final step toward noise mitigation.

During a design review, a reviewer questions your use of Eq. (6.8) rather than the following (from Chapter 3):

$$FFP_i = (FD_i - FD_0)/FD_0 \tag{3.8}$$

You explain that when the nominal value of a feature, FD_0, is less than 1, and especially when the functional-failure threshold is less than 1, it is more appropriate to use an alternative model:

$$FFP_i = (FD_i - FD_0) \tag{6.9}$$

6.6 Diagnostic and Prognostic Procedure

The specifications include a description of the diagnostic and prognostic procedures for an SMPS power supply and an EMA.

6.6.1 SMPS Power Supply

The diagnostic procedure for the power supply is straightforward: transform sensor data into an FFP signature, then into DPS data, and then into FFS data; and then input the FFS data to a prediction framework. When SoH is less than 100 and greater than 0, the power supply is degraded. When SoH first falls below 100, a diagnosis can be made. Issue alerts as specified in the node definition.

6.6.2 EMA

The sensors in the sensor framework produce three sets of measurements, one for each of the three phase currents – both positive and negative scalar values. Sensor measurement data is processed by computational routines (software and/or firmware) to produce three pairs of peak-RMS values: one pair of positive and negative values for each phase.

Fault-Tree Processing: Detection and Diagnosis
Check the amplitudes of the positive and negative special-peak RMS values for each phase current:

1. When both values for a phase exhibit decreasing amplitudes:
   - Fault mode: increasing on-resistance in the switching power transistors for that phase.
   - Fault isolation: the transistor in the H bridge circuit for the phase half that exhibits the largest decrease in amplitude.
2. When both values for all three phase currents exhibit increasing amplitudes:
   - Fault mode: increasing load on the EMA motor.
   - Fault isolation: either (or both) increasing friction or weight exerted on the EMA motor.
3. When both values for a single (or two) phase currents exhibit increasing amplitudes:
   - Fault mode: reduced impedance, such as that from shorting between turns in a motor winding.
   - Fault isolation: motor winding for the phase exhibiting increased current.
4. Otherwise, EMA status equals *no fault found*.

6.7 Specifications: Robustness and Capability

Your customer approves your initial design specifications and approves further design and development of a prototype PHM system to demonstrate robustness and capability by concurrently monitoring, detecting levels of degradation, and issuing appropriate alerts and prediction information for the system configuration shown in Figure 6.18: (i) a nondegraded power supply, (ii) a degraded power supply, (iii) a nondegraded EMA, (iv) an EMA with a degraded power MOSFET, and (v) an EMA with excessive loading on the motor shaft. Data saved in the output file labeled DS2 – |7| in Figure 6.29 – and the output file labeled DS4 – |24| in Figure 6.30 – provide input data to test, verify, and demonstrate the system.

A set of final specifications for the robust prototype of an exemplary PHM system includes the following: (i) special root mean square (RMS) method for processing phase currents; (ii) prognostic-enabling of the power supplies; and the EMAs for power-switching MOSFETs and motor loading. The final specifications need to support your claim that your team will develop a robust prototype of an exemplary PHM system that is capable of providing the required prognostics.

6.7.1 Node-Based Architecture

Your PHM system needs to define the polling schema for each node, to include scope, definition, and interval. Scope includes, for example, which program(s) is (are) to be used to process the collected data and, of course, requirements related to data conditioning and predictions. Those and other considerations lead you to design an online, node-based architecture such as the annotated diagram in Figure 6.52, which includes your PHM system and others. You are not designing, for example, a system to perform CBM, a system to perform autonomous logistic support, or a system to support and provide fault management.

Your PHM system is to include the following major functions and capabilities:

- System control and data flow.
- A system-node definition, and node definitions for two EMA subsystems, each comprising a power supply and two EMA assemblies.
- A library of data sets: input, output, and checkpoint/restart data sets to support interruptible PHM operations.
- Computational libraries containing program modules and routines you provide to perform data conditioning, fusing, transforming, and anything else required to support the PHM system.
- A library of system services to support and provide, for example, graphical and end-user interfaces, messages and alerts, information display and printing, and event handlers.

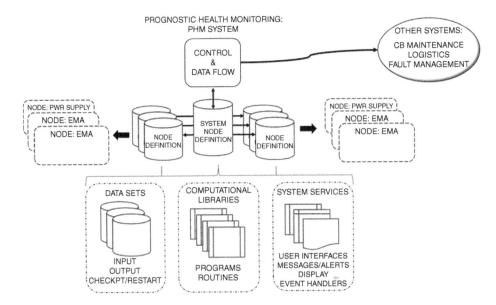

Figure 6.52 Block diagram for an example of a robust PHM system.

Scope of Effort

In general, a robust PHM system includes, but is not limited to, the following:

Sensor hardware, firmware, and drivers
- Primary sample rate, burst-mode (high frequency) sampling at a lower cyclic rate, and so on.
- Frequency-based versus amplitude-based, voltage versus current, and so on.
- Active filtering, analog-to-digital data conversion, and so on.

Vector program modules
- Number of inputs.
- Fusion – data and domain.
- Signature modeling, transform, and smoothing.

Prediction program
- Prognostic distance, accuracy, and convergence.

Services programs
- User interfaces such as command line, graphical input/output, and printing.
- Resource management: editing, selection, and modification.
- Messages, data logging, and reports.
- Alerts.
- Event handling.

Project Scope

For this phase of the project, the scope of effort for you and your team is the following:

- The design of a system node definition (SND).
- The design of application-specific node definitions (NDEF).
- The design of any required input, output, and checkpoint/restart data sets.
- The use of the computational program modules and routines previously designed and developed by your team as needed and desired to verify elements of the design. Design and development details of those modules and routines are not included.
- The use of the base set of service programs and routines previously designed and developed by your team as needed and desired to verify elements of the design. Design and development details of those services, programs, and routines are not included.
- Design information is to be provided as follows:
  - Use an end user interface (EUI, command line). Design of a graphical user interface (GUI) is deferred.
  - Assume there will be a message handler to contain all language-specific text.
  - Assume there will be a supporting platform for polling and sampling.
  - Asynchronous event handling is deferred for this phase of the project.
- Prediction services will include the following:
  - Prognostic information.
  - Prognostic accuracy: evaluation and reporting services.

6.7.2 Example Design

You decide that nodes in the prototype PHM system are logical nodes that may comprise more than one physical node. For example, each EMA has three physical nodes to which

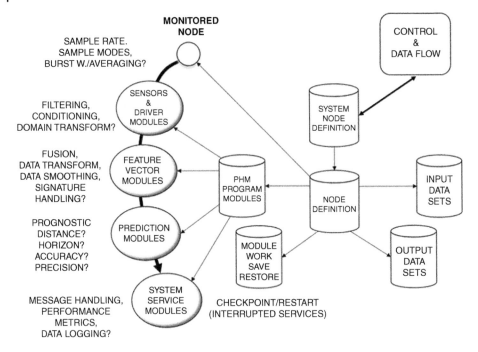

Figure 6.53 Architectural block diagram for a node definition.

an AC measurement sensor is attached. It is more practical to think of a single EMA node comprising three physical sensors and, therefore, three signals. Referring back to Figure 6.52, there are a total of six nodes: two for the power supplies and four for the EMAs. Your architecture for each node definition is shown in Figure 6.53.

An example design for your prototype PHM system includes the following: (i) a system node definition; (ii) specific node definitions; (iii) a set of program modules to be used to collect, condition, and process data, to produce prognostic information, and for system services; (iv) a library containing input data sets; (v) a library containing output data sets; and (vi) a library for saving and restoring operational information, including that for checkpoint/restart support.

Referring to Figure 6.53, these are the major frameworks of a node: (i) the physical nodes to which sensors are attached; (ii) the sensors and, as applicable, their driver modules and/or embedded firmware; (iii) feature-vector modules comprising computational programs and routines that process sensor data to produce input signatures to the prediction framework; (iv) prediction programs/algorithms that process input from a vector framework to produce prognostic information; (v) system services to support informational and alert types of messages, to perform data logging, to analyze and produce performance metrics, and so on; and (vi) the PHM system itself, which provides operational-flow and data-flow control.

Your design uses definition files, as shown in Figure 6.53 (i) they are maintained in clear-text format for ease-of-use and readability; (ii) they are flexible, to accommodate a wide range of support requirements; and (iii) they lend themselves to a table-driven design. There are two types of definition files: (i) system node definition and (ii) node definition.

6.8 Node Specifications

The example design calls for two sets of specifications for nodes: a single system node definitions (SND) and six node definition (NDEF) files.

6.8.1 System Node Definition

An SND is used to define the nodes the PHM system is to monitor. You could, for example, define and use an SND for functional operation; another SND for system testing; another SND for demonstrations; and so on. Figure 6.54 is an example SND to support the subsystem shown in Figure 6.23:

The first two lines identify a checkpoint library (*NCPTLIB*) and a file prefix (*NCPTFNP*):

- Used to check for an already-initialized system.
- Support inadvertent replacement.

The next three lines define the sampling clock:

- *NSMPINT*. The interval between cyclic samples in hours – one-hour interval specified.
- *NSMPRATE*. The cyclic sampling rate – four times in the specified interval.
- *NSMPCLK*. The initial sampling clock.

The next line, *NSMPMODE*, specifies the system mode: TEST, which includes demonstration or real-time:

- *TEST* specifies test/demonstration mode.
- Anything else specifies real time.

There are six nodes (*NDNUMID*) in the system:

- 50, 51, 52, 60, 61, and 62.
- Assigned node numbers are arbitrary and do not have to be in order – they are used for human readability.

Each system node points to a node definition file:

- *NDLIB* specifies the library.
- *NSFNAME* specifies the name of the file in the library.

6.8.2 Node Definition

An SND points to one or more NDEFs: one for each node to be processed. Each NDEF contains several subsections (blocks), as illustrated in Figure 6.55, each of which applies to a particular set of frameworks and/or specifications. You can use multiple

```
%*****************************************
% System Node Definition file
%*****************************************
NCPTLIB    = 'C:\DXPHMLIB\DXCPT\' ;      % Checkpoint-restart library
NCPTFNP    = 'SN_' ;                     % CPT filename prefix
NSMPINT    = '1' ;                       % Sampling interval, # hours
NSMPRATE   = '4' ;                       % Sampling rate, 4 times/NSMPINT
NSMPCLL    = '0' ;                       % Set sampling clock, hour
NSMPMODE   = 'TEST' ;                    % Test/demo, else real time
NSMPTSTP   = '1' ;                       % Test seconds per NSMPINT
%***
NDNUMID    = 50 ;                        % Assigned node number
NDLIB      = 'C:\DXPHMLIB\DXNDEF\' ;     % Node definition library
NDFNAME    = 'DXNDPWR1.txt' ;            % Node definition fname.typ
NDIDNAME   = 'SMPS: 1,EMA' ;             % Name ID
NDNUMID    = 51 ;                        % Assigned node number
NDLIB      = 'C:\DXPHMLIB\DXNDEF\' ;     % Node definition library
NDFNAME    = 'DXNDEMA1.txt' ;            % Node definition fname.typ
NDIDNAME   = 'EMA: 1,LEFT' ;             % Name ID
NDNUMID    = 52 ;                        % Assigned node number
NDLIB      = 'C:\DXPHMLIB\DXNDEF\' ;     % Node definition library
NDFNAME    = 'DXNDEMA2.txt' ;            % Node definition fname.typ
NDIDNAME   = 'EMA: 2,LEFT' ;             % Name ID
NDNUMID    = 60 ;                        % Assigned node number
NDLIB      = 'C:\DXPHMLIB\DXNDEF\' ;     % Node definition library
NDFNAME    = 'DXNDPWR2.txt' ;            % Node definition fname.typ
NDIDNAME   = 'SMPS: 2,EMA' ;             % Name ID
NDNUMID    = 61 ;                        % Assigned node number
NDLIB      = 'C:\DXPHMLIB\DXNDEF\' ;     % Node definition library
NDFNAME    = 'DXNDEMA3.txt' ;            % Node definition fname.typ
NDIDNAME   = 'EMA: 3,RIGHT' ;            % Name ID
NDNUMID    = 62 ;                        % Assigned node number
NDLIB      = 'C:\DXPHMLIB\DXNDEF\' ;     % Node definition library
NDFNAME    = 'DXNDEMA4.txt' ;            % Node definition fname.typ
NDIDNAME   = 'EMA: 4,RIGHT' ;            % Name ID
%***
ENDDEF     = -1 ;                        % end of node definition
```

Figure 6.54 Example of a system node definition.

instantiations of an NDEF in your PHM system: for example, one NDEF for the two power supplies and another NDEF for the four EMAs.

You prepare a set of NDEF specifications for each block: (i) they support the functional and operational requirements for a robust prototype PHM system, and (ii) you fully expect them to be updated in a stepwise fashion as design and development proceed. The first node definition you define is for NODENUMID = 50 (refer back to Figure 6.54).

Node Status

The node status block (Figure 6.56) contains three entries:

- *NDNUMID* is the numeric identifier of the node and contains the value as specified in the SND.
- *NODEINISW* is an initialization switch that is set to 1 when the node is initialized.

```
%*****************************************************
% XMPNDEF Example Node Definition - BLOCK DIAGRAM
%*****************************************************
NODE STATUS
SAMPLING SPECIFICATIONS
ALERT SPECIFICATIONS
SPECIAL FILE SPECIFICATIONS
FEATURE VECTOR FRAMEWORK
PREDICTION FRAMEWORK
PERFORMANCE & SERVICES
INPUT-OUTPUT LIBRARY & FILE SPECIFICATIONS
CHECKPOINT-RESTORE LIBRARISPECIFICATIONS
DEVICE DRIVER ID & UNITS OF MEASURE
OTHER PROGRAM IDENTIFIERS
```

Figure 6.55 Block diagram of an example node definition.

```
%***************************************************************************
% XMPNDEF Example Node Definition
%***************************************************************************
NDNUMID   = 0 ;      % node num  from SND (after initialization)
NODEINISW = 0 ;      % 0 = not initialized, 1 = initialized
NODESTAT  = 0 ;      % 0 = Operational  1 = Non-operational  2 = Suspended
%                      5 = Failed        9 = skip
```

Figure 6.56 NDEF: node status.

- *NODESTAT* is the status of the node – operational, non-operational, suspended, failed, or skip (do not monitor).

Sampling Specifications

The sampling specifications block (Figure 6.57) contains five entries:

- *SRDEF* defines the basic sampling period, which can be one of three that you design for: hourly, daily, or weekly. The specified period is hourly.
- *SRMONITOR* specifies the number of sampling periods between samples the node is to be monitored, in the absence of any detected degradation. In this case, the specification calls for a period of 36 hours between samples.
- *SRWATCH* specifies the number of sampling periods between samples when the node is degraded and SoH is over 50% – in this case, once every 12 hours.
- *SRTRACK* specifies the number of sampling periods between samples when the SoH of the node is between 25% and 50% – in this case, once every hour.
- *SRALERT* specifies the number of sampling periods between samples when the SoH of the node is less than 25% – in this case, every half hour.

Alert Specifications

The alert specifications block (Figure 6.58) has eight entries, four to specify when to issue alert types of messages and four switches for housekeeping:

- *AST1* and *AST1SW* are to issue an alert when degradation is first detected.
- *AST2* and *AST2SW* are to issue an alert when SoH falls below 75%.

```
%**Node monitoring rate - definition only
SRDEF      = 1 ;       % 1 = hourly 2 = daily, 3 = weekly
SRMONITOR = 36 ;       % Rate when SoH = 100         every 36 hours
SRWATCH    = 12 ;      % Rate when 100 > SoH > 50    every 12 hours
SRTRACK    = 1 ;       % Rate when SoH <= 50         every 1 hour
SRALERT    = 0.5 ;     % Rate when SoH <= 25         every 0.5 hour
```

Figure 6.57 NDEF: sampling specifications.

```
%**ALERT SoH Thresholds: when SoH falls below threshold level
%  -n means ignore ALERT threshold
AST1       = 100 ;     % Issue ALERT when SoH < threshold
AST2       = 75 ;      % Issue ALERT
AST3       = 50 ;      %     when
AST4       = 25 ;      %         SoH <= threshold
AST1SW     = 0 ;       % Warning issued when 1
AST2SW     = 0 ;       % else
AST3SW     = 0 ;       %    not
AST4SW     = 0 ;       %       issued
```

Figure 6.58 NDEF: alert specifications.

- *AST3* and *AST3AW* are to issue an alert when SoH falls below 50%.
- *AST4* and *AST3SW* are to issue an alert when SoH falls below 25%

Special Files Specifications
The Special Files specifications block (Figure 6.59) contains seven entries for the specifications for three special files used to log data buses:

- *NDFCLOSE* specifies whether the special files are to be closed after a data point is processed.
- *NDFID* is a file ID for the bus file containing the original data.
- *NDBUSFN* is the file name for the bus file containing the data from the sensor framework.
- *NVFID* is a file ID for the bus file containing the data from the vector framework.
- *NVBUSFN* is the file name for the bus file containing the data from the vector framework.
- *NPFID* is a file ID for the bus file containing the data from the prediction framework.
- *NPBUSFN* is the file name for the bus file containing the data from the prediction framework.

```
%**Special Files
NDFCLOSE  = 0 ;          % close files after data pt - 0 = no, 1 = yes
NDFID     = -1 ;         % file id if open
NDBUSFN   = 'none' ;     % file name after initialization
NVFID     = -1 ;         % file id if open
NVBUSFN   = 'none' ;     % file name after initialization
NPFID     = -1 ;         % file id if open
NPBUSFN   = 'none' ;     % file name after initialization
```

Figure 6.59 NDEF: Special Files specifications.

Feature Vector Framework

The Feature Vector block comprises three sections, – a primary section, a data-smoothing section, and an FFP-to-DPS transformation section:

FD = feature data = FD0*(1 + dP/P0)
FFP = (FD – FD0 – NM)/FD0 where NM = FNM*FD0/100 or
FFP = (FD – FD0 – NM) when FDNOMDIV = 0

The primary section contains six entries to specify transforming CBD to FD:

- *FNM* specifies the NM, a percentage of the nominal FD value (FD0)
- *FAILM* specifies the percent failure margin: 70 means failure when FFP = 0.70
- *FDNOM* specifies the data value considered to be the nominal FD value (FD0)
- *FDNOMV* specifies the percent expected variance in FD0
- *FDNOMMIN* specifies the minimum value to be used for FD0.
  - When measured FD0 is less than FDNOMMIN, use FDNOMIN.
  - Else use measured FD0.
- *FDNOMDIV* specifies whether division is (nonzero value) or is not (zero value) to be performed.

The data smoothing section contains four entries for data smoothing:

- *FD0PTS* specifies the number of CBD points to average, to calculate the value for FD0.
  - When specified as zero (0), FDNOM is used.
  - When variance in calculation exceeds FDNOMV, FDNOM is used.
- *FDPTS* specifies the number of FD points to average for smoothing.
- *FFPPTS* specifies the number of FFP points to average for smoothing.
- *FFSPTS* specifies the number of FFS points to average for smoothing.

The FFP-to-DPS transform section contains two entries for transforming FFP signature data to DPS data:

- *XDPS* specifies the FFP model for selecting the FFP-to-DPS transformation model:
  - 0 means do not transform.
  - 1 means use FFP = $FD0*(dP/P0)^n$.
  - 2 means use FFP = $FD0*([1/(1 - dP/P0)]^n - 1)$.
  - 3 means use FFP = $FD0*(1-[1/(1 + dP/P0)]^n)$.
  - 4 means use FFP = $FD0*((1 + dP/P0)^n - 1)$.
  - 5 means use FFP = $FD0*(1 - (1 - dP/P0)^n)$.
  - 6 means use FFP = $FD0*(\exp(dP/P0) - 1)$.
  - 7 means use FFP = $FD0*(1 - \exp(-dP/P0))$.
- *XDPSNV* specifies the value to use for n or lambda (lifetime).

Prediction Framework

The prediction framework section (Figure 6.61) contains two entries for specifying values used in producing prediction information:

- *PITTFF0* specifies a time to failure (TTF) value after the onset of degradation:
  - The default value is 200 sampling periods.
- *PIFFSMOD* specifies the expected shape of the input data:
  - 1 means a concave characteristic curve.

```
%**Feature Vector              (A)
FNM         = 2.5 ;      % Noise margin - % of FD0
FAILM       = 70 ;       % Failure margin - % above 0
FDNOM       = 0 ;        % Feature Data (FD), Nominal value
FDNOMV      = 3 ;        % FD Nominal variance (%)
FDNOMMIN    = 1 ;        % minimum nominal value: if less, perform DIV
FDNOMDIV    = 0 ;        % 0 = subtract only (no division) - small signal
                         % else FFP = (FD - DIV)/DIV

                              (B)
FD0PTS      = 20 ;       % # data points to average to calculate FD0
FDPTS       = 4 ;        % Rolling FD (CBD) points to average: up to 4
FFPPTS      = 4 ;        % Rolling FFP points to average: up to 4
FFSPTS      = 4 ;        % Rolling FFS points to average: up to 4

%**FFP to DPS Transform        (C)
XDPS            = 0 ;         % none
%               = 1 ;    % FFP = FD0*(dP/P0)^n
%               = 2 ;    % FFP = FD0*([1/(1 - dP/P0)]^n - 1)
%               = 3 ;    % FFP = FD0*(1-[1/(1 + dP/P0)]^n)
%               = 4 ;    % FFP = FD0*((1 + dP/P0)^n - 1)
%               = 5 ;    % FFP = FD0*(1 - (1 - dP/P0)^n)
%               = 6 ;    % FFP = FD0*(exp(dP/P0) - 1)
%               = 7 ;    % FFP = FD0*(1 - exp(-dP/P0))
XDPSNV          = 1 ;    % n or Lambda (life) value
```

Figure 6.60 NDEF: feature-vector framework – (a) primary, (b) smoothing, (c) FFP-DPS transform.

```
%**Prediction (Prognostic) Information
PITTFF0    = 4800 ;    % estimated time (hours) after onset of damage
PIFFSMOD   = 2 ;       % model 1=Concave, 2=Linear, 3=Convex, 4=complex
PIFFSNM    = 3 ;       % noise margin (%) for estimating begin degradation
```

Figure 6.61 NDEF: Prediction Framework.

- – 2 means a linear curve – this is the default value.
- – 3 means a convex curve.
- – 4 means a complex (irregular) curve.
- *PIFFSNM* specifies the NM (%) for estimating beginning of degradation (BD).

An operational note about PIFFSMOD: your PHM system is prepackaged with a basic set of prediction algorithms embedded in a software program called DXARULE (Adaptive Remaining Useful Life Estimator) at no additional charge, because it does not provide special handling of nonlinear input data (PIFFSMOD is ignored). Your customer has the option of acquiring and using an advanced version, ARULEAV.

Performance Services/Graphics

The Performance Services/Graphics section (Figure 6.62) contains three entries of specifications related to performance:

- *PGSW* is reserved for a future, planned capability (graphical user interface).
- *PGALPHA* specifies the alpha (accuracy) percent for evaluating predication accuracy.
- *PSSW* specifies whether performance metrics are (1) or are not (0) to be written to a logging file.

```
%**Performance Services/Graphics
PGSW      = 1 ;        % Future
PGALPHA   = 10 ;       % Prognostic alpha for Prognostic Horizon
PSSW      = 1 ;        %  Write PI results to file, 0 = no, 1 = yes
```

Figure 6.62 NDEF: Performance Services/Graphics.

```
%**Input & Output Files********************************
FDINLIB  = 'C:\DXPHMLIB\DXINDATA\' ; % Input data library
OUTLIB   = 'C:\DXPHMLIB\DXOUTDATA\'; % Output data library
INFILE   = 'DXDATFR1' ;              % File name for input data
OUTFILE  = 'DXDATFR1 ' ;             % File name for output data
INTYPE   = '.txt' ;                  % Input file type, text
OUTTYPE  = '.csv' ;                  % Output file type, text
```

Figure 6.63 NDEF: Input & Output Files.

Input & Output Files

The Input & Output Files section (Figure 6.63) contains six entries to specify the input and output files for data:

- *FDINLIB* specifies the input data library.
- *OUTLIB* specifies the output data library.
- *INFILE* specifies the name of the input-data file.
- *OUTFILE* specifies the name of the output-data file.
- *INTYPE* specifies the type of the input-data file (.txt is the default).
- *OUTTYPE* specifies the type of the output-data file (.csv is the default).

Checkpoint Library and File Names

The work files section (Figure 6.64) contains nine entries to specify the names of the library, the name prefixes, and the type to be used as work files to save and restore the operating environment to support a checkpoint and restart:

- *CPTLIB* specifies the checkpoint (and restore) library for work files.
- *CPTTYP* specifies the type of the work files.
- *DDWFN* specifies the name prefix of the work file for a device driver program.
- *FVWFN* specifies the name prefix of the work file for a feature vector program.

```
%**Checkpoint Library & File Name**************************
CPTLIB   = 'C:\DXPHMLIB\DXCPT\';      % Checkpoint restore
CPTTYP   = 'csv'                      % file name type
DDWFN    = 'DDW_' ; % Device Driver work
FVWFN    = 'FVW_' ; % Feature Vector work
PIWFN    = 'PIW_' ; % Prognostic Information work
PSWFN    = 'PSW_' ; % Prognostic Services work
VSWFN    = 'VSW_' ; % Visual Services work (future)
EFWFN    = 'EFW_' ; % End File Services work (future)
PAWFN    = 'PAW_' ; % Performance Analysis work
```

Figure 6.64 NDEF: Checkpoint Library & File Name.

- *PIWFN* specifies the name prefix of the work file for a prognostic information program.
- *PSWFN* specifies the name prefix of the work file for a prognostic services program.
- *VSWFN* specifies the name prefix of the work file for a visual services program.
- *EFWFN* specifies the name prefix of the work file for an end file program.
- *PAWFN* specifies the name prefix of the work file for a performance analysis program.

Device Driver Program ID and Data Units of Measure

The data units of measure section (Figure 6.65) contains three entries to define the units of measure for input data:

- *DDPID* specifies the program identifier for a device driver program.
- *DDUT* specifies the text to be used for units of time.
- *DDUA* specifies the text to be used for units of amplitude.

You choose this format because, in the future, you might decide to specify units of measure for other forms of data, such as that from a feature vector program (FVPID).

Program Identifiers

The program identifiers section (Figure 6.66) contains six entries to specify the identifiers (numeric values) of the programs, other than the device driver, to be used to process the node:

- *FVPID* specifies the numeric identifier for a feature vector program.
- *PIPID* specifies the numeric identifier for a prognostic information program.
- *PSPID* specifies the numeric identifier for a prognostic services program.
- *VSPID* specifies the numeric identifier for a visual services program.
- *EFPID* specifies the numeric identifier for an end file program.
- *PAPID* specifies the numeric identifier for a performance analysis program.

End of Definition

The program identifier section is followed by an End of Definition (*ENDDEF*) entry (Figure 6.67), as an explicit end rather than an implied end.

```
%**Device Driver Program ID and Units of Measure**************
DDPID     = 1 ;       % Required entry     DXDDTXT01
DDUT      = '[hr]' ; %    Units of time
DDUA      = '[kHz]' ;%    Units of amplitude
```

Figure 6.65 NDEF: Device Driver Program ID and Units of Measure.

```
%**Other Program IDs*****************************************
FVPID     = 1 ;       % Required entry     DXFV1S01
PIPID     = 2 ;       % Required entry     DXARULE
PSPID     = 1 ;       % Optional entry     DXPS01
VSPID     = 0 ;       % Place holder
EFPID     = 0 ;       % Place holder
PAPID     = 0 ;       % Optional entry     DXPA01
```

Figure 6.66 NDEF: Other Program IDs.

```
%**End of Definition***********************************
ENDDEF    = -9;        % end of node definition
```

Figure 6.67 NDEF: End of Definition.

6.8.3 Other Node Definitions for the Prototype PHM System

Because you have chosen a flexible, robust method – SND and NDEF files – for specifying how nodes are to be processed, and because you have used NDNUMID = 50 in Figure 6.54 for the first node definition, the definitions for the other five nodes are simplified:

- You edited the test file for the SMPS to change the time units of measure from days to hours for consistency.
- You selected reasonable and general threshold values.

Node NDNUMID = 60

For this node, the SMPS in the second EMA subsystem, you do the following: copy the first node definition, change the input file name, and save it with a new name (see Figure 6.68).

Node NDNUMID = 51

For this node, the first EMA in the first EMA subsystem, you do the following: copy the node definition for node 50, make the changes indicated in Figure 6.69, and save the updated file as DXNDEMA1.txt.

Node NDNUMID = 52, 61, and 62

For the other three EMA nodes, copy DXEMA10, and save as three new NDEFS: DXNDEMA20, DXNDEMA30, and DXNDEMA40. In those files, change the INFILE names to the following: DXDAT_CBD51.csv, DXDAT_CBD52.csv, DXDAT_CBD61.csv, and DXDAT_CBD62.csv; also change the OUTFILE names to the following: DXDAT_OUT51.csv, DXDAT_OUT52.csv, DXDAT_OUT61.csv, and DXDAT_OUT62.csv.

```
Copy DXNDPWR1.txt
Change INFILE from 'DXDATFR1' to 'DXDATFR2'
Change OUTFILE from 'DXDATFR1' to 'DXDATFR2'
Save NDEF as 'DXNDPWR2.txt'
```

Figure 6.68 NDEF updates to support node second SMPS (node 60).

```
Copy DXNDPWR1.txt and rename to DXNDEMA1.txt
Change the following entries as indicated
    INFILE    = 'DXEMA10' ;
    OUTFILE   = 'DXEMA10' ;
    INTYPE    = '.csv' ;
    FAILM     = 25       % FFP FAILURE (when FFP = 0.25)
    DDPID     = 2 ;      % DXDDCSV03
    DDUA      = '[mA]' ;
    FVPID     = 3 ;      % DXFV0303
```

Figure 6.69 NDEF updates to support EMA 1 (node 51).

For node 62, change NM from 3 to 1, because further experiments have revealed that 3% is too high an NM value.

6.9 System Verification and Performance Metrics

An important aspect of system verification for a PHM system is exemplified by the following: how accurate are the prognostic estimates? Important considerations for answering that question include the following (Celaya et al. 2010; Hofmeister et al. 2013, 2016, 2017, 2018a,b):

- *Time when degradation is detected, compared to the true time of the onset of degradation.* An offset type of error is unavoidable because of, for example, noise, discrete intervals between the times samples are taken, measurement uncertainty, data smoothing (which tends to move a signature downward and leftward), and rounding and loss of precision as data is fused and transformed from data types and domains to other data types and domains.
- *Time when functional failure is detected compared to the true time of functional failure.* An offset type of error is unavoidable, for the same reasons as for detection errors.
- *Time when all subsequent RUL estimates are within a specified accuracy* (α). A useful figure of merit is named *convergence efficiency* (χ).
- *Time when all subsequent prognostic horizon (PH) estimates are within a specified accuracy* (α). PH is defined as the relative time of a sample with respect to when degradation was detected plus the RUL estimate (see Figure 6.70).

Table 6.1 lists base acronyms and terminology used to measure and calculate performance metrics related to prognostics.

6.9.1 Offset Types of Errors

The magnitude of offset errors is primarily dependent on how well you have conditioned the data, especially with respect to eliminating and mitigating noise.

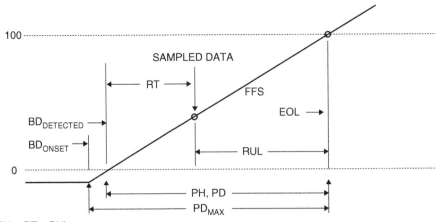

PH = RT + RUL
RT = TIME BETWEEN WHEN DATA IS SAMPLED AND WHEN DEGRADATION IS DETECTED

Figure 6.70 Illustration and relationship of PH to BD, sample time, RUL, and EOL.

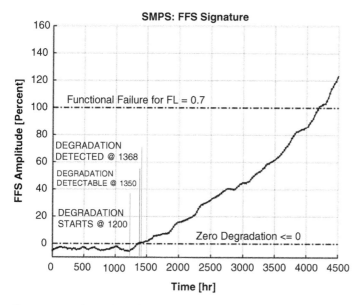

Figure 6.71 Relationship of degradation times and an FFS.

Degradation Detection

A degradation-offset error (refer back to Figure 6.43 and to Figure 6.71) is primarily due to noise, the noise mitigation you employ, and the sampling rate and precision you have chosen. Referring to Figure 6.71, you introduced degradation into an SMPS when the test time was 1200 hours; the earliest your system could detect that degradation is at time 1350 when the FFS becomes greater than 0; and degradation was detected at time 1368 (24-hour interval between samples). Thus the degradation-offset error is 168 hours (refer to Table 6.2).[2,3]

If the magnitude of this error does not meet requirements (the magnitude is unacceptable), you need to do one or more of the following:

- Reduce the amount of noise by, for example using noise filtering in your sensor hardware and/or firmware and/or by reducing discretization and rounding errors in your data converters and/or computational routines.
- Reduce the sampling period. For example, if the sampling period is 24 hours and you take a sample 1 second before degradation begins, the smallest degradation-offset error is 23 hours and 59 seconds. Generally, your system will experience a degradation-offset error equal to one-half of the sampling period.
- Reduce the magnitude of the NM you specify, using FNM in your NDEF (see Figure 6.60) – provided the margin is at least as large as the largest positive value of noise in your signal.

Reducing and/or mitigating degradation-offset errors becomes a compromise between cost (primarily sensor and data-converter cost) and loss of precision due to, for example, data smoothing and throughput/bandwidth (reduced because of higher sampling rates).

---

2 True EOL may also be referred to as *actual EOL* or *ground truth EOL*.
3 NASA, in June 2015, used the term *ground truth* instead of *ideal* or *actual*.

Table 6.2 Table of terms and definitions for performance metrics.

Accuracy	See RA.
Alpha, α	The plus/minus variance used to determine when accuracy of EOL is within specification.
BD	Beginning (onset) of degradation.
$BD_{DETECTED}$	Time when degradation is detected.
BD_{OFFSET}	Offset error difference in time between when degradation is detected and when the true onset of degradation began: $$BD_{OFFSET} = BD_{DETECTED} - BD$$
EOL	End of life. EOL is an estimate of the time (the time sample) that functional failure is likely to occur. The sample time at which EOL is first determined to have occurred is the final estimate. EOL might also be called time of failure, predicted end of life, or predicted time of failure.
True EOL, EOL_0	An estimate of the time that functional failure occurred. True EOL is an estimate because the times that sampling occur rarely occur at exactly the time of true EOL.
FL	Function-failure level. Value of an FFP signature at which a prognostic target is defined as being no longer capable of operating within specifications.
NM	Noise margin.
Noise	Any measurable feature in a signal other than the feature of interest.
OFFSET OFFSET error	Systemic difference between a true value in magnitude or time and the measured value. Generally used in terms of an error: for example, "A primary error related to NM is an offset error in detecting the time of the onset of degradation."
PD, PD_{MAX}, PDMAX	Prognostic distance. The interval in time between detection of degradation and EOL: a measure of the time to perform fault management, including repair and load balancing, to prevent failure. The maximum PD is calculated as follows: $$PD_{MAX} = EOL_0 - BD$$
PD_{FOM}	Figure of merit for prognostic distance. A measure of the effectiveness of detecting the onset of degradation: $$P_{DFOM} = 1 - BD_{OFFSET}/(EOL - BD)$$
PH, PHα	Prognostic horizon (or prediction horizon). The estimated, relative time of failure (EOL): the RUL estimate plus the current, relative time of that estimate $$PH = current\ time + RUL$$ where current time is calculated as (sample time − BD), a relative time (RT) value. PHα occurs when all subsequent PH estimates are within α (percent) of EOL.
PH convergence, $\chi\alpha$	Figure of merit for convergence of PH estimates to within a specified accuracy (α) of true EOL: $$\chi\alpha = PD/PD_{MAX} = 100^*(PH\alpha/(EOL_0 - BD))$$ $$\chi\alpha \approx 100^*(PH\alpha/(EOL - BD_{DETECTED}))$$

(Continued)

Table 6.2 (Continued)

PTOF	Predicted time of failure. The estimated time when a prognostic target is expected to functionally fail. It is calculated as the sum of the time when degradation is detected and the PH estimate:
	$$PTOF = BD + PH$$
RA	Relative accuracy. The accuracy of an estimated RUL value with respect to the actual RUL at the time of a data sample. RA is a performance metric having meaning only after functional failure occurs (TFF is determined):
	$$RA\ (\%) = 100^*(1 - RUL/RUL0)$$
	where RUL = estimated RUL at sample n and RUL0 = ideal RUL for sample n. Contrast RA with accuracy:
	$$RUL\ accuracy\ (\%) = 100^*RUL/RUL_0$$
	RA is applicable to any prognostic or predictive estimate such as SoH or EOL.
RT	Relative time between a sampled data point and the time when degradation was detected:
	$$RT = (data\ time) - BDDETECTED$$
TTF	Time to failure (see RUL). The time it takes a component, device, assembly, or system to fail – the end of its useful life.

Figure 6.72 Illustration of the uncertainty of determining exactly when functional failure occurs.

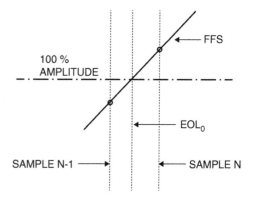

Time of Functional Failure: Uncertainty/Indeterminate

Functional failure (end of useful life) detection is subject to the same uncertainty as detection of the onset of degradation, because sampling almost always occurs before or after functional failure, as illustrated in Figure 6.72, rather than exactly at EOL_0. This is another offset type of error.

From a practical point of view, you should not be concerned about this offset error: remember that the primary purpose of your PHM system is to detect degradation far enough in advance to initiate and complete any remedial action necessary to prevent a future failure. For that purpose, there is a useful measurement called prognostic distance (PD): the period of time between when degradation begins and when functional failure occurs. The sampling interval needs to be much less than the PD (see Figure 6.73).

6.9.2 Uncertainty in Determining Prognostic Distance

There is an unavoidable error in determining PD because of the following (see Figure 6.73): (i) noise, especially in the output of a sensor; (ii) data conditioning, especially data smoothing; (iii) NM; and (vi) sampling, especially the sampling period.

Ideally, the sensor system in your PHM system would be absent all noise – no thermal noise, no harmonic distortion such as hum, no power-supply-switching noise, no coupling of signals, and no discretization and rounding errors between analog-to-digital and digital-to-analog conversions. But that is not going to be the case. Therefore, it becomes impossible for your PHM system to detect the exact time when degradation begins (true BD): you need to estimate BD.

Ideally, your PHM system would sample your sensor data at the onset of degradation – but that is not going to happen except by chance – even if you use an unreasonable sampling period (such as microseconds). Again, you need to estimate BD.

Ideally, the magnitude of the noise in your sampled data would be sufficiently low that you would not have to employ noise mitigation, such as data conditioning using a rolling average method, or the use of a noise margin (NM). Data conditioning employing noise mitigation results in shifting of the transfer curve (the FFS curves in Figure 6.73) of your sensing system, which comprises everything between the sensed node and the input to your prediction program, downward and to the left, which leads to offset type of errors.

Assuming you do not want to continue to use a degraded prognostic target until it actually fails, the best your PHM system can do is to estimate a future time when functional failure is likely to occur. The prediction programs you are using – DXARULE and ARULEAV – do just that: the PH outputs are estimates of the time between an estimated BD and when functional failure is likely to occur.

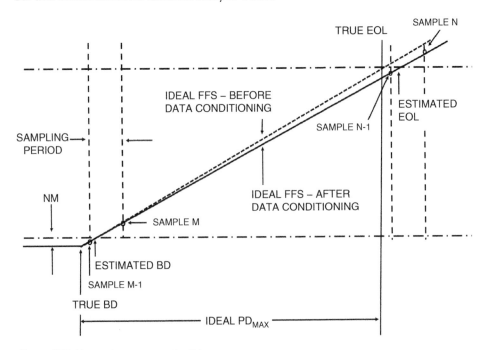

Figure 6.73 Uncertainty: prognostic distance.

Ideally, your PHM system would detect the exact moment when degradation begins and would then estimate exactly how much longer it would take for the prognostic target to functionally fail. Your PHM system would then provide a maximum lead time between onset of degradation and functional failure (PH = Ideal PD_{MAX} in Figure 6.73). But this is not going to happen, primarily because, as you learned in Chapter 1 (refer back to Figures 6.3 and 6.4 in our review of Chapter 1), you cannot know when a specific prognostic target will fail. This means you need to make design and implementation compromises, especially with regard to the following: (i) your sensing system and noise; (ii) your sampling rate and period to meet prognostic requirements for PD, accuracy, and precision; and (iii) your estimate as to how long it will take for functional failure to occur after the onset of degradation.

Sensing System and Noise

It is prohibitively costly to attempt to design and/or use a noiseless sensing system. So, you need to consider the required accuracy of your system: for example, if the requirement is that estimates are to be within 25% accurate within one-quarter of the period between detection of degradation and the time of functional failure, it would not be appropriate to design and/or use a sensing system with a maximum noise level of less than 1%. Similarly, for an estimate accuracy of 10% within three-quarters of the time between detection of degradation and functional failure, it would not be appropriate to design and/or use a sensing system with a maximum noise level of 25%.

Another aspect of sampling is noise. In Chapter 5, sampling was said to be a form of low-pass filtering. For investigative and verification purposes, you decide to sample your EMA data using hourly (SRDEF = 1 in Figure 6.57) and daily (24-hour) sampling (SRDEF = 2), which produces, for example, the respective plots shown in Figure 6.74. Your customer approves changing the sampling to once every 24 hours (daily).

Sampling: Rate and Period

It is expensive in terms of costs, implementation, and operation to sample at high rates. For example, if a PHM accuracy requirement stated that the estimated time of failure (such as a PH estimate) must be within 25% accurate within 100 hours of functional failure with a precision of 10 hours, it would be inappropriate to sample data at 1 MHz (a sampling period of one microsecond). Similarly, it would be inappropriate to sample data once every 24 hours. As previously discussed, the required estimate accuracy (25% in this case) and the estimate precision determine the maximum sampling period as follows:

Precision requirement. Ten-hour precision means you need to sample at least twice in that period.

Accuracy requirement. A 25% accuracy requirement means you need to quadruple the sample rate – in this case, to at least eight times in the 10-hour period. If, for example, you choose 10 times, the sampling period becomes 1 hour.

Estimating When Failure Is Likely

Chapter 2 introduced distribution modeling and failure curves having a failure distribution. As the independent variables change in distribution models, the rate of change of a given curve varies, which creates a family of failure curves having a failure distribution with a time-to-failure (TTF) as illustrated in Figure 6.75 (repeated from Chapter 2).

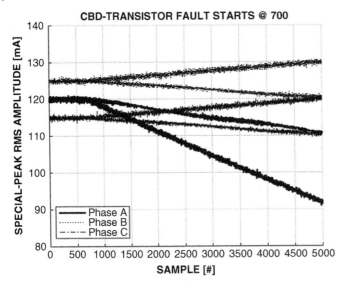

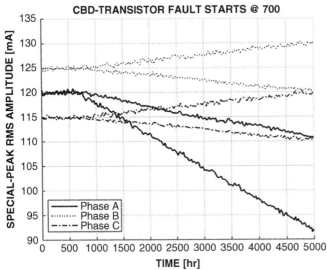

Figure 6.74 CBD at 1-hour sampling (top) and at 24-hour sampling (bottom).

The failure distribution is a probability density function (PDF), and the TTF is the expectation or the 0.50 value of the cumulative density function (CDF) of that PDF.

Estimations are based on assumptions and likelihood that include the following with respect to PD (refer to Figures 6.73, 6.75, and 6.76):

- Assume that a manufacturer's TTF value (or TTF value from another source) applies to and is reasonable for the prognostic target: therefore, set PITTFF0 equal to that value (see Figure 6.61).
- Assume a certain probability, such as 95%, that the fastest failures occur on or before half that time.

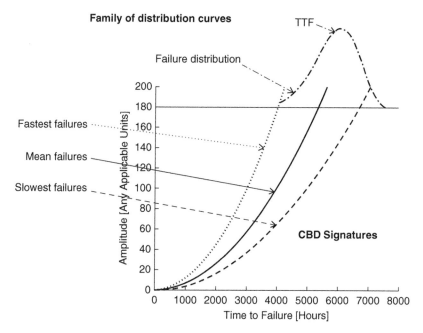

Figure 6.75 Family of failure curves, failure distribution, and TTF.

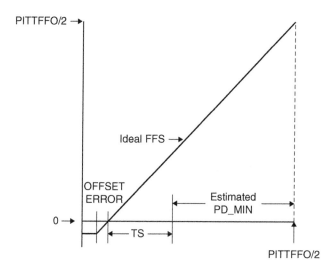

Figure 6.76 Diagram of an initial estimate for PD.

Therefore, an initial estimate for a minimum value for prognostic distance is calculated as follows:

Let PD_{EST_MIN} represent that minimum value, and let $OFFSET$ represent an error in time of detection due to NM, and let TS represent the designed-for sampling period. Then

$$PD_{EST\_MIN} = (PITTFF0/2) - OFFSET - TS \tag{6.10}$$

where OFFSET is calculated as follows:

Let NM_{FFS} represent the NM expressed as an FFS value.
Let FL_{FFS} represent the value used to convert either FFP or DPS data into FFS data.
Let NM_{FD} represent the NM expressed as a FD value.
Let FNM represent the value you specified in your NDEF (refer back to Figure 6.60).
Let FD_0 represent the nominal FD value in the absence of degradation.
Let $FDNOMMIN$ represent the value you specified in your NDEF (refer back to Figure 6.60).

Then

$$NM_{FD} = (FNM/100) \text{ when } FD_0 \geq FDNOMMIN$$

else

$$NM_{FD} = FNM$$

Then

$$NM_{FFS} = (NM_{FD}/FL_{FFS})$$

$$OFFSET = NM_{FFS}(PITTFF0/2) \tag{6.11}$$

Example 6.2   For example, when PITTFF0 is 2000 hours and NM_{FFS} is 0.030 (3.0%), then OFFSET is estimated as 30 hours (0.03 times 1000 hours); and if TS is 24 hours, then an estimated value for PD_{EST_MIN} using Eqs. (6.10) and (6.11) becomes 946 hours. If that value does not meet requirements, then assuming PITTFF0 is sufficiently correct, you need to either reduce OFFSET by reducing noise so you can reduce the NM and/or you need to increase the sampling rate to decrease TS.

Example 6.3   Suppose you have been given the following prognostic requirements and specifications statements:

"Degradation must be detected at least 1000 hours before failure, and the RUL estimates need to be within 10% at least 240 hours ±24 hours before failure."

This is interpreted as follows:

At least 1000 hours means PD_{EST_MIN} needs to be at least as large as 1000 hours.
"Within 10% at least 240 hours" means PH_{10} must be at least as large as 240 hours.
24 hours and "within 10%" means you need a minimum sampling rate of (2 samples /24 hours)*(100/10): 20 samples every 24 hours. This leads you to choose a designed-for-sampling rate of once per hour; so TS, in this case, is 1 hour.

But wait, what effect does PH_{10} have on your design, and how do you estimate whether your design is likely to meet requirements? That question is answered in the next section.

6.9.3   Estimating Convergence to Within PHα

Convergence of prognostic estimates to within PH_α is primarily dependent on the following factors, which you can change: (i) the efficiency of your prediction program; (ii) the noisiness of the input to your prediction program; (iii) the linearity (or lack of it) of the input to your prediction program; and (iv) the sampling rate and, therefore, the sampling period of your design.

Convergence Efficiency

Suppose the prediction program you use in your PHM system employs a Kalman-like algorithm coupled with a random-walk algorithm that treats a data point as a particle that has weight and momentum (a data point tends to keep moving at a given velocity and direction). Then, for example, one prediction program might converge from an initial amplitude error at a rate of 20% per data point and another at 10%.

Example 6.4 You have, by design, chosen a PITTFF0 value that you believe is likely to be 50% higher than an expected fastest failing time. Assume the input to a prediction algorithm is close to being ideal. As part of an evaluation of your design, you create a table that lists the expected number of data points that need to be processed to converge to within 25%, 10%, and 5% accuracy. When asked, the supplier of your prediction program, ARULEAV, asserts that the program corrects initial offset errors at a convergence of about 20% per data point. More important, the supplier asserts that convergence is also dependent on the noise and the degree of nonlinearity in the input signature. So you decide to evaluate the performance of your design using a prediction program that converges at 20% correction per data point versus another that converges at 10% per data point. This evaluation is summarized in Table 6.3, which lists the variation in number of data points and time-to-converge for two prediction programs. The range is rather large – from 11 hours to 264 hours in the example.

6.9.4 Performance Metrics

Refer to Figure 6.77 to relate prognostic distance (PD) and another performance metric called PH convergence to the FFS input and the RUL and PH output.

Prognostic Distance

PD is the period of time between when degradation begins and when functional failure occurs. You can define a performance metric called PD figure of merit (PD_{FOM}), as follows:

$$PD_{FOM} = (PD/PD_{MAX}) = (EOL - BD_{DETECTED})/(EOL - BD)$$

From the definition of BD_{OFFSET}, $BD_{DETECTED} = BD_{OFFSET} + BD$, so that

$$PD_{FOM} = (EOL - BD - BD_{OFFSET})/(EOL - BD) = (1 - BD_{OFFSET}/(EOL - BD))$$

Table 6.3 SMPS example: table of number of data points and times to converge to α.

| | | Initial error = 50% | | | | | |
| | | $\alpha = 25\%$ | | $\alpha = 10\%$ | | $\alpha = 5\%$ | |
		# Points	Time	# Points	Time	# Points	Time
Program 1	TS = 12 h	4	48 h	8	96 h	11	132 h
20%	TS = 1 h	4	4 h	8	8 h	11	11 h
Program 2	TS = 12 h	7	84 h	16	192 h	22	264 h
10%	TS = 1 h	7	7 h	16	16 h	22	22 h

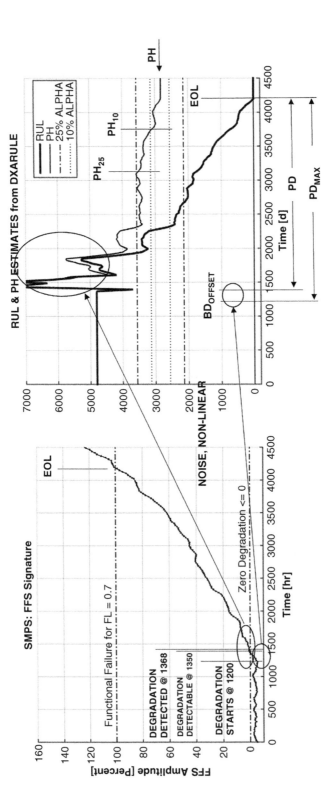

Figure 6.77 Comparison: FFS and plots of RUL and PH estimates.

and the performance metric becomes

$$PD_{FOM} = 1 - BD_{OFFSET}/PD_{MAX} \tag{6.12}$$

For your SMPS degradation test (refer back to Figure 6.71), BD_{OFFSET} was calculated to be 168 hours, and PD_{MAX} is 3000 hours (4200 – 1200), which means the calculated value of PD_{FOM} is 0.944. Your interpretation is that there is about a 6% offset error, due primarily to noise and NM. You can reduce the NM to improve the figure of merit, but to make a significant reduction, you need to reduce the noise in the data. PD_{FOM} is primarily related to how well you have conditioned the data and mitigated noise.

PH Convergence

As you have discovered, a very important performance metric, if not the most important, is PH convergence to within accuracy α (PHα): all subsequent PH estimates are within α percent of the actual PH. PH convergence is primarily due to how the prediction algorithms handle perturbations (noise) and nonlinearity in data.

So, how well did the prediction program, DXARULE, perform? Fortunately, The PHM system you designed has a log-file capability: you can write an output log file of the prognostic file (see Figure 6.78) to support post-processing of prognostic information. The last three entries in the header file are visual aids for analysis purposes: the PGALPHA value (10) specified in the NDEF file, the node ID (50), and the file name of the output file.

From the output bus file, you obtain the following measurement values: $BD_{DETECTED}$ occurred at 1368 hours, EOL (functional failure) occurred at 4200 hours, PH_{25} occurred at 3120 hours, and PH_{10} occurred at 3768 hours, which results in the following calculated values for the convergence performance metric: χ: $\chi_{25} = 38\%$ and $\chi_{10} = 16\%$, which might or might not be acceptable to your customer.

6.9.5 Prognostic Information: RUL, SoH, PH, and Degradation

Still referring to Figure 6.78, you see the last four columns in the output bus file contains the following prognostic information: RUL, SoH, PH, and BD_D – the latter is a

[hr]	[kHz]	[hr]	FSS	[hr]	RUL	SoH	PH	BD		10	50 DXDATFR1.csv
0.00	2667.85	0.00	-5.21	0.00	4800.00	100.00	4800.00	0.00			
*	*	*	*	*	*	*	*	*			
1344.00	2755.40	1344.00	-0.58	1344.00	4800.00	100.00	4800.00	0.00			
1368.00	2768.71	1368.00	0.13	1368.00	4793.03	99.87	4797.40	1363.62		Degradation Detected	
*	*	*	*	*	*	*	*	*			
*	*	*	*	*	*	*	*	*			
3096.00	3635.65	3096.00	46.02	3096.00	1844.07	53.98	3576.45	1363.62			
3120.00	3662.21	3120.00	47.42	3120.00	1769.43	52.58	3525.80	1363.62		PH within 25%	
3144.00	3687.97	3144.00	48.79	3144.00	1700.55	51.21	3480.92	1363.62			
*	*	*	*	*	*	*	*	*			
*	*	*	*	*	*	*	*	*			
3744.00	4187.77	3744.00	75.24	3744.00	730.41	24.76	3110.79	1363.62			
3768.00	4232.96	3768.00	77.64	3768.00	646.43	22.36	3050.81	1363.62		PH within 10%	
3792.00	4267.31	3792.00	79.45	3792.00	586.49	20.55	3014.87	1363.62			
*	*	*	*	*	*	*	*	*			
*	*	*	*	*	*	*	*	*			
4176.00	4631.54	4176.00	98.73	4176.00	34.00	1.27	2846.38	1363.62			
4200.36	4675.50	4200.36	101.06	4200.36	-12.18	0.00	2824.56	1363.62		EOL detected	
4224.00	4685.51	4224.00	101.59	4224.00	0.00	0.00	2824.56	1363.62			
*	*	*	*	*	*	*	*	*			

Figure 6.78 Prognostic bus (log file).

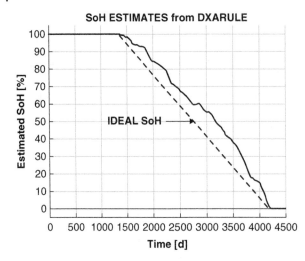

SoH ESTIMATES from DXARULE

Figure 6.79 SoH plot from DXARULE.

calculated value for detected beginning of degradation. You plotted the RUL and PH columns to create Figure 6.77, you used BD_D to calculate BD_{OFFSET}, and you used the SoH column to create Figure 6.79.

6.10 System Verification: Advanced Prognostics

After reviewing the verification results of your SMPS function tests, you and your customer decide to rerun the function testing using an advanced version of the prediction program (ARULEAV) instead of the base program (DXARULE). Both you and your customer are extremely pleased with the results.

6.10.1 SMPS: FFP Signature Directly to FFS

Input Changes
Referring back to Figure 6.66, the only alteration you make is to change PIPID from 2 to 1, the ID for program ARULEAV. The same input data set is used, and the same conditioning is performed as before.

Prognostic Information Results: Bus Output
Examining the output bus file using ARULEAV (Figure 6.80) and comparing the output bus file using DXARULE (Figure 6.78) reveals that the detection of degradation occurs at the same time (BD_D at 1368 hours) and detection of EOL occurs at the same time (PH is 0 at 4200 hours). However, there is a significant improvement in PH estimates: PH_{25} is reached at 2328 hours instead of 3120 hours ($\chi_{25} = 67\%$ instead of 38%), and PH_{10} is reached at 3696 hours instead of 3768 hours ($\chi_{10} = 18\%$ instead of 16%).

Plots: RUL, PH, and SoH
When you plot the RUL, PH, and SoH estimates (Figure 6.81) using data in the output bus file, the first thing you notice is that there is significantly less RUL and PH noise. Next, you notice that the plots are more linear compared to those using DXARULE: the

[hr]	[kHz]	[hr]	FSS	[hr]	RUL	SoH	PH	BD_D	10	50 DXDATFR1.csv
0	2667.85	0	-5.2104	0	4800	100	4800	0		
*	*	*	*	*	*	*	*	*		
1344	2755.4	1344	-0.5761	1344	4800	100	4800	0		
1368	2768.71	1368	0.1285	1368	4788.468	99.9087	4792.845	1363.624		Degradation detected
*	*	*	*	*	*	*	*	*		
*	*	*	*	*	*	*	*	*		
2304	3249.198	2304	25.5623	2304	2666.072	74.4301	3581.981	1388.091		
2328	3282.208	2328	27.3096	2328	2510.77	72.7616	3450.68	1388.091		PH within 25%
2352	3307.971	2352	28.6733	2352	2388.819	71.25	3352.729	1388.091		
*	*	*	*	*	*	*	*	*		
3672	4100.302	3672	70.6141	3672	802.5878	25.9694	3090.517	1384.07		
3696	4123.131	3696	71.8225	3696	763.5184	24.8263	3075.448	1384.07		PH within 10%
3720	4153.425	3720	73.426	3720	708.1646	23.2636	3044.094	1384.07		
*	*	*	*	*	*	*	*	*		
*	*	*	*	*	*	*	*	*		
4176	4631.537	4176	98.7341	4176	29.7124	1.053	2821.642	1384.07		
4200.36	4675.505	4200.36	101.0614	4200.36	-12.18	0	2804.11	1384.07		EOL detected
4224	4685.512	4224	101.5912	4224	0	0	2804.11	1384.07		
*	*	*	*	*	*	*	*	*		

Figure 6.80 Output file for SMPS using ARULEAV.

PD and PD_{MAX} are unchanged, and, therefore, PD_{FOM} is unchanged. Compare the plots in Figure 6.81 to those in Figures 6.77 and 6.79.

6.10.2 SMPS: FFP Signature to DPS to FFS

You and the customer approve another design change: transforming the FFP signature to DPS and then to a FFS transform, as described in Chapters 4 and 5.

Input Changes
From your PoF and FMEA analyses and the characteristic curve, you decide to use a square root–based FFP to DPS transform, but with n = 0.45 instead of 0.5. Your node-definition changes are as follows:

- XDPS from 0 to 3 (refer back to Figure 6.60).
- XDPSNV from 1 to 0.45 (refer back to Figure 6.60).
- PIFFSMOD from 2 to 4 (refer back to Figure 6.61).

Prognostic Information Results: Bus Output
Examining the output bus file using ARULEAV (Figure 6.82) and comparing it with the output bus file using DXARULE (refer back to Figure 6.78) reveals that the detection of degradation occurs at the same time (at 1368 hours), and detection of EOL occurs at the same time (PH is zero at 4200 hours). There is significant improvement in PH estimates: PH_{25} is reached at 1560 hours compared to the previous 2328 and 3120 hours ($PH_{25} \sim$ 94% instead of 67% or 38%), and PH_{10} is reached at 2760 hours compared to the previous 3696 hours and 3768 hours ($PH_{10} \sim$ 53% instead of 18% or 16%). See Table 6.4 for a list of the performance measurements and metrics.

Plots: RUL, PH, and SoH
When you plot the RUL, PH data, and SoH estimates using data in the output bus file, you notice a further decrease in noise, and the plots are even more linear. Compare the plots in Figure 6.83 to those in Figure 6.81.

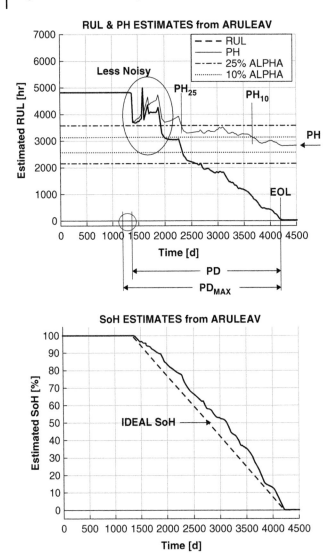

Figure 6.81 Plots of the RUL, PH, and SoH estimates produced by ARULEAV.

Final Design Specifications

The customer approves your changes to the design specifications for prognostic-enabling of the SMPS for degradation of the capacitance in the output filter: (i) use the advanced version, ARULEAV (PIPMOD = 1); (ii) use FFP to DPS signature transform (XDPS = 3); (iii) use 0.45 as a power value (XDPSNV = 0.45); and (iv) use a complex signature model in ARULEAV (PIFFMOD = 4).

Continue to use the remaining design specifications: (i) NM (FNM = 2.5); (ii) FFP failure threshold (FAILM = 70); (iii) number of points to calculate nominal FD value (FD0PTS = 20); (iv) number of rolling FD points for data smoothing (FDPTS = 4); (v) number of rolling FFP points for data smoothing (FFPPTS = 4); and (vi) number of rolling FFS points for data smoothing (FFSPTS = 4).

[hr]	[kHz]	[hr]	FSS	[hr]	RUL	SoH	PH	BD_D	10.00	50.00 DXDATFR1.csv
0.00	26.68	0.00	-5.21	0.00	4800.00	100.00	4800.00	0.00		
*	*	*	*	*	*	*	*	*		
1344.00	27.55	1344.00	-0.58	1344.00	4800.00	100.00	4800.00	0.00		
1368.00	27.69	1368.00	0.13	1368.00	4786.05	99.91	4790.43	1363.62		
*	*	*	*	*	*	*	*	*		
*	*	*	*	*	*	*	*	*		Degradation detected
1536.00	28.17	1536.00	2.71	1536.00	3660.78	96.10	3809.15	1387.62		
1560.00	28.31	1560.00	3.40	1560.00	3338.29	95.09	3510.66	1387.62		PH within 25%
1584.00	28.41	1584.00	3.97	1584.00	3222.19	94.26	3418.57	1387.62		
*	*	*	*	*	*	*	*	*		
*	*	*	*	*	*	*	*	*		
2760.00	35.34	2760.00	40.62	2760.00	1119.08	44.92	2491.23	1387.85		
2784.00	35.32	2784.00	40.55	2784.00	1146.64	45.03	2546.35	1384.29		PH within 10%
2808.00	35.30	2808.00	40.40	2808.00	1149.24	44.56	2579.10	1378.14		
*	*	*	*	*	*	*	*	*		
*	*	*	*	*	*	*	*	*		
4176.00	46.32	4176.00	98.73	4176.00	0.32	0.01	2817.47	1358.85		
4200.36	46.76	4200.36	101.06	4200.36	-24.04	0.00	2817.47	1358.85		EOL detected
4224.00	46.86	4224.00	101.59	4224.00	0.00	0.00	2817.47	1358.85		
*	*	*	*	*	*	*	*	*		

Figure 6.82 Output file for SMPS using DPS-based FFS and ARULEAV.

Table 6.4 Performance measurements and metrics.

Prognostic target	PI specifications	BD	EOL	PD	PH	
Name	PITTFF0 FFS NM	At time estimated BD	At time estimated	Maximum FOM	$PH_{ERROR} = (PITTFF0/PD_{MAX})$ 25% @ time χ [%] – pts. [#]	10% @ time χ [%] – pts. [#]
SMPS	4800 [h] 3%	1368 [h] 1261 [h]	4200 [h] 4176 [h]	2939 [h] 96.4%	Initial PH error = 63% 1560 [h] 93.5% – 9 pts.	2760 [h] 52.6% – 23 pts.
EMA load	4800 [h] 2%	504 [h] 441 [h]	3168 [h] 3164 [h]	2727 [h] 97.7%	Initial PH error = 76% 576 [h] 97.3% – 3 pts.	1368 [h] 68.3% – 19 pts.
EMA winding	4800 [h] 2%	1104 [h] 1067 [h]	2760 [h] 2745 [h]	1693 [h] 97.8%	Initial PH error = 184% 1320 [h] 87.2% – 10 pts.	1680 [h] 66.0% – 25 pts.
EMA power Transistor	4800 [h] 1%	960 [h] 958 [h]	4440 [h] 4434 [h]	3476 [h] 99.9%	Initial PH error = 38% 984 [h] 99.3% – 2 pts.	1512 [h] 84.1%

6.11 PHM System Verification: EMA Faults

You propose, and your customer agrees, to always use the advanced version of ARULE (ARULEAV) because even though EMA signatures are linear, ARULEAV has demonstrated that it provides faster convergence from an initial error (determined

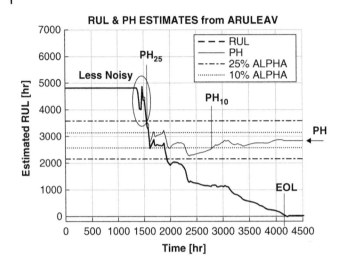

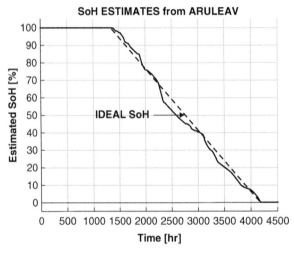

Figure 6.83 Plots of the RUL, PH, and SoH estimates using DPS-based FFS and ARULEAV.

by PITTFF0) compared to DXARULE. Accordingly, you change the PIPID from 2 (DXARULE) to 1 (ARULEAV) in the NDEF files for all four EMA nodes.

Since the FFP signatures for the three types of EMA faults are already linear (Figures 6.41, 6.43, and 6.51), it is not necessary to perform FFP-to-DPS-to-FFS transforms. However, because your customer requests that sampling be changed from a 1-hour to a 24-hour interval, you should rerun your functional EMA tests, data conditioning, and transforms, and then input those FFS signatures into ARULEAV; then verify and accept the RUL, PH, and SoH estimates; and produce, verify, and accept the performance metrics.

6.11.1 EMA: Load (Friction) Type of Fault

Your function test collects CBD, transforms that to FD, and creates FFP signature data, which is directly transformed into FFS data (Figure 6.84). This data is input to ARULEAV

Figure 6.84 CBD, FFP, and FFS for EMA node 51 (friction/load).

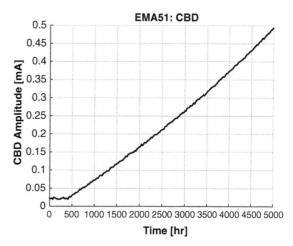

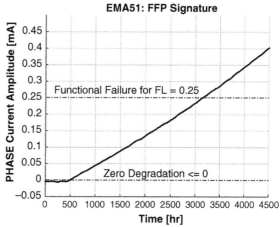

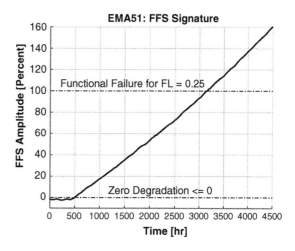

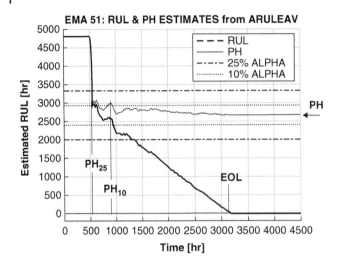

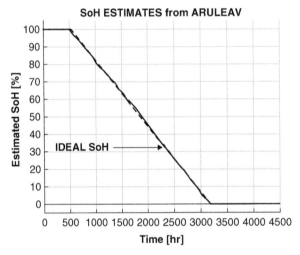

Figure 6.85 EMA (load) plots: RUL, PH, and SoH estimates using DPS-based FFS and ARULEAV.

one data point (time and amplitude) at a time, which results in data and performance metrics: Figure 6.85 and Table 6.3.

From the logged output file, you collect and calculate the following information (in hours): BD was estimated to occur at 441 – a difference of 3 hours from the 444 hours when degradation actually began; BD_{DET} at 504, BD_{OFF} is calculated as 63 using the estimated BD; EOL was detected at 3168 – just prior to functional failure; EOL was estimated at 3164; PD_{MAX} is calculated as 2727 (3148 minus 441); PD_{FOM} is calculated as 97.7%; PH_{25} was reached at 576; and PH_{10} was reached at 1368. From that information, you calculate the following: χ_{25} is 97.3% from $((2727 - (576 - 441))/2727)$ and χ_{10} is 68.3%, which are excellent results and exceed our recommended targets of 70% and 35%. See Table 6.4 for a summary of results.

The plots and performance metrics confirm that an FFP-to-DPS transform is not necessary: the FFP signature is only slightly curvilinear. In fact, after plotting (Figure 6.85) and other analyses, you and the customer are ecstatic with the RUL, PH, and SOH estimates and the performance metrics (Table 6.4). To better appreciate the results, compare and contrast the bottom plot in Figure 6.85 to the top plot in Figure 6.77 (DXARULE), to the bottom plot to that in Figure 6.81 (ARULEAV), and to the right-hand plot in Figure 6.83.

6.11.2 EMA: Winding Type of Fault

After data collection and processing, your function test produces FFS data (Figure 6.86) for node 61, which is input to ARULEAV one data point (time and amplitude) at a time. The very accurate RUL, PH, and SoH estimates are plotted in Figure 6.87 and listed in Table 6.4.

You carefully document the calculations you used – including those used in an initial version of a Prognostic Analysis program and those you hand-calculated. You also document potential additions and changes that were discovered during verification testing; all such additions and changes need to be submitted, reviewed, and approved before being officially incorporated in the design of the PHM system.

6.11.3 EMA: Power-Switching Transistor Type of Fault

Your function test produces the CBD, the FFP data, and the FFS data (Figure 6.88) for node 62, which is input to ARULEAV one data point (time and amplitude) at a time. The very accurate RUL, PH, and SoH estimates are plotted in Figure 6.89 and listed in Table 6.4.

The plot of the FPP signature and FFS data in Figure 6.88 verifies the design change from an NM of 3 to 1 (percent of the nominal FD value). You document that it is necessary to handle the input data as multiple inputs (multivariate data) that required each to be normalized to compensate for the differing levels of reference between the phase currents (as seen in the top plot in Figure 6.88).

6.12 PHM System Verification: Functional Integration

To verify the design of the PHM system, you need to include the prototypes you have already designed and developed. Additionally, you need to design and develop a rapid prototype of control and data flow of your PHM system that interfaces with the nodes, those programs, and your prognostic-related programs already in hand: DXARULE (basic prognostic support), ARULEAV (advanced prognostic support), and DXPA01 (prognostic performance, analysis). A high-level procedure for a node-based architecture for your PHM control and data flow is shown in Figure 6.90.

6.12.1 Functional Integration: Control and Data Flow

You need to verify the initial design of your PHM system by doing the following: (i) verify functional operation – process the node data and produce prognostic information; and (ii) verify prognostic accuracy – process the prognostic information using PAPID (DXPA01) to produce performance metrics and obtain approval of those metrics.

You have already completed an initial functional test for node 50 (degraded); completed a test for node 60 (not degraded) and verified a no-degradation detected

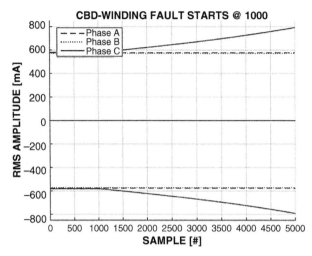

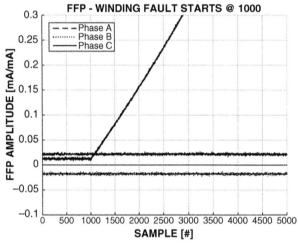

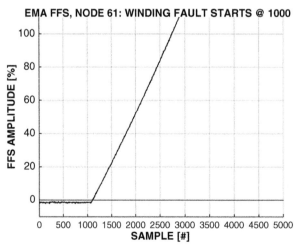

Figure 6.86 CBD, FFP, and FFS for EMA node 61 (winding).

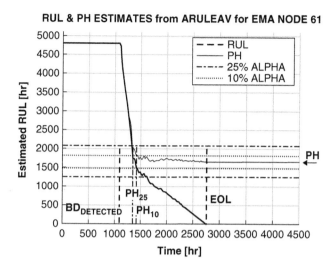

Figure 6.87 EMA (winding) plots: RUL, PH, and SoH estimates; DPS-based FFS and ARULEAV.

state; and completed tests for node 51 (degraded, loading/friction type of fault), node 52 (not degraded), node 61 (degraded, motor-winding type of fault), and node 62 (degraded, power-switching transistor fault).

Next you perform a concurrent test of all six nodes using your prototype PHM system: an integrated function test. The system-produced information is shown in Figure 6.91 (startup initialization), Figure 6.92 (node alerts, part 1), and Figure 6.93 (node alerts, part 2). At the end of the tests, you plot and examine the output files and verify that the results are as expected – they have not changed.

6.12.2 System Performance Metrics: Summary

Table 6.4 lists selected performance measurements and metrics for the functional tests performed during your system verification.

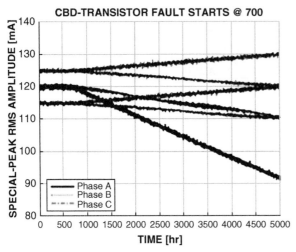

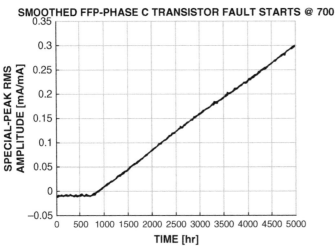

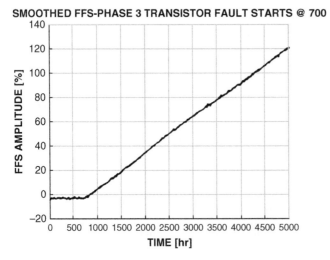

Figure 6.88 CBD, FFP, and FFS for EMA node 62 (power transistor).

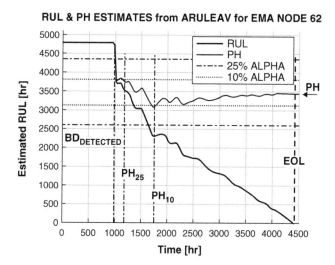

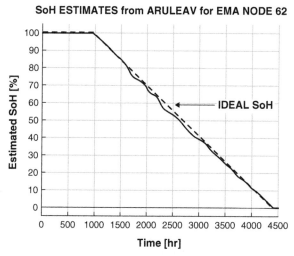

Figure 6.89 EMA (transistor) plots: RUL, PH, and SoH estimates; DPS-based FFS and ARULEAV.

6.12.3 PHM System: Plans

You have successfully verified the initial design of your robust, exemplary PHM system; demonstrated that design; prepared and submitted that report to your customer; and scheduled and held a phase type of wrap-up review. You and your customer agree in principal that the next phase in the project is to prepare an initial design for the following:

- Fault management to handle detected degradation at each of the alert levels. In the initial design, the response to alerts was to continue operation.
- Variable sampling times. In the initial design, the same sampling rate was used prior to and after detection of degradation, despite the node-monitoring specifications shown in Figure 6.57.

```
%**************************************************************************
%** PHM CONTROL & DATA FLOW
%**************************************************************************
    Housekeeping and initialization
        Read SND
        For each defined node (NDEF) in SND
            Read, parse, and evaluate the NDEF
            Initialize operating parameters, variables, and constants
            Initialize work areas for checkpoint/restart
    Main
        When Node active
            Restore (or initialize) operating environment
                Call DDPID to get data (output DD bus)
                Call FVPID to process data (output FV bus)
                Call PIPID to produce prognostic information (output PI bus)
            Save operating environment
            When node functionally fails or is made inactive
                Call PAPID (if specified) to analyze performance
            Point to next node (in the ring)
        Otherwise
            Point to next node
    end
```

Figure 6.90 PHM: high-level control and data flow.

```
>> DXPHM01('-1','DXSNDIVP1','-1')
I4: AN INITIALIZED SND (DEFINITIONS) FILE HAS BEEN FOUND
I19: RECEIVED RESPONSE: REPLACE
I1: SAVED NODE DEFINITION FILE IS REPLACED: C:\DXPHMLIB\DXCPT\ND_50.mat
I1: SAVED NODE DEFINITION FILE IS REPLACED: C:\DXPHMLIB\DXCPT\ND_51.mat
I1: SAVED NODE DEFINITION FILE IS REPLACED: C:\DXPHMLIB\DXCPT\ND_52.mat
I1: SAVED NODE DEFINITION FILE IS REPLACED: C:\DXPHMLIB\DXCPT\ND_60.mat
I1: SAVED NODE DEFINITION FILE IS REPLACED: C:\DXPHMLIB\DXCPT\ND_61.mat
I1: SAVED NODE DEFINITION FILE IS REPLACED: C:\DXPHMLIB\DXCPT\ND_62.mat


I18: INITIALIZED NODE 50: OPENING FILE C:\DXPHMLIB\DXINDATA\DXDATFR1.txt
I18: INITIALIZED NODE 51: OPENING FILE C:\DXPHMLIB\DXINDATA\DXDAT_EMA10.txt
I18: INITIALIZED NODE 52: OPENING FILE C:\DXPHMLIB\DXINDATA\DXDAT_EMA20.txt
I18: INITIALIZED NODE 60: OPENING FILE C:\DXPHMLIB\DXINDATA\DXDATFR2.txt
I18: INITIALIZED NODE 61: OPENING FILE C:\DXPHMLIB\DXINDATA\DXDAT_EMA30.txt
I18: INITIALIZED NODE 62: OPENING FILE C:\DXPHMLIB\DXINDATA\DXDAT_EMA40.txt
```

Figure 6.91 Initialization: system nodes.

- Upgrade the current Prognostic Analysis program to include all the measurements, estimates, and calculations shown in Table 6.4, which includes some hand-calculated values.
- Visual services support: a GUI rather than the EUI you used. An example is shown in Figure 6.94.

A next step is to prepare a set of updated requirements, objectives, and specifications. You are likely to discover that the next phase requires working closely with the customer, which is likely to be time-consuming and, at times, frustrating:

- *Fault management.* You need to understand and account for the operation and logistics of the customer's current, and possible future, maintenance and supply

Figure 6.92 System alerts, part 1.

```
W19: SOH for NODE: 51 IS 99.8663 PERCENT
I44: RUL IS 4776.1 [hr] AT DATA TIME: 504 [hr]
I43: RECEIVED RESPONSE: CONTINUE
W19: SOH for NODE: 62 IS 99.9713 PERCENT
I44: RUL IS 4795.8 [hr] AT DATA TIME: 960 [hr]
I43: RECEIVED RESPONSE: CONTINUE
W19: SOH for NODE: 61 IS 99.8078 PERCENT
I44: RUL IS 4695.3 [hr] AT DATA TIME: 1104 [hr]
I43: RECEIVED RESPONSE: CONTINUE
W19: SOH for NODE: 50 IS 99.9087 PERCENT
I44: RUL IS 4786.1 [hr] AT DATA TIME: 1368 [hr]
I43: RECEIVED RESPONSE: CONTINUE
W19: SOH for NODE: 61 IS 74.4394 PERCENT
I44: RUL IS 1312.3 [hr] AT DATA TIME: 1536 [hr]
I43: RECEIVED RESPONSE: CONTINUE
W19: SOH for NODE: 51 IS 74.4327 PERCENT
I44: RUL IS 2114.7 [hr] AT DATA TIME: 1224 [hr]
I43: RECEIVED RESPONSE: CONTINUE
```

system. For example, the customer might wish all of the following to be supported as responses to alerts:

- CONTINUE operation.
- SUSPEND operation.
- SCHEDULE maintenance, removal and replacement, and inspection.
- MONITOR: update/change sampling rate and/or alert levels.
- OFFLINE: end operation.
- ONLINE: resume operation of a suspended or offline prognostic target.
- NODE: add, update, or remove node definitions (NDEFs) for a system.

- *Variable sampling.* You are likely to discover that each set of LRUs has different sampling requirements based on how and where it is used. For example, an EMA used to adjust the seat in a cockpit is likely to be less critical compared to that used to position a wing surface.
- *Prognostic Analysis program.* Even though you might believe the design of this program is primarily in your zone of expertise, you are likely to find that your customer has other ideas. For example, should the program analyze the specifics for an alpha accuracy of 5%, 10%, and 25%? Your customer might want your program to do any or all of the following:
  - Analyze and display the accuracy statistics for a prognostic target as specified by PGALPHA (refer back to Figure 6.62).
  - Analyze and display the estimated accuracy of the estimated RUL values when alerts are issued.
  - Plot measurements and metrics only when specifically asked for.
- *Visual services.* The design of a GUI for a PHM system is probably the most difficult, time-consuming, and costly aspect of the system, especially because of the following:
  - The look and feel you prefer may be disliked by your customer.
  - The GUI must be compatible with and integrate with existing diagnostic and/or prognostic solutions.
  - The breadth and depth of the functionality need to cover all aspects of the design system: definitions, programs, operational specifications, and so on. Further, every instance where you add, delete, or update functionality is likely to require design, review, testing, and approval costs and time.

```
W19: SOH for NODE: 62 IS 74.3818 PERCENT
I44: RUL IS 2373.2 [hr] AT DATA TIME: 1776 [hr]
I43: RECEIVED RESPONSE: CONTINUE
W19: SOH for NODE: 51 IS 49.7792 PERCENT
I44: RUL IS 1386.1 [hr] AT DATA TIME: 1896 [hr]
I43: RECEIVED RESPONSE: CONTINUE
W19: SOH for NODE: 61 IS 49.9565 PERCENT
I44: RUL IS 854.1 [hr] AT DATA TIME: 1944 [hr]
I43: RECEIVED RESPONSE: CONTINUE
W19: SOH for NODE: 50 IS 74.4301 PERCENT
I44: RUL IS 1471.8 [hr] AT DATA TIME: 2304 [hr]
I43: RECEIVED RESPONSE: CONTINUE
W19: SOH for NODE: 61 IS 24.9286 PERCENT
I44: RUL IS 392.3 [hr] AT DATA TIME: 2352 [hr]
I43: RECEIVED RESPONSE: CONTINUE
W19: SOH for NODE: 51 IS 24.5532 PERCENT
I44: RUL IS 658.2 [hr] AT DATA TIME: 2520 [hr]
I43: RECEIVED RESPONSE: CONTINUE
W19: SOH for NODE: 62 IS 49.7584 PERCENT
I44: RUL IS 1665.2 [hr] AT DATA TIME: 2640 [hr]
I43: RECEIVED RESPONSE: CONTINUE
W19: SOH for NODE: 61 IS 4.8343 PERCENT
I44: RUL IS 79.7 [hr] AT DATA TIME: 2664 [hr]
I43: RECEIVED RESPONSE: CONTINUE
W20: NODE: 61 HAS FUNCTIONALLY FAILED
I44: AT DATA TIME: 2760 [hr]
I43: RECEIVED RESPONSE: CONTINUE
W19: SOH for NODE: 51 IS 4.2125 PERCENT
I44: RUL IS 112.2 [hr] AT DATA TIME: 3048 [hr]
I43: RECEIVED RESPONSE: CONTINUE
W19: SOH for NODE: 50 IS 49.5872 PERCENT
I44: RUL IS 1011.9 [hr] AT DATA TIME: 3120 [hr]
I43: RECEIVED RESPONSE: CONTINUE
W20: NODE: 51 HAS FUNCTIONALLY FAILED
I44: AT DATA TIME: 3168 [hr]
I43: RECEIVED RESPONSE: CONTINUE
W19: SOH for NODE: 62 IS 24.7536 PERCENT
I44: RUL IS 825.9 [hr]AT DATA TIME: 3528 [hr]
I43: RECEIVED RESPONSE: CONTINUE
W19: SOH for NODE: 50 IS 24.8263 PERCENT
I44: RUL IS 402.2 [hr] AT DATA TIME: 3696 [hr]
I43: RECEIVED RESPONSE: CONTINUE
W19: SOH for NODE: 50 IS 4.5852 PERCENT
I44: RUL IS 60.9 [hr] AT DATA TIME: 4128 [hr]
I43: RECEIVED RESPONSE: CONTINUE
W20: NODE: 50 HAS FUNCTIONALLY FAILED
I44: AT DATA TIME: 4200.36 [hr]
I43: RECEIVED RESPONSE: CONTINUE
W19: SOH for NODE: 62 IS 4.9946 PERCENT
I44: RUL IS 172.9 [hr] AT DATA TIME: 4248 [hr]
I43: RECEIVED RESPONSE: CONTINUE
W20: NODE: 62 HAS FUNCTIONALLY FAILED
I44: AT DATA TIME: 4440 [hr]
I43: RECEIVED RESPONSE: CONTINUE
```

Figure 6.93 System alerts, part 2.

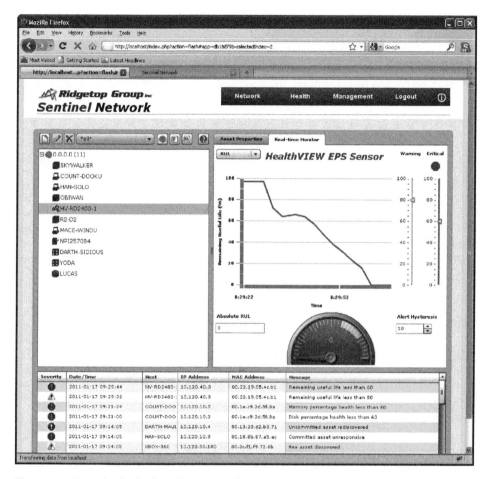

Figure 6.94 Example of a GUI for a PHM system. Source: Ridgetop (2018).

6.13 Summary: A Robust Prototype PHM System

This chapter presented topics related to the design of a robust prototype of an exemplary PHM system, in addition to the approaches and methods presented in the previous chapters. For example, we used and expanded on the following:

- Modeling, frameworks, and prognostic information – introduced in Chapter 1
- Failure distribution, prognostic approaches (statistics based), machine learning, and, especially, modeling of CBD signatures – introduced in Chapter 2
- Conditioning and transforming signatures (CBD, FFP, FFS) – introduced in Chapter 3
- Heuristic-based handling of signatures, noise, and NM – introduced in Chapter 4
- Non-ideal data: effects and conditioning such as noise, uncertainty, sampling, delays, feedback, and so on – introduced in Chapter 5

This chapter was essentially a case study, in terms of both material content and presentation style, with you in the role of a lead architect/designer of PHM solutions. The study was a system comprising multiple inputs and nodes, and univariate and multivariate inputs from two different types of units (power supply and EMA). The units were configured in two subsystems of three nodes each: a power supply node and two EMA nodes; and the study include four different failure modes.

A clear-text, node-based architecture was used to describe the PHM system, its programs, and its control and data flow. The study illustrated that accuracy is dependent on many factors, including but not limited to noise; sampling rates, modes, and periods; input operating specifications (such as PITTFF0 and NM); and the framework programs used (such as DXARULE versus ARULEAV).

The next chapter is devoted to miscellaneous topics related to prognostic enabling: electronic health solutions – evaluation of suitability and comprehensive and coverage, such as considerations related to selecting a prognostic target, which is typically an equipment asset; the relationship of failure distribution, TTF, mean time to failure (MTTF), and mean time before/between failure (MTBF); a comparison study of a some PHM approaches – conservative usage-based, conservative CBD-based (diagnostic), practical CBD-based (diagnostic), and practical CBD-based (prognostic); and reliability, bathtub curve, and diagnostic triggers for prognostics.

References

Celaya, J.R., Saxena, A., Wysocki, P. et al. (2010). Towards prognostics of power MOSFETs: accelerated aging and precursors of failure. Annual Conference of the Prognostics and Health Management Society, Portland, Oregon, US, 10–14 Oct.

Hofmeister, J., Wagoner, R., and Goodman, D. (2013). Prognostic health management (PHM) of electrical systems using conditioned-based data for anomaly and prognostic reasoning. *Chemical Engineering Transactions* 33: 992–996.

Hofmeister, J., Goodman, D., and Wagoner, R. (2016). Advanced anomaly detection method for condition monitoring of complex equipment and systems. 2016 Machine Failure Prevention Technology, Dayton, Ohio, US, 24–26 May.

Hofmeister, J., Szidarovszky, F., and Goodman, D. (2017). An approach to processing condition-based data for use in prognostic algorithms. 2017 Machine Failure Prevention Technology, Virginia Beach, Virginia, US, 15–18 May.

Hofmeister, J.P., Goodman, D.L., and Szidarovszky, F. (2018a). Transforming condition-based data signatures into functional failure signatures. IEEE 2018 Aerospace Conference, Big Sky, Montana, US, 3–9 March.

Hofmeister, J.P., Goodman, D.L., and Szidarovszky, F. (2018b). Heuristic-based approach: degradation signatures and CBD signatures. IEEE 2018 Aerospace Conference, Big Sky, Montana, US, 3–9 March.

IEEE. (2017). Draft standard framework for prognosis and health management (PHM) of electronic systems. IEEE 1856/D33.

Medjaher, K. and Zerhouni, N. (2013). Framework for a hybrid prognostics. *Chemical Engineering Transactions* 33: 91–96. DOI: 10.3303/CET1333016.

National Science Foundation Center for Advanced Vehicle and Extreme Environment Electronics at Auburn University (CAVE3). (2015). Prognostics health management for electronics. http://cave.auburn.edu/rsrch-thrusts/prognostic-health-management-for-electronics.html (accessed November 2015).

Pecht, M. (2008). *Prognostics and Health Management of Electronics.* Hoboken, NJ: Wiley.

Ridgetop Group. (2013). Phase 2 final report: physical modeling for anomaly diagnostics and prognostics. NASA Contract NNX11CA04C, NASA Ames RC, Moffett Field, California, US, 7 June.

Ridgetop Group. (2018). Sentinel network, view of the graphical user interface for an electronic power supply (EPS). Courtesy and permission of Ridgetop Group, Inc., 3560 West Ina Road, Tucson, AZ, 85741.

Further Reading

Filliben, J. and Heckert, A. (2003). Probability distributions. In: *Engineering Statistics Handbook*. National Institute of Standards and Technology. http://www.itl.nist.gov/div898/handbook/eda/section3/eda36.htm.

Jenq, Y.C. and Li, Q. (2002). Differential non-linearity, integral non-linearity, and signal to noise ratio of an analog to digital converter. Portland, Oregon: Department of Electrical and Computer Engineering, Portland State University.

O'Connor, P. and Kleyner, A. (2012). *Practical Reliability Engineering*. Chichester, UK: Wiley.

Tobias, P. (2003a). Extreme value distributions. In: *Engineering Statistics Handbook*. National Institute of Standards and Technology. https://www.itl.nist.gov/div898/handbook/apr/section1/apr163.htm.

Tobias, P. (2003b). How do you project reliability at use conditions? In: *Engineering Statistics Handbook*. National Institute of Standards and Technology. https://www.itl.nist.gov/div898/handbook/apr/section4/apr43.htm.

7

Prognostic Enabling: Selection, Evaluation, and Other Considerations

7.1 Introduction to Prognostic Enabling

As you have already learned, a key objective of deploying a prognostic-enabled system is to monitor prognostic targets to provide advanced warning of failures in support of condition-based maintenance (CBM). There are criteria associated with, for example, equipment availability and other metrics, test coverage, and confidence levels. To meet the criteria, the various sensing, signal-processing, and computational (algorithms) routines in a prognostics and health management/monitoring (PHM)[1] system need to be factored into the entire design. CBM methods and approaches – especially those using condition-based data (CBD) signatures that are ultimately transformed into functional failure signature (FFS) data that is processed by a very good prognostic information program – provide significant advantages over (i) a system based on statistical or other methods applicable to populations rather than a specific instantiation of a population and (ii) a system based on using CBD to detect damage without prognosing when such damage will result in the system no longer operating within specifications.

7.1.1 Review of Chapter 6

Chapter 6 presented a design of an exemplary prototype of a PHM system that prognostic-enabled multiple instantiations of systems and prognostic targets with excellent results (see Table 7.1).

The design supported multiple subsystems (two), each comprising prognostic targets: a power supply and two electro-mechanical actuators (EMAs). The design included monitoring each of the two power supplies for a single mode of failure and monitoring each of the four EMAs for three failure modes. The monitoring, conditioning, and processing were all based on CBD signatures, and the prognostic approaches and methods provided excellent, if not superior, results:

- Degradation was detected at least 96% of the distance in time before functional failure occurred.
- Estimate accuracy was within 25% at least 93% of the distance in time before functional failure occurred.

1 There are other names for PHM. A common one is integrated vehicle health management (IVHM); another is prognostic health management.

Prognostics and Health Management: A Practical Approach to Improving System Reliability Using Condition-Based Data, First Edition. Douglas Goodman, James P. Hofmeister and Ferenc Szidarovszky.
© 2019 John Wiley & Sons Ltd. Published 2019 by John Wiley & Sons Ltd.

Table 7.1 Performance measurements and metrics.

Prognostic target	PI specifications	BD	EOL	PD	PH		
Name	PITTFF0	@ time	@ time	Maximum	$PH_{ERROR} = (PITTFF0/PD_{MAX})$		
	FFS NM	Estimated BD	Estimated	FOM	25% @ time	10% @ time	
					χ [%]–pts [#]	χ [%]–pts [#]	
SMPS	4800 h	1368 h	4200 h	2939 h	Initial PH error = 63%		
	3%	1261 h	4176 h	96.4%	1560 h	2760 h	
					93.5%–9 pts	52.6%–23 pts	
EMA load	4800 h	504 h	3168 h	2727 h	Initial PH error = 76%		
	2%	441 h	3164 h	97.7%	576 h	1368 h	
					97.3%–3 pts	68.3%–19 pts	
EMA winding	4800 h	1104 h	2760 h	1693 h	Initial PH error = 184%		
	2%	1067 h	2745 h	97.8%	1320 h	1680 h	
					87.2%–10 pts	66.0%–25 pts	
EMA power	4800 h	960 h	4440 h	3476 h	Initial PH error = 38%		
Transistor	1%	958 h	4434 h	99.9%	984 h	1512 h	
					99.3%–2 pts	84.1%	

- Estimate accuracy was within 10% at least 52% of the distance in time before functional failure occurred.
- A time to failure (TTF) value of 9600 hours was used to specify a prognostic TTF (using PITTFF0) value of 4800 hours:
  - All four of the prognostic targets that failed did so prior to that time: at estimated times of 2745 hours, 3164 hours, 4176 hours, and 4434 hours.
  - The actual TTF was 3630 hours.

7.1.2 Electronic Health Solutions

Electronic health solutions, such as those described in this book, become part of a PHM system. They are sometimes referred to as a *prognostic ecosystem* (see Figure 7.1), within which such solutions can be categorized at levels as shown in Figure 7.2: die, component, board, module, and system (Ridgetop Group 2018). Other levels could be added, such as an assembly of boards and a collection of modules into a replaceable unit.

Ecosystems

An ecosystem can be described as prognostic models within a system that includes descriptions of data, quantification of uncertainty, justification and validation of model selection, and limitations of application (Astfalck et al. 2016). The locations in the broader view of an ecosystem shown in Figure 7.1 are the following: location 1 is a system or subsystem comprising one or more line-replaceable units (LRUs) that are prognostic enabled (monitored for damage and/or degradation); location 2 is a PHM system that acquires, manipulates, manages, and processes data to produce prognostic

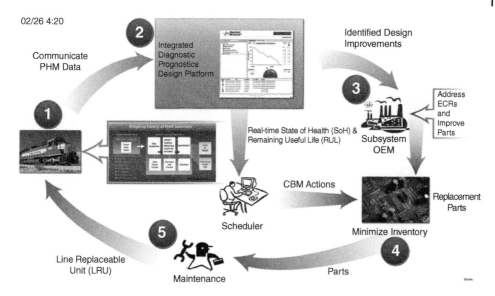

Figure 7.1 Example of a broad view of an ecosystem.

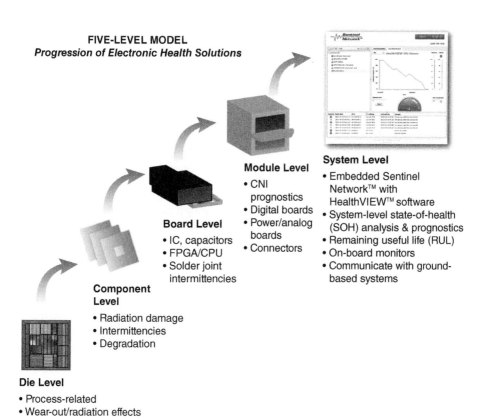

Figure 7.2 Example of a five-level model for health solutions.

information that is used to initiate a service and maintenance action; location 3 is where failures are analyzed and products are improved by a supplier; location 4 are repositories of LRUs, assemblies, components, and devices used for service and repair; and location 5 are maintenance personnel who perform service and maintenance.

Levels of Solutions

A complex PHM system contains devices, components, boards, subassemblies, and so on. A sensor is attachable to any node within a system, and therefore health solutions that process sensor data can be categorized in accordance with the node to which the sensor is attached, as exemplified by the five-level model of health solutions shown in Figure 7.2 (Ridgetop 2018).

7.1.3 Critical Systems and Advance Warning

Critical systems are considered vital to the ongoing operation of everyday life, and criticality is a key consideration when evaluating and selecting a node for prognostic enabling (a prognostic target). For example, a power system in an aircraft, or a gear box in a wind turbine, would be considered critical, since their operation is essential to meet design objectives of the overall system. Another example of criticality is the safety of life and health: prevention and avoidance of loss of life and injury is a primary objective of a system. Fault severity and fault propagation also play a role in the definition of systemwide criticality.

Advance warning, such as an alert, of any impending failure of a mission-critical or safety-critical prognostic target is vital: a properly designed PHM system will provide detection of anomalies that affect the ability of the system to operate within specifications and issue appropriate alerts. For example, Chapter 6 included examples of messages and alerts issued by an exemplary prototype PHM system. A PHM system will issue alerts (health monitoring) and/or initiate appropriate actions (health management) such as soft shutdowns, load shedding, and scheduling maintenance. There might also be various levels of alerting where threshold levels and fault models can be used to prioritize what information is available to an operator of an aircraft, or a seagoing ship, or a machine tool on the manufacturing floor. This brings in the notion of fault severity, access to the information, and what is done to mitigate an issue that results in an alert.

7.1.4 Reduction in Maintenance

To save money and resources, maintenance intervals can be optimized based on actual evidence of degradation. For example, a system might have components that fail after 250 hours of operation, some that fail after about 600 hours, and others that fail at about 500 hours. A usage-based PHM system might be designed to do one of the following:

1. Schedule repair and/or replace maintenance for all units on or before 200 hours of operation – a 20% safety margin with respect to failure.
2. Schedule repair and/or replace maintenance for all units on or before 400 hours of operation.

In a first design of a PHM system, an objective might be to avoid all unexpected failures, but at increased maintenance costs: for example, an average 300 hours of lost usage for each instance of avoidance, and cost increases due to increased maintenance actions (more frequent replacement). A second design might focus on reducing sustainment costs by increasing the time between maintenance actions – but unexpected failures would increase. Typically, even disregarding mission and safety issues, the cost of an unexpected failure is higher than an early repair-and-replace action.

So, instead of a usage-based PHM system, we advocate CBM using a PHM system that is CBD-based; uses signature-based detection and prognostic approaches and methods; and employs a fast, highly accurate set of data-conditioning, prediction, and computational routines. The advocated approach of using CBD signatures is an effective method for handling variability introduced by the operational environment: operating in the desert of Arizona is very different than operating equipment in a rainy, cold environment such as that in Puget Sound, Washington.

7.1.5 Health Management, Maintenance, and Logistics

This book is focused on prognostic enabling to monitor the health of a system of nodes: the prognostic targets chosen because those nodes have signals that change in response to degradation of devices, components, and so on that, when they fail, have a critical effect on the operation of the system. They may cause mission-critical functions to cease or otherwise operate out of specifications, or they may create a hazardous threat to the safety of the system or life.

But monitoring, per se, does not avoid unexpected outages, does not repair anything, and does not prevent loss of life. Refer back to Figure 7.1: a PHM system needs to provide services for health management, maintenance, and logistic support to schedule maintenance, locate and deliver parts and equipment, and dispatch a maintenance team.

Management

Given the accuracy of the prognostic information produced by the PHM system in Chapter 6, it would not be unreasonable to defer maintenance until a detected state of health (SoH) value at or below a specified level, such as 25%. PHM management support might be designed to act on alerts, such as those shown in Figure 7.3, which are

```
. . .
W19: SOH for NODE: 61 IS 24.9286 PERCENT
I44: RUL IS 392.3 [hr] AT DATA TIME: 2352 [hr]
. . .
W19: SOH for NODE: 51 IS 24.5532 PERCENT
I44: RUL IS 658.2 [hr] AT DATA TIME: 2520 [hr]
. . .
W19: SOH for NODE: 62 IS 24.7536 PERCENT
I44: RUL IS 825.9 [hr]AT DATA TIME: 3528 [hr]
. . .
W19: SOH for NODE: 50 IS 24.8263 PERCENT
I44: RUL IS 402.2 [hr] AT DATA TIME: 3696 [hr]
. . .
```

Figure 7.3 Example of alerts for SoH at or below 25%.

. . .

W19: DAMAGE IS DETECTED AT NODE: 51 AT DATA TIME: 504 [hr]

. . .

W19: DAMAGE IS DETECTED AT NODE: 62 AT DATA TIME: 960 [hr]

. . .

W19: DAMAGE IS DETECTED AT NODE: 61 AT DATA TIME: 1104 [hr]

. . .

W19: DAMAGE IS DETECTED AT NODE: 50 AT DATA TIME: 1368 [hr]

. . .

Figure 7.4 Example of alerts for a damage-detection approach.

excerpted from Chapter 6. In contrast, PHM management support might be designed to act on damage-detected alerts, such as those shown in Figure 7.4.

Example 7.1   For the example prognostic alerts shown in Figure 7.3 from a PHM system with prediction accuracy within 10% for better or SoH estimates at or below 25%, and maintenance scheduled 72 hours after an alert, a cost-effective set of maintenance actions might be the following:

- Repair the EMA at node 61 at 3324 hours: at least 280 hours before failure.
- Repair the EMA at node 51 at 2592 hours: at least 510 hours before failure.
- Repair the EMA at node 62 at 3692 hours: at least 675 hours before failure.
- Repair the switch-mode power supply (SMPS) at node 50 at 3762 hours: at least 500 hours before failure.

Example 7.2   For the example detection alerts shown in Figure 7.4 from a PHM system using a conservative approach to maximize system health and minimize unscheduled outages, maintenance might be scheduled to occur 72 hours after a detection alert:

- Repair the EMA at node 51 at 576 hours: more than 3324 hours before failure.
- Repair the EMA at node 62 at 1032 hours: more than 2592 hours before failure.
- Repair the EMA at node 61 at 1176 hours: more than 2692 hours before failure.
- Repair the SMPS at node 50 at 1440 hours: more than 3762 hours before failure.

Maintenance

A PHM system needs to alert users when maintenance is required – based on physical evidence of degradation, and not on an arbitrary number of elapsed hours. In addition to alerts, a PHM needs to provide for and support maintenance-related services to avoid unnecessary replacements, increase usage of systems, decrease downtime, and reduce sustainment costs. The approaches and methods used for maintenance are application specific, need be integrated with health management and logistic services, and are beyond the scope of this book.

Logistics

A critical function of a PHM system is logistics support. Parts and equipment must be located and delivered to the service and repair site, and a service and maintenance team needs to be dispatched to arrive on or after, but not before, the arrival of needed parts and equipment. Additionally, maintenance and inventory records need to be updated; and suppliers, vendors, and manufacturers must be notified per contractual obligations. The latter is especially true when dealing with a government agency such as the Department of Defense. Logistic support might also be required to arrange for and record the outcome of ancillary activities such as cause-and-effect review of repairs.

7.1.6   Chapter Objectives

The previous chapters were focused on PHM aspects deemed critical to the design of a PHM system, including approaches and methods not suitable for CBD-based prognostics. The overall objective of this book is to provide you with the knowledge to understand, evaluate, design (at least at a high level), and verify health monitoring. The introduction of this chapter has briefly introduced you to the concept of ecosystems, critical systems and warnings, reduction in maintenance, and health management. The remainder of this chapter is devoted to the evaluation, selection, and specifications of prognostic targets: nodes to be monitored to detect damage and provide prognostic information, to avoid unscheduled outages of critical functions and loss of safety in a system.

7.1.7   Chapter Organization

The remainder of this chapter is organized to present and discuss topics related to prognostic enabling:

7.2 Prognostic Targets: Evaluation, Selection, and Specifications
  This section includes descriptions of the meaning and relationship of TTF, time before/between failure (TBF), prognostic distance (PD), and prognostic horizon (PH); distributions of the onset of degradation and functional failure; mean time to failure (MTTF); and mean time before/between failure (MTBF).
7.3 Example: Cost-Benefit of Prognostic Approaches
  This section is devoted to the cost-benefit analysis of prognostic approaches and includes example comparisons of no PHM, two usage-based approaches, CBD-based detection, and CBD-based prognostics.
7.4 Reliability: Bathtub Curve
  This section is devoted to the bathtub curve, prognostic triggers, and the relationship of the bathtub curve to failure rate and MTBF.
7.5 Chapter Summary and Book Conclusion
  This section summarizes and ends both the chapter and the book.

7.2   Prognostic Targets: Evaluation, Selection, and Specifications

Selecting a target to be prognostic enabled probably seems pretty straightforward: collect and analyze historical records pertaining to maintenance and repair to identify those targets having high rates of failure and/or failure of mission-critical and/or safety-critical parts regardless of failure rate. Prepare a cost-benefits business case: cost to replace or repair, cost associated with unplanned failure, savings due to prolonged time in use, savings due to reduction in sustainment costs and unplanned downtime, and so on. But you also need to factor in a hard-to-quantify cost related to criticality of mission and/or safety (refer back to Section 1.8).

7.2.1 Criteria for Evaluation, Selection, and Winnowing

Because you are designing a PHM system to prognostic enable an operational system, your team will not perform any traditional failure mode and effect analysis (FMEA) or failure mode effect and criticality analysis (FMECA) (DAU 2018). Instead, your team will review the existing FMEA and FMECA data; the historical failure, service, and repair data; and any other data related to failure. The focus is on identifying, selecting, and winnowing prognostic targets. To select and winnow a list of prognostic candidates, including those identified as candidates by FMEA/FMECA, you need to know the following:

- *TTF or TTFF*. An estimate of the time to failure (alternatively, time to functional failure). This is a primary focus of this chapter.
- *PD (estimated)*. An initial estimate of the distance in time between the onset of degradation and functional failure from which a prediction algorithm converges upward or downward to a true PH.
- *Failure mode to be detected*. The failure mechanism that causes the characteristic shape of the signature captured by one or more sensors.
- *Severity classification*. A qualitative assessment of the consequences of functional failure.
- *Cost of failure*. An estimate of the cost of an unplanned failure versus the cost of service and repair before failure.
- *Cost of prognostic enabling*. An estimate of, for example, the cost of the sensor, PHM support, design, development, testing, qualification, and fielding.
- *Cost-benefit analysis*. An estimate of, for example, the change in sustainment cost plus time in use due to prognostic enabling, the effective savings achieved, and the change in service actions because of prognostic enabling.

7.2.2 Meaning of MTBF and MTTF

PHM systems are typically referenced to either MTBF or MTTF, and therefore you should know and understand the difference between them (Speaks 2005):

- *MTBF*. Calculated mean time between (or before – context dependent) failures of a repairable system.
- *MTTF*. Expected TTF of a nonrepairable system.

As you can see, the meaning of (and therefore the use of) these terms is dependent upon the definition of *failure* and the definition of *repairable*. To achieve an understanding of those terms, recollect that Chapter 2 introduced failure in time (FIT) in Eq. (2.15):

$$\text{Failure rate} = \lambda = \frac{(\text{number of failures})}{(\text{number of tested parts})(\text{hours of test}) \, AF} \, 10^9 \, \text{FIT}$$

where AF is the value of an acceleration factor for a specified test. Refer to Tables 2.3 and 2.4 for examples.

Example 7.3 Suppose a particular system has a unit (device, component, assembly, and so on) that is tested for a vendor according to a required test regime that calls for 40 units to be tested for 2500 hours. The AF value for the required test is 10 000 hours per highly accelerated life test (HALT) cycle, and a HALT cycle has a duration of 1 hour. When the HALT is run, 1 of the 40 fails. The failure rate would be calculated as

$$\text{Failure rate} = \lambda = \frac{(1)}{(40)(2{,}500 \text{ cycles})(10{,}000 \text{ hours/cycle})} \; 10^9 = 1 \text{ FIT}$$

But you are not given a failure rate: instead, you are told that the FIT number is 50. Now you need to know the following to relate that FIT number to a failure rate (Ellerman 2012; NIST 2018):

$$\text{FIT} = 1/\lambda \tag{7.1}$$

where 1 FIT $= 1/10^9$ hours.

Example 7.4 Given a FIT number of 50, you calculate the failure rate:

$$\lambda = 1/(50 \; FIT) = 10^9/50 = 20{,}000{,}000 = 20 \; \textit{million hours!}$$

Even though your research confirms that your calculation is correct, because the calculated rate of failure is so large, you decide to calculate an MTBF (mean time before failure) value using Eq. (7.1) (Abernethy 2006; RAC 2005; Speaks 2005; Weibull 2008):

$$MTBF = (\textit{total time})/(\textit{number of failures}) \tag{7.2}$$

But this does not help, because you do not know the total time, which you know is calculated using Eq. (7.3):

$$\textit{Total time} = (\text{\# of tested units}) * (\textit{test time}) \tag{7.3}$$

You also don't know the number of tested units or the test time. So, you find another expression for MTBF

$$MTBF = 10^9/FIT \tag{7.4}$$

which, for FIT $= 50$, is equal to the Example 7.4 calculated value for λ: 20 000 000 hours. Literature research reveals the following:

- In 1993, MTTF was defined as "mean time to first" failure (Seymour 1993), and the expression to calculate that value is given as

$$MTTF = (\textit{total time})/(\textit{number of failures})$$

- So, MTTF (circa 1993) is calculated the same way as MTBF in 2018, when *B* means *before*.
- You discover that MTBF is also defined as "mean time between failures" for a repairable product.
- MTTF is defined as "mean time to failure" (no reference to first) for a nonrepairable product.
- A relationship of MTBF to MTTF is defined as follows (Ellerman 2012):

$$MTBF = MTTF + MTTR \tag{7.5}$$

where MTTR is defined as "mean time to repair."

- Ergo, you have a multiplicity of definitions and acronyms that are ambiguous and cause uncertainty regarding use and meaning.

Regardless of meaning and/or definition, neither MTTF nor either of the two definitions of MTBF is useful for prognostic enabling. MTTF and MTBF should be limited to a classical definition of reliability (Section 1.6): without intervention, there is a 63% probability the system will fail prior in time to the value of MTTF or MTBF.

7.2.3 MTTF and MTBF Uncertainty

Figure 7.5 illustrates the relationship of a failure distribution (density of failures/time), the MTBF (failure rate *between* failures), the MTBF (mean time *before* a first failure), and MTTF: MTTF and MTBF were originally defined for an exponential distribution having a constant, low failure rate: for example, solid-state (integrated circuit) devices. Those devices are subjected to one or more accelerated tests, such as a HALT, with test results extrapolated to normal life using an AF (Ellerman 2012; O'Connor and Kleyner 2012; NIST 2018; RAC 2005; Speaks 2005; Wilkins 2002).

Typical commercial FIT values for solid-state devices are in the 50–1000 range, with FIT values in the range of 1–10 for space applications (Johnston 2010). There are simulators that calculate values called MTTF and MTBF using simulated failure times (Weibull 2008), which adds even more uncertainty as to the meaning of and/or calculation of a particular MTTF and/or MTBF value.

Even worse, different failure distributions, different CBD signatures, and so on, can result in identical (or nearly identical) reliability metrics such as MTTF: compare Figures 7.5 and 7.6. Finally, this book asserts that attempts to use a TTF value of hundreds of thousands (or larger) of hours for CBD-based prognostics is nonsensical: an MTTF value of 100 000 hours is equivalent to more than 11 years.

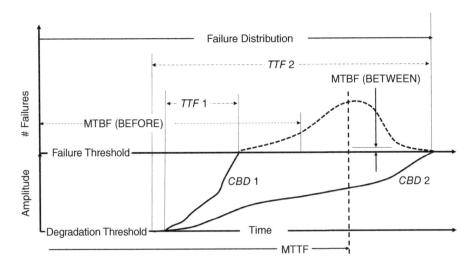

Figure 7.5 MTTF, TTF, and PITTFF0: CBD signature and failure distribution.

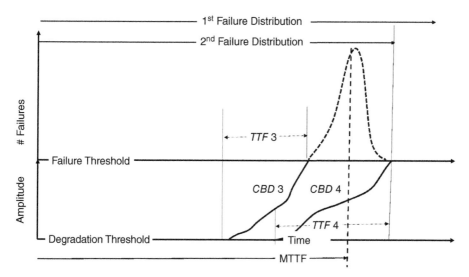

Figure 7.6 Same MTTF for different failure distributions and signatures.

7.2.4 TTF and PITTFF

We need to know how to determine, calculate, and/or estimate a TTF value that begins when degradation begins and ends when functional failure occurs (see TTF 1 and TTF 2 in Figure 7.5). But at the time when degradation is first detected, there is no a priori knowledge of that future time of failure; yet our prediction program needs to converge from an initial estimate of that time to a very accurate estimate of the time of failure. Research into reliability metrics such as MTTF, MTBF, and FIT indicates that they are not really close in value to what we need for TTF.

The prediction program we are using, ARULEAV, provides a parameter called PIT-TFF0 (introduced in Chapter 6) as a means to specify an initial value for TTF. TFF is not a value that is looked at or specified by manufacturers and/or vendors of products; failures in the field are usually due to either an anomalous event such as a lightning strike or degradation. Degradation typically is not caused by a part entering what is referred to as the *wear-out region* of a bathtub curve; rather, degradation is typically due to an accumulation of fatigue damage caused by cyclic stresses and strains (such as thermal and mechanical) during operation (Hofmeister et al. 2006).

We can estimate an initial value for TTF using a number of methods: a service-life determination, an end-use test method, or an MTTF-based method. But be aware that the supplier of the prediction program advises that, in general, that program converges to within 25% accuracy in less time when the initial estimate is higher, rather than lower, in comparison to the true time of functional failure.

TTF: Service Life Determination
Instead of using MTTF or MTBF values for an extremely low failure rate, you might use end-use values based on service-life values from vendors. For example:

- A tire is warranted to last 5 years or 40 000 miles, and on or about that amount of usage, the tire is deemed "worn out" and is replaced: the tire is not repairable. The new tire (same brand and type) is also warranted to last 40 000 miles. The tire can be said to have a TTF of 40 000 miles.

- A fuel pump is warranted to last 5 years or 40 000 miles, whichever occurs first. When the fuel pump fails, it is replaced with a remanufactured pump that is warranted to last 2 years or 25 000 miles, whichever occurs first: the fuel pump is deemed repairable. Since the repaired pump has a smaller expected lifetime, the fuel pump can be said to have a time before failure (TBF) of 2 years or 25 000 miles: less than the original expected lifetime. In this situation, choose the larger value.

Set the PITTFF0 parameter to twice the value you have of the service-life determined value for TTF, for three reasons:

- The supplier advises that the prediction program generally converges to a solution from a high initial error.
- Reliability theory states that a mean lifetime, TTF, means 63% of failures are likely to occur before that mean.
- Human factors: a low, initial error means that the prediction program is likely to produce increasing values of remaining useful life (RUL) for decreasing values of SoH. For example, Figure 7.9 plots the prediction results for the same data using a high value for PITTFF0 (left-hand plot) and a low value for PITTFF0 (right-hand plot): at best, the results are perplexing to an operator; at worst, they may cause distrust and loss of confidence in the reliability of the PHM system.

TTF: End-Use Test Method

You can calculate TBF and TTF values in the same way as MTBF and MTTF values, respectively, using the following (Weibull 2008):

$$TBF = (total\ operating\ hours)/(\#\ failures)\ \text{as for MTBF} \tag{7.6}$$

$$TTF = (total\ failure\ hours)/(units\ tested)\ \text{as for MTTF} \tag{7.7}$$

Example 7.5 In Chapter 6, an example system of six LRUs was used: two power supplies and four EMAs, of which four of the six LRUs failed. Use that data to calculate the values for TBF and TTF, and tabulate the results (see Table 7.2). For example, at the end of 5000 hours of testing six LRUs and four failures:

- TBF = (6*5000)/4 = 7500 hours
- TTF = (2745 + 3164 + 4176 + 4434)/4 = 3630 hours

There is a significant difference between the TTF and TBF values, as shown in Table 7.2. The difference is due to TTF values being calculated using independent failure times, whereas TBF values are calculated using dependent failure times, because of a common understanding of the word *between*. Since it seems reasonable to use TTF rather than TBF, and since the TTFs are more consistent, you decide to use TFF: any question of whether to include the time of the first failure when calculating TBF becomes moot.

Again, set PITTFF0 to twice the value you calculate.

Table 7.2 Summarized list of TBF and TTF calculations for various periods of operation.

TBF and TTF for various periods of operation

Operating time (h)	Total LRUs	Failed LRUs	TBF (h)	TTF (h)
3 000	6	1	$18\,000/1 = 18\,000$	$2745/1 = 2745$
3 500	6	2	$21\,000/2 = 10\,500$	$5909/2 = 2955$
4 000	6	2	$24\,000/2 = 12\,000$	$5909/2 = 2955$
4 500	6	4	$27\,000/4 = 6750$	$14519/4 = 3630$
5 000	6	4	$30\,000/4 = 7500$	$14519/4 = 3630$

TTF: MTTF-Based Method

Referring to Figure 7.5, a simplistic MTTF-based method would be to set TTF equal to MTTF, but that method works well when the spread of the majority of the failures is wide compared to the value of MTTF. In such cases, simply setting PITTFF0 to twice the value of MTTF suffices. But if you are fairly confident that the situation is more like that illustrated in Figure 7.6, you need to specify a lower value for PITTFF0. However, it might be the case that your PHM system supports the same type of LRUs in two distinct operating environments: one that induces earlier-than-expected failures (akin to a situation like that shown in Figure 7.5) and a second that is less variable and causes failures to be more closely bunched together (akin to a situation like that shown in Figure 7.6). Further, to avoid misunderstanding and/or for procedural reasons, suppose you must always set PITTFF0 to a value (such as MTTF) specified by a manufacturer, vendor, or governmental agency. In such situations, you need a method to cause the prediction program to adjust the specified PITTFF0 value. The supplier of your prediction program agrees, and changes are made to provide a node-definition parameter, PITTFADJ, to allow you to adjust how the value of PITTFF0 is handled:

- When PITTFADJ is not specified or is less than 1, the value of PITTFF0 is used as an initial value for calculating the next RUL and PH values.
- When PITTADJ $\geq$ 1, the initial value is calculated as $[1 - \exp(-\text{PITTADJ})]$ (PITTFF0).

Example 7.6 From Table 7.1, TTF was calculated at 3630 hours; and from Eq. (7.7), PD(EST) = 2722 hours. So, you decide to set PITTFF0 = 2722 and do the following (Figure 7.7):

- Compare this calculated result against the original result of using PITTFF0 = 4800 (see Table 7.3).
- Create comparison plots of the maximum PD versus the estimated PD using the calculated value of PITTFF0 (see Figures 7.8 and 7.9).
- Create comparison plots of RUL and PH using the original value and the calculated value for PITTFF0 (see Figure 7.10).
- Calculate and tabulate the percentage difference between the estimated, initial PD using the new calculated PITTF0 value and the original value of 4800 hours.

The results indicate that there is a reduction in the difference, and the plots confirm that the prediction program converges from either a high or low initial estimate.

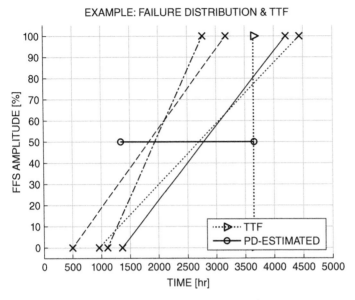

Figure 7.7 Failure plots with average values for TTF and PD = PITTFF0.

Table 7.3 Tabulated calculations for PD(EST).

PD(EST) calculations and comparison to measured values							
LRU type	BD detected	Failure detected	PITTFF calculated	PD(EST) measured	PITTFF-PD(EST)	% Difference Calculated	Table 7.1
EMA	504	3168	2732	2664	−370	−14%	61%
EMA	960	4440	2732	3480	−1186	−34%	10%
EMA	1104	2760	2732	1656	638	39%	123%
SMPS	1368	4200	2732	2832	−538	−19%	21%

Example 7.7 As another test, you set the new parameter to 2.0 (PITTFADJ = 2.0), recalculate the values, and run tests to create new plots. The before and after results are tabulated in Table 7.4 and plotted in Figures 7.11–7.14. There are now both positive and negative differences, which is what you would expect from using an estimated initial value; and, with the exception of Figure 7.11, convergence to 25% is improved or about the same.

7.3 Example: Cost-Benefit of Prognostic Approaches

We will use the SMPS-EMA examples from Chapter 6 as a base platform to construct an example situation to illustrate the cost-benefit for various approaches. You and your customer agree that cost-benefit analyses are not to include unavoidable catastrophic failures (such as, for example, being hit by another vehicle) and that, because of criticality

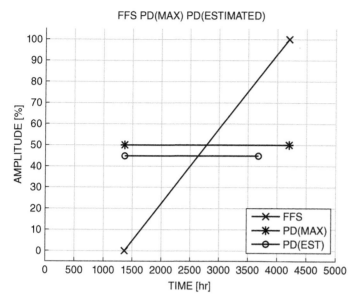

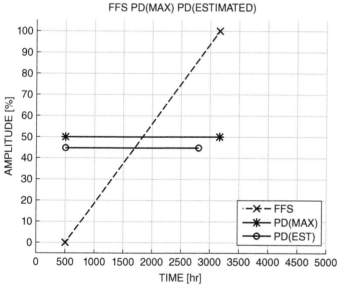

Figure 7.8 Estimated PD and actual PD: power supply (top) and EMA load (bottom).

considerations, repair-and-removal activity due to unexpected functional failures will be held to less than 5% of the total number of repairs and removals.

7.3.1 Cost-Benefit Situations

The situations are the following: (i) none; (ii) usage-based, MTTF; (iii) usage-based, 2/3 MTTF; (iv) CBD-based, replace when damage is detected; and (v) CBD-based, replace within 720 hours after estimated SoH becomes 75% or less. Although the first situation

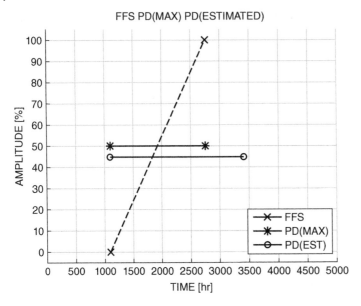

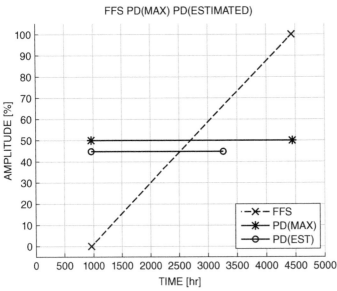

Figure 7.9 Estimated PD and actual PD: EMA winding (top) and EMA transistor (bottom).

fails to meet the requirement that unexpected failures will be less than 5%, it establishes a baseline estimate of cost.

Your customer arranged for special delivery of 6 power supplies and 12 EMAs. Those 18 units were subjected to end-use tests similar to HALTs (refer back to Section 2.2). Your team assisted in the design of the experiment and building of the test beds; the tested units (power supplies and EMAs) needed to be prognostic enabled, as described in Chapter 6. For analysis purposes, test failures are to be evaluated as though all of

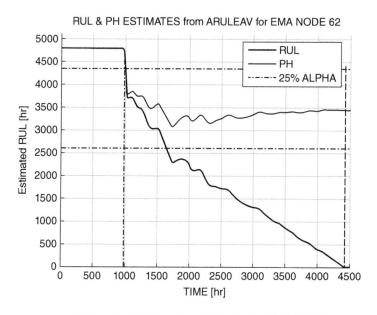

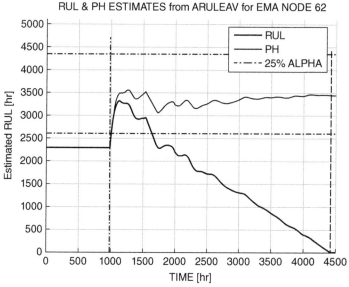

Figure 7.10 RUL and PH plot for PITTFF0 = 4800 and for PITTFF0 = 2290.

the tested units were installed at the same time and failed in the sequences and times indicated by the test.

The numbers to be used in a cost-benefit analysis are provided by the customer and listed in Tables 7.1 and 7.5. After examining the rest results (see Figure 7.15 and Table 7.6), your customer concludes the cost-benefit analysis for the power supply scenario is sufficient for evaluation of the five approaches.

Table 7.4 Tabulated calculations using enhanced program with PITTFF0 = 4800.

				Comparison of PD(EST) before and after ARULEAV is enhanced			
				PITTFF0 = 4800 (h): Initial PD (PD0) = PITTFF0*PITTFADJ			
	BD	Failure	PD	PD0 - PD		% Difference	
LRU type	detected	detected	measured	Before	After	Before	After
EMA	504	3168	2664	4296	3646	61%	37%
EMA	960	4440	3480	3840	3190	10%	−8%
EMA	1104	2760	1656	3696	3046	123%	84%
SMPS	1368	4200	2832	3432	2782	21%	−2%

7.3.2 Cost Analyses

No PHM Approach
In this scenario, the system runs until a unit fails – a "do nothing until failure occurs" approach. When failure occurs, the failed unit is removed and replaced. The system is restarted and runs until the next unit fails, and so on, until the end of a sustainment period of 14 400 hours (600 days of operation):

- On average, #1 power supplies would fail and be replaced: 14 400/4200 = 3.4 times:
  - Cost = 3.4*10000 + 0*2000 + 3.4*4000 = 47 600

Repeating this for the remaining five power supplies results in 24 removals and replacements because of unexpected outages due to degradation failure, at a baseline cost of $336 000.

Usage-Based MTTF Approach
In this scenario, all power supplies are replaced when usage equals 3592 hours (MTTF). Examination of the data in Table 7.6 shows that power supplies #3, #4, and #6 would functionally fail before they are replaced. In the sustainment period, a total of 27 power supplies would be removed and replaced, at a total cost of $354 000 – an increase of $18 000 per system during the sustainment period to reduce the number of unexpected outages from 24 to 14. However, 52% of all repair and removal actions would be attributable to degradation failures. This approach fails to meet requirements.

Usage-Based 2/3 MTTF Approach
In this scenario, all power supplies are replaced when usage equals 2395 hours. There would be no unexpected outages, but 36 power supplies would be replaced at a cost of $432 000 – an increase in cost of $96 000 and a reduction of unexpected failures to zero. That $96 000 becomes $96.0 million for a population of 1000 systems.

Damage-Detection Approach
In this scenario, whenever damage is detected in a power supply, it is removed and replaced within 720 hours, which would also, for these supplies, result in zero unexpected outages. This approach is seemingly a good one because there would be

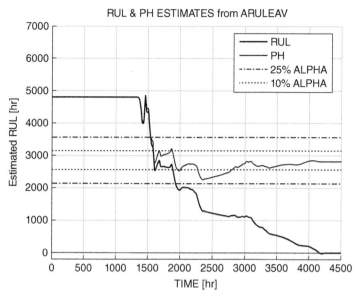

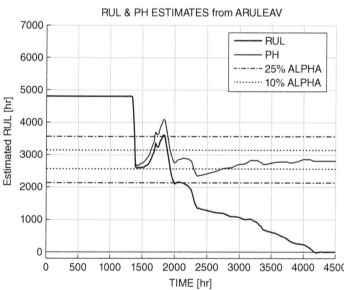

Figure 7.11 RUL and PH plots for SMPS: before and after PITTFADJ = 2.0.

no unplanned outages due to degradation leading to failure. But there would be a large increase in removal and replacement activity: a total of 43 power supplies in the 14 400-hour sustainment period at an estimated cost of $516 000 per system – an increase of $180 000 more than the baseline cost per system, which becomes $180.0 million for a population of 1000 systems.

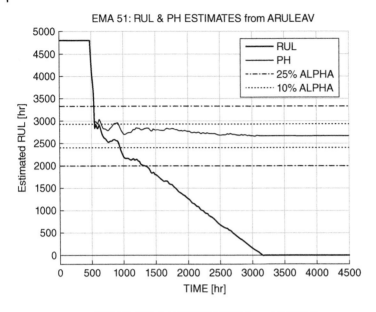

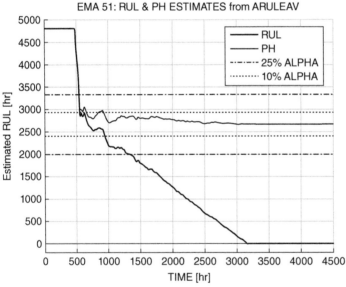

Figure 7.12 RUL and PH plots for EMA 51: before and after PITTFADJ = 2.0.

SoH at 75% or Less Approach

In this scenario, whenever a prognostic SoH estimate is 75% or less for a power supply, it is removed and replaced within 120 hours: again, for this approach, there would be zero unexpected outages. This approach would result in the removal and replacement of 28 power supplies during the sustainment period at an estimated cost of $336 000 per system – neither a savings nor an increase in cost compared to the baseline cost of a "do nothing until failure" approach.

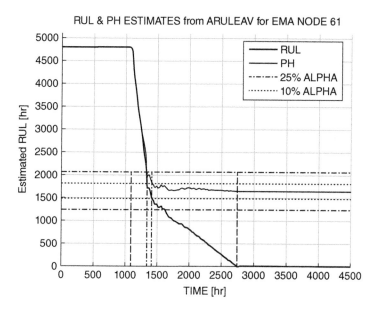

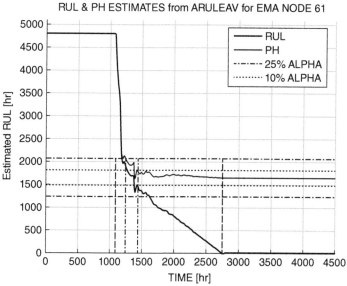

Figure 7.13 RUL and PH plots for EMA 61: before and after PITTFADJ = 2.0.

Yes, there is the cost of the sensors and PHM systems, but the SoH approach has the following advantages:

- It reduces unexpected outages due to degradation failure to zero.
- It is less expensive than the next-most-expensive approach (usage-based 2/3 MTTF).
- It has the fewest repair and replace activities (28) compared to the usage-based 2/3 MTTF approach (36) or the damage-detection approach (43).

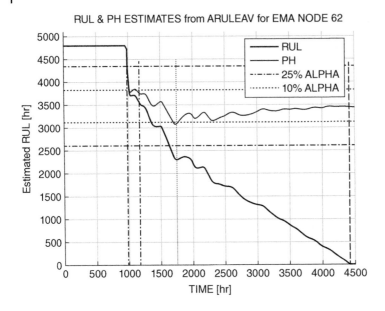

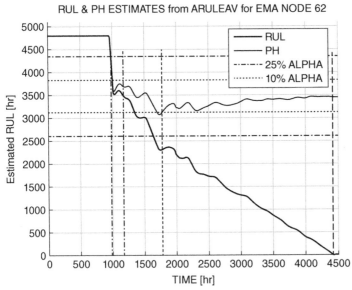

Figure 7.14 RUL and PH plots for EMA 62: before and after PITTFADJ = 2.0.

- It requires the smallest number of repair hours and thereby increases mission-availability time.
- It does not have the drawback of unexpected early onset of degradation and/or faster-than-expected time from onset to failure – which are drawbacks of the usage-based 2/3 MTTF approach.
- It saves $180 million compared to the damage-detection approach for a 1000-system network over a 14 400-hour sustainment period.

Table 7.5 Cost estimates for benefits evaluation of prognostic enabling.

| LRU name | Estimated costs: use for benefits analysis | | | | Sustainment (Use) |
	Acquisition	Scheduled R&R	Unplanned failure	Expected life (h)/LRU	Period (h)/LRU
Power supply	$10 000	$2 000	$4 000	3 500	14 400
EMA	$25 000	$3 000	$6 000	3 500	14 400

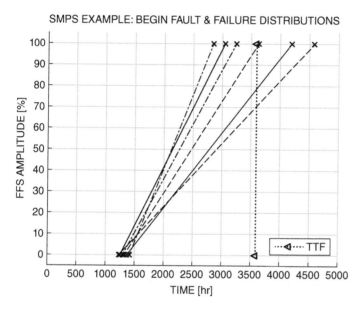

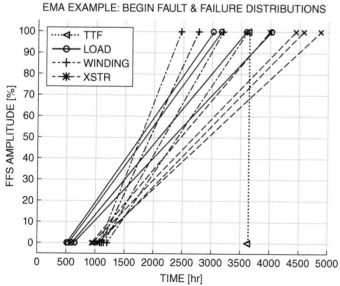

Figure 7.15 Plots: test results for six power supplies (top) and 12 EMAs (bottom).

Table 7.6 Summarized list of test results.

Power supply	Detect degradation	Detect failure	EMA	Detect degradation	Detect failure
Supply #1	1368	4200	EMA #1	540	3168
Supply #2	1320	3624	EMA #2	516	3024
Supply #3	1296	3240	EMA #3	588	3600
Supply #4	1248	3048	EMA #4	636	4032
Supply #5	1224	4584	EMA #5	1104	2760
Supply #6	1416	2856	EMA #6	1080	2472
			EMA #7	1152	3192
			EMA #8	1200	3624
			EMA #9	960	4440
			EMA #10	936	4008
			EMA #11	1008	4872
			EMA #12	1056	5304

Sustainment costs can be further reduced when your PHM system is sufficiently accurate and reliable to let your customer defer maintenance until SoH estimates fall below 50% or even lower.

CBD-Based Approaches to CBM

Of the five approaches in the cost analyses, two are based on CBD: the damage-detection approach and the SoH approach. The damage-detection approach is diagnostic in nature: it processes CBD and detects damage, and maintenance is scheduled. The SoH approach is prognostic in nature: it processes CBD, detects damage, and provides estimates of SoH that are used to trigger scheduling of maintenance. The prediction program, ARULEAV, also provides RUL and PH estimates for use in health management.

7.4 Reliability: Bathtub Curve

A bathtub curve, shown in Figure 7.16, is a statistical depiction of the failure rate over the lifetime of a population of electronic products. There are three distinct regions involved, where the curve depicts the failure rate versus time. Beginning on the left, and moving to the right:

- The first region is the infant mortality region, where burn-in cycles can be used to weed out those products susceptible to early failure due to material or manufacturing flaws.
- The constant-failure region is where the failure rate is very low. Even so, fielded products are susceptible to fatigue damage.
- The wear-out of end-of-life region is marked with increasingly high levels of failure rates as the products wear out. However, electronic products rarely fail due to

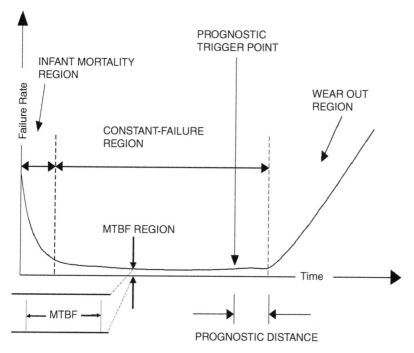

Figure 7.16 Bathtub curve showing three regions, MTBF, and a prognostic trigger point.

wear-out; instead, failure is most often due to cyclic stresses and strains that cause fatigue damage that accumulates (Hofmeister et al. 2006). As fatigue damage accumulates, one or more measurable signals changes, which can be captured as signature data.

So, there is really nothing about a bathtub curve that can be used to enable or to support CBD-based prognostics.

7.4.1 Bathtub Curve: MTBF and MTTF

As you can see in Figure 7.16, MTBF (between failures) is not a time-axis value: it is a failure-rate value. Neither MTTF nor MTBF is seen in a typical view of a bathtub curve, perhaps because of the relationship suggested in Figure 7.17 (Seastrunk 2016).

7.4.2 Trigger Point and Prognostic Distance

Also shown in Figure 7.16 is a conceptual diagram intended to convey the notion that it is possible to employ a prognostic trigger to provide advance warning of a probable failure within time PD while, at the same time, conveying a notion that useful life does not extend into the wear-out region. Figure 7.18 conveys a more practical view of prognostic trigger points:

- Envision multiple sensors attached to a complicated product such as an EMA with a built-in power supply, as already described in this book.

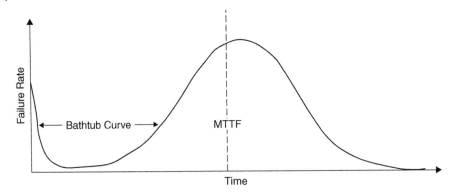

Figure 7.17 Possible relationship of bathtub curve to failure distribution and MTTF.

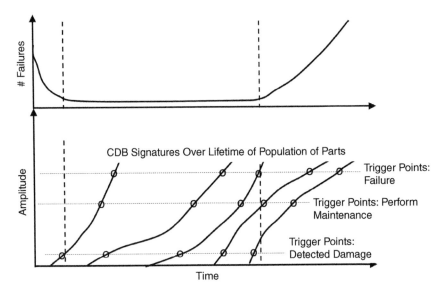

Figure 7.18 Multiple instances of CBD signatures and trigger points.

- When a population of those EMAs are deployed and put into use, they become damaged due to cyclic stress, weak/defective components, and random events.
- Continued use increases the level of damage to the point that the attached sensors detect the damage, trigger alerts, then trigger maintenance activity, and then trigger alerts that failure is imminent or has occurred.

7.5 Chapter Summary and Book Conclusion

This is the final chapter in this book. We discussed topics related to the selection, evaluation, and other considerations of prognostic enabling. The introduction briefly touched on critical systems, advance warning, and health management. The bulk of this chapter presented a rationale for not using reliability metrics such as MTTF and MTBF; instead, a rationale was presented for using the time between the onset of degradation and the time when such degradation results in functional failure: TTF. Methods to determine or calculate a value for TTF include service life, end-use testing, and MTTF-based methods. A section on cost-benefit analysis of prognostic approaches included example approaches for the following approaches: (i) no PHM; (ii) usage-based MTTF; (iii) usage-based 2/3 MTTF; (iv) damage-detection; and (v) SoH at 75% or less. The section after that focused on the bathtub curve and how it relates to failure distributions, MTBF, MTTF, trigger points, and CBD signatures.

By no means does this book cover the entirety of information related to PHM and CBD – conditioning, modeling, and processing for CBM. On the other hand, this book contains a wealth of information dealing with basic approaches and, importantly, CBD signatures and how to process and linearize those signatures, which lessens the burden on prediction programs and improves the accuracy of prognostic information. Chapter 6 presented the design of a hypothetical prototype PHM system to illustrate the challenges a designer might face and to demonstrate the application of the approaches discussed in this book.

References

Abernethy, R.B. (2006). *The New Weibull Handbook*, 5e, 2006. Berringer and Associates.

Astfalck, L., Hodkiewicz, M., Medjaher et al. (2016). A modelling ecosystem for prognostics. Annual Conference of the Prognostics and Health Management Society.

DAU (2018). Failure modes & effects analysis (FMEA) and failure modes, effects & criticality analysis (FMECA). Acquipedia, Defense Acquisition University, 9820 Belvoir Road, Fort Belvoir, VA 22060. https://www.dau.mil/acquipedia.

Ellerman, P. (2012). *Calculating Reliability Using FIT & MTTF: Arrhenius HTOL Model*, MicroNote™ 1002, Rev 0. MicroSemi Corp. https://www.microsemi.com/document-portal/doc_view/124041-calculating-reliability-using-fit-mttf-arrhenius-htol-model (accessed 2018).

Hofmeister, J.P., Lall, P., and Graves, R. (2006). In-situ, real-time detector for faults in solder joint networks belonging to operational, fully programmed field programmable gate arrays (FPGAs). In: *Proceedings of the IEEE AUTOTESTCON, Anaheim, CA, 18–21 Sept. 2006*, 237–243. IEEE.

Johnston, A. (2010). *Reliability and Radiation Effects in Compound Semiconductors*, 117–132. Singapore: World Scientific Publishing Co. Pte. Ltd.

O'Connor, P. and Kleyner, A. (2012). *Practical Reliability Engineering*. Chichester, UK: Wiley.

RAC (2005). *Reliability Toolkit: Commercial Practices Editions*. Reliability Analysis Center, Rome Laboratory. https://www.dsiac.org/sites/default/files/journals/2Q2005.pdf (accessed 2018).

Ridgetop Group. (2018). Sentinel Network, view of the graphical user interface (GUI) for an electronic power supply (EPS). Courtesy and permission of Ridgetop Group, Inc., 3580 West Ina Road, Tucson, AZ, 85741.

Seastrunk, B. (2016). Reliability, warranty, and why nothing lasts as long as it used to. http://opinionbypen.com/reliability-warranty-and-why-nothing-lasts-as-long-as-it-used-to (accessed 2018).

Seymour, B. (1993). MTTF, failrate, reliability and life testing. Application Bulletin, AB-059, Burr-Brown Corporation. Tucson, AZ.

Speaks, S. (2005). Reliability and MTBF overview. Vicor Reliability Engineering. http://www.vicorpower.com/documents/quality/Rel_MTBF.pdf (accessed August 2015).

Tobias, P. (2003). Assessing product relability. In: *Engineering Statistics Handbook*. National Institute of Standards and Technology. https://www.itl.nist.gov/div898/handbook/apr/apr.htm.

Weibull. (2008). MTTF, MTBF, mean time between replacements and MTBF with scheduled replacements. HotWire 94. https://www.weibull.com/hotwire/issue94.

Wilkins, D.J. (2002). The bathtub curve and product failure behavior, part two – normal life and wear-out. HotWire 22. https://www.weibull.com/hotwire/issue22/hottopics22.htm (accessed 2018).

Further Reading

Filliben, J. and Heckert, A. (2003). Probability distributions. In: *Engineering Statistics Handbook*. National Institute of Standards and Technology. http://www.itl.nist.gov/div898/handbook/eda/section3/eda36.htm.

Tobias, P. (2003a). Extreme value distributions. In: *Engineering Statistics Handbook*. National Institute of Standards and Technology. https://www.itl.nist.gov/div898/handbook/apr/section1/apr163.htm.

Tobias, P. (2003b). How do you project reliability at use conditions? In: *Engineering Statistics Handbook*. National Institute of Standards and Technology. https://www.itl.nist.gov/div898/handbook/apr/section4/apr43.htm.

Index

Prognostics and Health Management: A Practical Approach to Improving System Reliability Using Condition-Based Data,
First Edition. Douglas Goodman, James P. Hofmeister and Ferenc Szidarovszky.
© 2019 John Wiley & Sons Ltd. Published 2019 by John Wiley & Sons Ltd.